Pain and Touch

Handbook of Perception and Cognition
2nd Edition

Series Editors
Edward C. Carterette
and **Morton P. Friedman**

Pain and Touch

Edited by
Lawrence Kruger
Department of Neurobiology
University of California, Los Angeles
Los Angeles, California

Academic Press
San Diego   London   Boston
New York   Sydney   Tokyo   Toronto

Copyright © 1996 by ACADEMIC PRESS

All Rights Reserved.
No part of this publication may be reproduced or transmitted in any form or by any means, electronic or mechanical, including photocopy, recording, or any information storage and retrieval system, without permission in writing from the publisher.

Academic Press, Inc.
525 B Street, Suite 1900, San Diego, California 92101-4495, USA
http://www.apnet.com

Academic Press Limited
24-28 Oval Road, London NW1 7DX, UK
http://www.hbuk.co.uk/ap/

Library of Congress Cataloging-in-Publication Data

Pain and touch / edited by Lawrence Kruger.
        p.   cm. -- (Series, Handbook of perception and cognition, 2nd ed.)
    Includes bibliographical references and index.
    ISBN 0-12-426910-9 (alk. paper)
    1. Touch.   2. Pain—Psychological aspects.   I. Kruger, Lawrence.
    II. Series: Handbook of perception and cognition (2nd ed.)
    BF275.P35   1996
    152.1'82--dc20   96-22523
                                                            CIP

PRINTED IN THE UNITED STATES OF AMERICA
96   97   98   99   00   01   BC   9   8   7   6   5   4   3   2   1

Contents

2   The Psychophysics of Tactile Perception and Its Peripheral Physiological Basis

Joel D. Greenspan and Stanley J. Bolanowski

⚡3   *Somatosensory Cortex and Tactile Perceptions*

Harold Burton and Robert Sinclair

⚡4   *Nociception and Pain: Evolution of Concepts and Observations*

Edward R. Perl and Lawrence Kruger

5   *Afferent Mechanisms of Pain*

Bruce Lynn and Edward R. Perl

6   *Measurement of Pain Sensation*

Richard H. Gracely and Bruce D. Naliboff

Contributors

Numbers in parentheses indicate the pages on which the authors' contributions begin.

Stanley J. Bolanowski (25)
Department of Bioengineering and
  Neuroscience
Institute for Sensory Research
Syracuse University
Syracuse, New York 13244

Harold Burton (105)
Department of Anatomy and
  Neurobiology
Washington University School of
  Medicine
St. Louis, Missouri 63110

C. Richard Chapman (315)
Departments of Anesthesiology,
  Psychiatry, and Behavioral Sciences
University of Washington
Seattle, Washington 98195

Richard H. Gracely (243)
Neuropathic Pain and Pain Assessment
  Section
Neurobiology and Anesthesiology
  Branch
National Institute of Dental Research
National Institutes of Health
Bethesda, Maryland 20892

Barry G. Green (1)
Monell Chemical Senses Center
Philadelphia, Pennsylvania 19104

Joel D. Greenspan (25)
Departments of Physiology and
  Neurosurgery
State University of New York Health
  Science Center
Syracuse, New York 13210

xi

Lawrence Kruger (179)
Department of Neurobiology
University of California, Los Angeles
Los Angeles, California 90024

Bruce Lynn (213)
Department of Physiology
University College London
London WC1E 6BT, United Kingdom

Bruce D. Naliboff (243)[1]
University of California, Los Angeles
Los Angeles, California 90024

Edward R. Perl (179, 213)
Department of Physiology
University of North Carolina—Chapel Hill
Chapel Hill, North Carolina 27599

Russell K. Portenoy (343)
Department of Neurology
Memorial Sloan–Kettering Cancer Center
New York, New York 10021

Robert Sinclair (105)
Department of Anatomy and Neurobiology
Washington University School of Medicine
St. Louis, Missouri 63110

Joseph C. Stevens (1)
John B. Pierce Laboratory and Yale University
New Haven, Connecticut 06519

Mark Stillman (315)
Department of Psychology and Behavioral Sciences
University of Washington
Seattle, Washington 98195

---

[1] Current address: West Los Angeles Veterans Administration Medical Center, Los Angeles, California 90073.

Foreword

The problem of perception and cognition is in understanding how the organism transforms, organizes, stores, and uses information arising from the world in sense data or memory. With this definition of perception and cognition in mind, this handbook is designed to bring together the essential aspects of this very large, diverse, and scattered literature to give a précis of the state of knowledge in every area of perception and cognition. The work is aimed at the psychologist and the cognitive scientist in particular, and at the natural scientist in general. Topics are covered in comprehensive surveys in which fundamental facts and concepts are presented, and important leads to journals and monographs of the specialized literature are provided. Perception and cognition are considered in the widest sense. Therefore, the work will treat a wide range of experimental and theoretical work.

The *Handbook of Perception and Cognition* should serve as a basic source and reference work for those in the arts or sciences, indeed for all who are interested in human perception, action, and cognition.

Edward C. Carterette and Morton P. Friedman

Preface

The title of this volume and the determination of its contents were significantly influenced by illuminating discussions with Edward Carterette, the editor who has taken large responsibility for organizing and selecting editors for the volumes dealing most directly with the sensory systems. In the precedent established in the *Handbook of Perception* (the volume titled *Feeling and Hurting*), the sense data derived from all sense organs, including those essential for *pain*, were subsumed under the ambiguous banner of *feeling*. Given the richness of the English language and thereby its capacity for greater precision, *feeling* evokes too many superfluous ideas to be a suitable synonym for tactile experience or touch, and *hurting* might be construed as suffering removed from sensory experience. Attracted by and more concerned with epistemology than semantics, this volume attempts to examine current knowledge and a modern perspective of touch and pain that incrementally builds on the surveys of these subjects appearing in the earlier *Handbook of Perception* published in 1978. Most importantly, the change in title of the entire *Handbook* series to now encompass *Perception and Cognition* suggests that simply reiterating the efforts of the past would not suffice for this next-generation handbook. Thus, we aimed to include the cognitive aspects of sensory experience to the extent that one can ever delve into the properties of awareness and judgment underlying cognition.

Chapter 1 on the historical survey of research on touch by Joseph Stevens and Barry Green derives from my challenging them to improve upon and extend their 1978 review of this subject for the previous *Handbook*. To the delight of the editors, their effort brought the fresh perspective that comes with maturity and from following a subject with avid passion and intensity for a long time. The deliberate objective was to acknowledge that the principal audience for these volumes would logically be groups of neuroscientists who take pride in identifying themselves as psychologists. This chapter sets the stage for the entire volume, thereby attempting to achieve something distinct from the conventional coverage of sensory physiology contained in standard textbooks.

Chapter 2 on peripheral cutaneous tactile information processing by Joel Greenspan and Stanley Bolanowski brings the broad sweep of psychophysics and experimental psychology applied to tactile perception into the modern era in a manner that is not available elsewhere; in any case, it bears little resemblance to the coverage of the subject in the 1978 *Handbook*, largely because of its outlook as well as its recency.

The central processing of tactile information has been reviewed so extensively in so many places that it might seem hard to justify still another recounting of what by now has become an almost traditional narrative. The challenge was met by the most versatile and polymathic researcher in this field, Harold Burton, whose experimental interests in recent years have been extended beyond sensory mapping into spheres examining the nature of tactile exploratory behavior and the new explosion of imaging techniques for examining regional changes in the human brain in the context of tactile discrimination and performance. Active touch was barely recognized as a subject worthy of serious discussion until quite recently, but Burton and Sinclair not only bring the subject up to date, they also provide a quite original perspective.

Covering the subject of pain in a parallel fashion is more difficult, and indeed the concepts of hurting and suffering need not be associated with activation of a specific, segregated set of sense organs. In the past three decades, the obfuscatory misuse of the term *pain* and the debate about the very existence of specific afferent nerves uniquely implicated in sensory reports of pain required a historical account of how nociception and nociceptors have entered current parlance. Edward Perl has been the leading pioneer in championing the notion of specific nociceptors and has devoted much of his career to following the development of the field. He agreed to cover the history and the basic neurophysiology of this subject with the assistance (and combat) of co-authors. The more difficult problem was finding someone sufficiently courageous to tackle the central representation of pain, the only sensory pathway that has largely eluded detailed topographic mapping using electrophysiological methods. I agreed to assist him in recounting the detailed modern history about nociceptors in a broader historical setting, which includes a focused summary of the elusive information on central representation of pain. Chapter 5 also attempts to explain the evolution of terminology leading to the distinction between nociception and pain as a proper introduction to a modern account and review of peripheral neural mechanisms with emphasis on the nociceptor literature, a task for which Ed Perl and Bruce Lynn are ideally suited.

The psychophysics of pain and the emerging systematization of clinical scaling methods were nicely reviewed by Richard Sternbach in the 1978 *Handbook of Perception* and in his more comprehensive subsequent monographs. Richard Gracely, who was completing his dissertation research on pain psychophysics at that time, has become a leading exponent of the employment of various categories of scaling methods for suprathreshold magnitude estimation as well as threshold measurement. He and Bruce Naliboff provide a most useful, modern, and original account in Chapter 6.

C. Richard Chapman covered pain pathology and therapeutics in the 1978 *Handbook,* but the explosion of information and interest in this field clearly indicated a need for a new and broader approach. Accordingly, he agreed to review pain pathology in a quite distinctive chapter with the able collaboration of Mark Stillman. It was agreed that an able and knowledgeable clinician should be recruited to provide a modern account of pain control covering the complexities and pitfalls of the myriad therapies in current use and favor. This presented the most difficult editorial decision, and after many colleagues were consulted, the consensus was that Russell Portenoy was the optimal choice. He has broad clinical experience and knowledge and met the challenge of presenting this subject organized and proportioned for the nonclinician psychologist audience as well as for the pain therapist. Portenoy's chapter serves as an excellent example of how rapidly psychological sophistication can permeate and empower the strategies underlying therapeutic selection. The nexus between basic research and clinical practice becomes progressively more evident in the last three chapters, and it is hoped that the entire volume, rather than serve merely as a reference work, can be a guide and a source of incentive to further inquiry into the compelling subject matter so enthusiastically presented by the authors of this book.

The production of this volume was ably conveyed to completion by Ms. Nikki Fine Levy of Academic Press and propelled through the final stages of copy editing and indexing by Ms. Eileen Favorite, a superb professional possessed of exceptional ability, patience, and wit. Their contributions are much appreciated.

Lawrence Kruger

CHAPTER 1

History of Research on Touch

Joseph C. Stevens
Barry G. Green

I. INTRODUCTION

By *touch* we refer here primarily to sensations aroused through stimulation of receptors in the skin—pressure, warmth, cold, and various blends of these attributes. It must be pointed out that in the broad sense touch includes more than just these sensations. Two other notable aspects are pain (see chapter 4, this volume, for a separate historical chapter) and sensations aroused by stimulation of receptors deeper within the body, such as in the lining of various internal organs (organic sensibility) and in the muscles and joints (proprioception).

When one palpates an object, such as a piece of sandpaper, a tennis ball, or an ice cube, one ordinarily fails to break down the experience into various attributes that simultaneously make up the total, unitary experience of touching. Among the potential attributes are roughness, warmth, cold, pressure, size, location, and weight. The perceived weight of an object can sometimes involve proprioception, but most attributes of touch come about from stimulation of receptors in the skin and just beneath the skin.

The more or less unitary character of the touch experience may help to explain why from the time of Aristotle to the middle of the 19th century touch nearly always was classified as one of the five senses. Even Aristotle

Pain and Touch
Copyright © 1996 by Academic Press, Inc. All rights of reproduction in any form reserved.

recognized that touch can have several different attributes, but before the mid-19th century the belief that the skin might house a variety of sense modalities had won only a handful of converts. Among the exceptions was Erasmus Darwin (1796), who posited several senses beside the traditional five. In the present context, the most interesting of these postulated senses was a separate *heat sense,* which Darwin reported remained intact in a patient who had lost sensitivity to pricking and pinching in his food but could still feel the heat radiated from a hot poker.

II. E. H. WEBER (1795–1878)

It was Weber, an influential physiologist at Leipzig, who initiated a trend toward the differentiation of touch with his 1846 publication of an essay on touch titled *Der Tastsinn und das Gemeingefühl* ("Touch and Common Sensibility") (see Weber, 1834, 1846, 1978). Thereafter, further differentiation (and opposition to it) came to lie at the heart of research on touch, reaching its classical expression a half-century later, also at Leipzig, in the publications of M. von Frey, about whom more will be said later on.

Weber's research led him to distinguish touch *(Tastsinn),* a nerve-receptor system housed solely by the skin itself, from a common sensibility *(Gemeingefühl)* made up of receptors that reside in nearly all human tissues and that mediate a wide variety of sensory experiences. *Gemeingefühl* comprised pain (including that originating in the skin), fatigue, hunger, thirst, feelings of well-being, and sexual pleasure—"Weber's scrap basket into which he cast a miscellany of sensory remnants," as Dallenbach (1939, p. 336) later put it.

Unlike common sensibility, touch could be functionally differentiated by its ability to distinguish a pair of sensations by means of differences in their pressures, their temperatures, or their locations on the body. No anatomical status, as to receptor or nerve specificity, was therein implied; on the contrary, Weber argued for the interdependence of these tactile "qualities." As often passed over as it is cited in the literature on the skin senses is his curious observation that a cold silver *Thaler* (a 19th-century German silver coin) placed on the forehead feels heavier (as much as two or more times) than a neutral one! This "Thaler illusion," as it is usually dubbed, can be very large and compelling. Weber conjectured that cooling an object makes it feel heavier and warming makes it feel lighter. More recent and much more thorough investigations (J. C. Stevens, 1979; J. C. Stevens & Green, 1978) comparing subjects' magnitude estimations of the heaviness of objects warmed to various degrees above and below what feels neutral to the skin demonstrates that both cooling and warming can intensify the impression of heaviness. Intensification of this kind can take place almost anywhere on the body surface and is much more impressive for cooling than for warming.

As mentioned, Weber gave "localization" an important status in touch. In order to study it he invented two methods, both of which have been widely used right up to the present. One was the "compass" test, which he used to determine the smallest discriminable distance between two points of contact, the so-called two-point limen. The other test yielded the *error of localization*, a measure of the accuracy with which a person can identify the locus of a point of contact. Application of these methods led to important findings regarding the spatial acuity of the skin. Mapping of the whole body surface revealed to the 19th-century scientist great variation of spatial acuity from one body part to another. The region of the hand, especially the fingertips, and the face, especially the lips and tongue, exhibit exquisitely fine resolution, compared, say, to the back, upper arm, or leg, which have only the most rudimentary power of spatial resolution. Another famous generalization became known as Vierordt's (1870) law of mobility. This law states that the two-point limen improves, up to about 20-fold, from the shoulder toward the more mobile end of the arm and hand (see Figure 1). Such measurements have generally withstood the test of time (Weinstein, 1968), though we shall see later that refinements of these old methods and

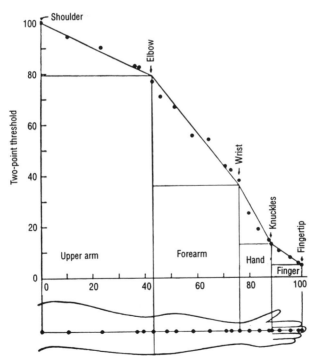

FIGURE 1   An illustration showing Vierordt's law of mobility. The two-point limen (ordinate) is plotted as the percentage of its size at the shoulder. (From Boring, 1942, using Vierordt's, 1870, data; reproduced by permission of Lucy D. Boring.)

invention of new ones for studying the skin's spatial acuity were to appear about a century after Weber introduced the subject.

Even more famous experiments than those on spatial acuity had to do with the discrimination of pressure sensation on the skin. Over a sizable range of stimulation, the just-perceptible increase in pressure (say, to a weight placed on the forearm) is a constant *fractional* increment of the weight. This fundamental generalization spread to one sense modality after another, eventually earned the title of *Weber's law,* and despite its frequent failure at weak stimulus levels, also withstood the test of time—it became "psychology's law of relativity," as S. S. Stevens (1951, p. 34) phrased it. In this regard Weber's influence on the whole domain of sensory psychophysics and physiology became, and remains, immense.

III. M. VON FREY (1852–1932)

After Weber, two important observations gave impetus to further differentiation of the skin senses. The first came from the clinic, the second from the laboratory. Spinal injuries and pressure blocks on nerves were frequently seen to affect pain, pressure, and thermal sensation differentially. From the laboratory, or rather from three laboratories independently within a span of only 3 years (1882–1885), came the discovery of *sensory spots* by Blix (1884) in Sweden, Goldscheider (1884) in Germany, and Donaldson (1885) in America. A sensory spot means a tiny area of the skin that elicits a sensation when touched, say, by a needle (pain), a hair (pressure), or by the tip of a temperature-controlled brass cone (warmth or cold). With this technique one could construct a punctiform map of any region of the skin for any of the four types of sensation. Always it turned out that the punctiform maps for the four types were independent of each other. Furthermore, it turned out that the density of spots for each type varied with the body locus mapped. Warmth spots are notable for their scarcity. For example, one expects to find on the average only one spot in a map of 5 cm² of the upper arm.

The neural meaning of sensory spots has drawn debate (Head, 1920; Jenkins, 1951; Melzack, Rose, & McGinty, 1962; J. C. Stevens, Marks, & Simonson, 1974), but for von Frey the spot maps furnished the chief cornerstone for what has become the classical theory of cutaneous sensitivity (Boring, 1942).

Two aspects of the theory deserve mention. The first regards *function;* where Weber treated warmth, cold, and pressure as interdependent attributes of the same sensation, von Frey regarded the four "attributes"—pain, pressure, warmth, and cold—as independent sense modalities. The second regards *structure;* von Frey claims to identify four specific receptor types that correspond to the four sensory qualities. This identification characterized a growing tendency in sensory physiology to extend Johannes Müller's "doctrine of specific nerve energies" by relating every sensory quality to its own specific nerve type. The anatomy of the day had uncovered a variety of encapsulated nerve endings as well as free nerve endings in the skin. Unfortunately, von Frey's educated guesses about which ones did which jobs turned out, on more careful scrutiny, to be mostly wrong or at best unsubstantiated (Dallenbach, 1929), although many a biology student has had to learn that Krause bulbs are for cold, Ruffini cylinders for warmth, Meissner corpuscles for pressure, and free nerve endings for pain. In von Frey's defense, the specific functions of the nerve-ending types known to the anatomist continue to be debated by physiologists. Moreover, von Frey himself acknowledged the tentative nature of his correlations between sensations and end organs; later generations converted hypothesis into dogma.

Despite the shakiness of the anatomy and fierce opposition from certain quarters (notably from Goldscheider, whose main objection related to the separate classification of pain), von Frey's theory seemed for a time to settle the question of differentiation of the skin senses (Boring, 1942). Its functional aspect long outlived its anatomical shortcomings, theory became textbook "fact," and thereafter the study of each modality tended to proceed as if the others did not exist. Even the study of touch "blends" was taken to support von Frey's view. At stake here was the nature of such complex touch perceptions as *wet, smooth, sticky, tickle,* and *oily*. Some of these percepts have yielded to analysis into the more basic sensations. "Wet," for example, yields to excellent synthesis by mingling pressure with cold (Bentley, 1900). More on this subject below.

IV. RESEARCH ON WARMTH AND COLD

Until about 1930, speculation about the nature of thermal sensibility outdistanced factual information, no doubt in large part because of the difficulty of producing and controlling thermal stimuli. The following phenomena, though, had come under laboratory investigation to one degree or another.

A. Punctiform Mapping of Warm and Cold

Maps of warmth and cold spots, mentioned earlier, furnished the main evidence for considering warmth and cold as two modalities. Although the maps tend to be unstable, when constructed with care they prove reliable enough to establish that the locations of warmth and cold spots are uncorrelated and that the concentration of cold spots far exceeds that of warmth spots (Figure 2). Also, the concentrations vary from one body locus to another; the lip, for example, has some six times as many cold spots as the sole (Strughold & Porz, 1931), and the finger has some nine times as many

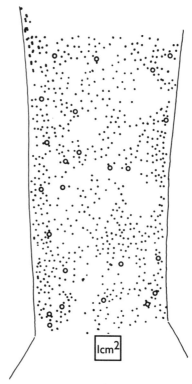

FIGURE 2   Map of warmth and cold spots over an area of about 100 cm² of the dorsal side of the right forearm; small dots = cold spots, open circles = warmth spots. The number of warmth spots in this area is 24 (density = .24 cm⁻²); the number of cold spots is about 700 (density = 7 cm⁻²). Thus, in this area there are about 29 times as many cold as warmth spots. (Redrawn from Strughold & Porz, 1931.)

warmth spots as the dorsal side of the upper arm (Rein, 1925). Some body parts have so few warmth spots as practically to negate the notion, so tempting to the early workers, that warmth receptors bear a one-to-one relation to warmth spots. If receptors exist only where there are spots, then large portions of the skin should be insensitive to warmth stimulation—but this is unlikely, as can be demonstrated by stimulating with larger contactors (1–2 cm² instead of the usual 1–2 mm² used for spot mapping). The larger stimuli will evoke warmth sensations reliably, even in regions of low spot concentration, like the upper arm.

It would seem in the light of this observation that there may exist many more receptors than spots, but that it usually requires the simultaneous stimulation of several of them to arouse a sensation reliably. This property of a sensory system, called *spatial summation,* comes under further discussion later.

Another complicating fact, first noted by von Frey in 1895 and cited in favor of specific nerve energies, is *paradoxical cold,* that is, the elicitation of a cold sensation from a cold spot using a contact stimulator heated to a temperature a little below the thermal pain threshold (Boring, 1942). All this leaves the meaning of sensory spots cloudy. They could represent high concentration clusters of receptors, a theory proposed by Jenkins (1951), or they could signal the location of a single receptor having greater than average sensitivity, or they could signal the recruitment of a second type of temperature receptor, which Head (1920) called a *protopathic* receptor (punctiform) as opposed to an *epicritic* (nonpunctiform) receptor system. Yet it is amazing how the simplistic hypothesis of von Frey continues to thrive.

Whatever their meaning, to many the temperature spots have seemed to call decisively for a separation of warmth and cold into two modalities.

B. Adaptation

Adaptation features prominently in thermal sensibility. As an easy demonstration, immerse the left hand in mildly cool water, the right hand in mildly warm water; the resulting thermal sensations will gradually fade in intensity and eventually vanish. Both hands then go into a third bath (32–33°C), which would normally feel neutral but now feels warm to the left hand and cool to the right hand. Eventually, of course, these sensations, too, will vanish.

To the German physiologist Hering (1880), this kind of observation supported the idea of a single thermal sense. The qualities (warmth and cold) relate to each other as *opponent processes* (analogous to the opponent processes in Hering's famous theory of color vision). Adaptation merely alters the null point (called *physiological zero*) on the stimulus continuum, allowing the same temperature to feel warm at one time and cool at another. Two more recent reversions of Hering's one-modality approach have been offered by Nafe (1929) and by Jenkins (1938).

Given the salience of thermal adaptation, the phenomenon has failed to get the full parametric treatment it deserves. When, for example, the rate or limits of adaptation have come under study, the interpretations have had to be tenuous or qualified because other parameters, such as areal extent and body locus of stimulation, have been left unexplored. Problems of stimulus control abound, especially in the earlier work. One can easily confound neural adaptation with the effects of vasoconstriction and vasodilation—as when the hands are immersed in water for some time. Despite a revival of interest in adaptation (Hensel, 1950; Kenshalo, Holmes, & Wood, 1968), with technological and methodological innovations and fresh measurements, Boring's (1942) evaluation of the state of the quantitative studies generally still holds: "They furnish little more than handbook data" (p. 500).

C. Quantity and Quality

Before 1930 there was relatively little interest in the problem of relating the magnitude of warmth and cold sensations to the stimulus magnitude. This may reflect the inability to test Weber's law. That any of the then current scales of temperature constituted only an interval scale (i.e., one with an arbitrary zero) rather than a ratio scale (i.e., one with a true zero) might seem to render the very idea of a Weber fraction illogical. Fechner (1860/1966), who wanted to extend his famous logarithmic "law" to thermal sensations, contrived to bypass this inconvenience by arbitrarily fixing zero on the temperature scale at 18.5°C—halfway between the freezing point of water (0°C) and body core temperature (37°C)!

The thermal stimulus, though forming a quantitative continuum, can arouse a variety of sensory qualities—pain, cold, warmth, and heat, which provokes a kind of stinging quality short of painful. Alrutz (1898) proposed the ingenious theory that this heat quality results from stimulation by temperatures sufficiently high to arouse simultaneously warmth and paradoxical cold but insufficiently high to arouse pain. This theory makes heat a fusion of warmth and cold.

Evidence for Alrutz's theory comes from touching a *thermal grill*, an apparatus that consists of alternate warm and cool tubes. One experiences a synthetic heat that resembles heat per se (although some have asserted that the quality lies more in the domain of pressure than of temperature). Polemic (often semantic), rather than physiological search and discovery, has characterized the subsequent history of the problem of heat as a separate quality. Nonetheless, synthetic heat demonstrates a genuine interaction between the warmth and cold senses, which more recent studies suggest may occur because both types of thermal sensations tend to be "referred" to the site of nearby mechanical stimulation (Green, 1977, 1979).

D. After 1930

Research on thermal sensation received a fresh boost after World War I with the development of practical precision radiometry in the infrared region of the electromagnetic spectrum. Electromagnetic radiation makes an excellent stimulus for the study of warmth and pain (less so for cold, because of the difficulty of creating a heat sink massive enough to attract sufficient radiation density from the body). For one thing, radiation solves the problem of concomitant mechanical stimulation. More important, it provides for variation of several parameters of interest to the psychophysicist and the physiologist: level, areal extent, duration, and wavelength composition. Even microwaves, which penetrate the body tissues readily, have come under study as a stimulus for warmth (Eijkman & Vendrik, 1961; Hendler, Hardy, & Murgatroyd, 1963; J. C. Stevens, 1983).

Although radiation had earlier found users from time to time, in the hands of J. D. Hardy and various associates (e.g., Hardy & Oppel, 1937, 1938; Herget, Granath, & Hardy, 1941; Oppel & Hardy, 1937a,b) it spurred the development of a new thermal psychophysics, tackling finally the more traditional psychophysical problems posed by, for example, students of vision and hearing. These new endeavors included the measurement of absolute and differential thresholds as functions of other variables such as the stimulus intensity, field size, body region, duration and wavelength; the thermal analog of the critical-flicker frequency in vision; and the scaling of apparent warmth level by a method remarkably prophetic of the method of magnitude estimation invented a couple of decades later by S. S. Stevens (1957) and, since about 1960, utilized in numerous investigations of the properties of thermal sensation (Marks, 1974).

Of the many phenomena studied by Hardy and his associates, perhaps the most revealing to the understanding of the workings of the thermal senses was *spatial summation*. Spatial summation means that the larger the stimulated area of the receptor surface, the greater the magnitude of the sensory response. To illustrate this, one need only immerse in cool or warm water first a finger, then the whole hand. This may seem too "obvious" to require an explanation, but some sensory systems behave differently. For example, as the areal size of a light increases, one usually sees simply an increase in area; apparent brightness level remains the same.

Just the opposite occurs with warmth sensations aroused by radiation. A person has only the feeblest capacity to judge the areal extent of a radiant field; increases in area generally register as increases in warmth level. Because of summation one can have relatively stable perception of the thermal environment; one is not at the mercy of the gross irregularity, seen in the fine grain, of the skin's sensitivity to heating, and, to a lesser degree, cooling.

One quantitative way employed by Hardy and Oppel (1937, 1938) to assess spatial summation is to measure the absolute threshold for each of several areal extents of the skin. In the case of warmth, area and irradiation level can be traded practically in proportion; that is, when the area is doubled, the irradiance necessary for threshold falls by nearly half. In the case of cold thresholds produced by negative radiation, summation clearly takes place but to a lesser degree, in the sense that a larger percentage decrease in area is needed to offset a given percentage increase in radiance. Such studies have stood up under repetition by others (Hensel, 1950; Kenshalo, Decker, & Hamilton, 1967; J. C. Stevens et al., 1974) and have expanded to include the study of supraliminal sensations of warmth from weak to strong in many different regions of the body. (For a review of the literature on thermal studies since Hardy see J. C. Stevens, 1983). Summation also governs the simple reaction time to thermal stimuli: the larger the area of stimulation (or its level), the faster the speed of reaction (Banks, 1976; Wright, 1951).

Other experiments have shed light on the neural locus of spatial summation. Summation takes place freely across the body midline and dermatome boundaries in general (even the two hands will sum their stimulation to a degree); it is, therefore, partly, and perhaps wholly, a central neural process (Hardy & Oppel, 1937).

On the neurophysiological front, the greatest progress has taken place not in the morphological and electrophysiological study of thermal receptors themselves, but rather in the recording of neural impulses from individual fibers in the cutaneous nerves of a variety of species, including the human species (for reviews, see Hensel, 1973 and Hallin, Torebjörk, & Wiesenfeld, 1981). On the whole, these recordings seem to favor at least one kind of specificity of the thermal senses, in that two types of nerve fibers have been discovered that respond specifically to temperature stimulation (and not to pressure or temperature associated with pain) and that differ from each other in their dynamic operating characteristics. One type increases its firing rate sharply when the skin is warmed and shows a transient inhibition on cooling; the other type increases its firing rate sharply when the skin is cooled and shows a transient inhibition on warming. Other fiber types show up that respond to mechanical-plus-warmth stimulation, and still others that respond to mechanical-plus-cold stimulation. The full implication for the classical specificity theory of somesthesis remains unclear, though the occurrence of the Thaler illusion implies that such bimodal fibers likely code the sensation of pressure rather than warmth or cold.

V. THREE THEORIES OF SOMESTHESIS

A. Classical Theory

Whatever its shortcomings, von Frey's four-element theory of somesthesis seemed to give order and direction to tactile research. Von Frey's association of a neural structure with a sensory quality provided an attractive simplification and scored heavily in the cause of reductionism. He had placed the atom of tactile sensation beneath the touch spot.

The von Frey hair stimulator soon established itself as a standard tool of the tactile psychophysicist. By probing the skin with hairs, von Frey discovered that the adequate stimulus for punctiform touch was the force exerted on the skin divided by the diameter of the hair (Boring, 1942). Kiesow extended Weber's law to touch spots, finding that the Weber fraction for pressure equals one-seventh over the middle range of intensities (Metcalf, 1928). Von Frey's theory proved consistent with Weber's interpretation of the limits of tactile spatial acuity as measured by the two-point limen, namely that the two-point limen reflected the distance between *sensory circles*. To some, touch spots gave concrete meaning to sensory circles, although in doing so one had to explain away evidence that the two-point limen can vary considerably with both practice and individual criterion (Boring, 1942; Friedline, 1918).

Tactile research has never enjoyed (or suffered through) an era in which one far-reaching theory regulated entirely the flow of psychophysical investigation. For many, the classical theory fell short of explaining common experience. What relevance, Katz was to ask, has probing the skin with horse hairs when "most men die without having experienced the stimulation of an isolated touch spot?" (Zigler, 1926, p. 327). Besides, when larger areas come under test, von Frey himself found that force divided by the diameter of the stimulus no longer equaled a constant at threshold (Boring, 1942). Also, Holway and Crozier (1937) later found that the difference limen (DL) for macroscopic (nonpunctiform) stimuli related more closely to hydrostatic pressure than to force divided by diameter. The importance of movement in tactile perception also found no obvious place in von Frey's theory, although distinctions had long been made between active and passive touch, and between sensations of impact (caused by rapidly loading the skin) and sensations of pressure (Griffing, 1895).

Valiant attempts were nevertheless exerted by some of Titchener's students to describe complex tactile sensations in terms of the four classic skin senses. Liquidity, solidity (Sullivan, 1923), roughness, smoothness (Meenes & Zigler, 1923), clamminess (Zigler, 1923) and, as mentioned earlier, wetness (Bershansky, 1923) were among the so-called touch blends analyzed into the four elements of warmth, cold, pressure, and pain. But such analysis frequently failed: For example, the perception of roughness and smoothness requires movement of the stimulus over the skin or of the skin over the stimulus. Some percepts seemed to contain sensations called *subcutaneous pressure,* and others called for the use of descriptive terms like *compact, clear,* and *sharp.* In retrospect, the inadequacy of a simple four-element analysis seems obvious, because even the quality of pressure itself varies experientially along dimensions like *firmness, brightness,* or *bluntness.*

B. Head's Theory of Dual Sensibilities

To some, the emergence in 1908 of Head's theory of dual cutaneous sensibilities held fresh promise for understanding somesthetic function (Head, 1920). Following peripheral section of cutaneous nerves, the recovery of sensitivity, Head observed, came in two phases. First, a primitive sensitivity appeared that enabled appreciation of heavy pressures, extremes of temperature and pain, but permitted no fine discrimination of intensity and quality or good spatial acuity (as measured by the two-point limen). This he called *protopathic sensibility.* Later on, discrimination improved, and sensations of light touch, pressure, and moderate temperature returned. This more refined perception he called *epicritic sensibility*. Head also postulated the existence of a subcutaneous *deep sensibility* that remained intact after section of cutaneous nerves.

Head's basic finding that sensation returned in stages received support from other workers (Boring, 1916, 1942; Trotter & Davies, 1909), but it was less obvious to them that the stages were as distinct as Head had described them and that the stages had to arise from two distinct afferent systems. Further investigation foundered, perhaps in part because the necessity of inflicting nerve damage tended to displace the investigation from the psychophysics laboratory into the clinic. Only a rare Spartan investigator practiced nerve section. Thus, Head's theory never gained clear acceptance or rejection and seemed to lose utility as a working hypothesis. Nevertheless, it raised interesting and important questions that outlived commitment to the theory as a whole. For example, the finding that protopathic temperature sensibility seemed to be punctiform, whereas epicritic temperature sensibility never seemed to be punctiform, may have relevance to the way spatial summation works in the thermal senses. In other words, the student of the skin senses can ill afford to ignore Head's observations.

C. Nafe's Pattern Theory of Feeling

As Head's theory drew less attention, and as the periodic failures of the introspectionists to analyze touch experience in terms of the classical modalities became evident, a new theory evolved that sought to explain somesthetic sensations without reference either to specialized neural structure or to varieties of sensory quality. Nafe (1929) introduced what he called "a quantitative theory of feeling" in which he pronounced that specialized receptors have "no factual basis at all" and that "particular experiences, for example, wet, cold or pressure, depend, for their similarities and differences, upon the 'pattern' or 'arrangement' of neural discharges" (p. 231). The patterns were thought to comprise variations in the frequency of impulses, the length of time the impulses continued, the area of skin over which the impulses arose, and the relative number of fibers activated. The theory also stressed that the concept of sensory quality, previously tied firmly to Müller's doctrine of specific nerve energies, be abandoned; somesthetic experiences were rather portrayed as lying on a single continuum of "brightness." In later writing, Nafe (1942) moderated his stand on quality, conjecturing that quality derives from the pattern of excitation, a "moment by moment representation of events at the periphery" (p. 14).

Nafe's theory struck at classical theory in almost every way. Where classical theory was specific, Nafe's theory was holistic; where classical theory was atomic, Nafe's was molecular. The quantitative theory obviated the problem of touch blends, since in principle any complex sensation could be produced by initiating the correct neural pattern. To some, however, this unbounded flexibility tended to deprive the theory of its power as a research tool. The underlying mechanisms never received clear statement. It remained for later workers, with the advantages of neurophysiological advances, to refine pattern theory and suggest plausible mechanisms (Bishop, 1946; Melzack & Wall, 1962; Weddell, 1955).

VI. RESEARCH ON TOUCH

Despite some of the inadequacies of the classical theory with respect to the relationship between receptor specificity and quality of sensation, the theory had a potent influence on tactile research in the 20th century. This influence becomes strongly evident in two areas: spatial acuity and vibration. Two other areas (adaptation and active touch) turned out to evolve largely independent of classical theory. We turn now to a brief account of each of these four areas.

A. Spatial Acuity

Following Weber's and Vierordt's 19th-century mapping of the body surface for spatial acuity, little substantive advance in this field occurred until a revival of interest took place beginning in the 1960s. Weinstein (1968) investigated pressure sensitivity (measured with von Frey hairs) together with the classical two-point limen and the error of localization over the body surface. His results agreed well with those of Weber nearly a century and a half earlier and confirmed Vierordt's law of outward mobility (Fig. 1). Areas of accurate localization also turned out to correspond well with areas where the two-point limen was small, although in most areas localization is finer by three or four times than is two-point resolution. But perhaps of greatest interest from the point of view of possible underlying mechanisms, Weinstein found that good two-point discrimination (and point localization) did not necessarily mean good sensitivity to pressure. The maps for pressure and acuity differed strikingly from each other in various ways. For example, pressure sensitivity varies over the body surface much less than does spatial acuity. A later mapping of the body surface for texture sensitivity (J. C. Stevens, 1990), by means of grooved bars described in a landmark study of texture psychophysics (Lederman & Taylor, 1972), revealed that the texture map had more in common with pressure sensitivity than with acuity. For example, although the fingertip greatly outperforms the forearm when it comes to two-point resolution, both regions give nearly equal impressions of degree of texture.

The 1960s saw the beginnings of fresh forays into the measurements of spatial acuity. Vierck and Jones (1969) and Jones and Vierck (1973) saw the two-point limen as a measure that can erroneously lead to the conclusion that the skin has poor spatial acuity. They found that the discrimination of area and length of touch stimuli on the forearm was nearly ten times as fine as the classical two-point threshold. It is unclear why this should be so, but perhaps there are multiple kinds of tactile spatial acuity, just as there are multiple types of visual acuity. Later Loomis and Collins (1978) showed a similar "hyperacuity" (i.e., 10 to 30 times finer than two-point threshold) when the stimulus was a moving shift in the locus of stimulation.

The two-point limen, a traditional darling of the tactile scientist, had been known for some time to pose serious methodological difficulty. Earlier workers, notably Friedline (1918), complained that the size of the two-point limen shrank unrealistically with greater and greater practice. It remained for Johnson and Phillips (1981), taking advantage of criterion-free methods associated with signal-detection theory, to diagnose the nature of the trouble and suggest alternative methods of assessing resolution. With sufficient practice a person is able to discriminate two points with very small or even zero separation from a single point. Such discrimination cannot represent true spatial resolution and must therefore invoke a different sensory cue, such as intensity, configuration, or orientation of the field of contact. These latter cues may be of interest on their own, of course, but they should not be confused with spatial resolution. In place of two-point measurement Johnson and Phillips offered a series of psychophysical measurements on the perception of gaps, of spatial gradients, and of alphabet letters impressed on the skin (others have used Braille letters, too). These methods all seem to reflect true spatial resolution, while avoiding the false cues that the compass test can evoke.

B. Vibration

When von Frey postulated the four skin senses he included sensitivity to vibration within the pressure sense. Katz disagreed, however, arguing instead for a separate vibration sense (Geldard, 1940a). To provide support for his position, von Frey undertook a series of experiments in which he demonstrated that points of greatest vibratory sensitivity on the skin corresponded well with pressure spots. This meant to him that vibration and pressure were mediated by the same receptors, and there must, therefore, be only a single pressure sense. Katz remained adamantly unconvinced, as did the many clinicians who observed that the sensitivities to vibration and pressure often suffered differentially in nerve injury, suggesting that the two types of stimulation act on different nerves (Gordon, 1936). It remained for Geldard (1940a,b,c) to restate and solidify von Frey's position in a series of meticulous experiments using punctiform pressure and vibratory stimuli. Finding the same correlation von Frey had observed 25 years earlier, Geldard concluded that the sensation of vibration was "a perceptual pattern of feeling of which pressure is but another temporal expression" (Geldard, 1940b, p. 279).

Vibration has since come into its own as a subdivision of tactile psychophysics (thanks in large measure to the efforts of Geldard and his students), and has even come to dominate the study of touch. Much of the research has involved phenomena that, at a glance, lack obvious relevance to theories of touch per se. The effect of frequency of vibration on sensitivity, for example, provided the subject for numerous studies (Geldard, 1940c; Gilmer, 1935; Knudsen, 1928; Sherrick, 1953; Verrillo, 1962). Another example has been the measurement of the Weber fraction for vibration frequency (Goff, 1967; Knudsen, 1928).

As often happens though, pursuit of specific and seemingly unrelated questions can lead to the discovery of facts that turn out to have more general relevance. Examination of sensitivity to various frequencies of vibration is an example. Geldard (1940c) brought to light a difference between the frequency function (at threshold) for punctate stimuli on touch spots (threshold was independent of frequency) and the frequency function for larger stimuli (threshold was typically a U-shaped function with a minimum in the neighborhood of 200 Hz). A further study of this difference (Verrillo, 1968) furnished evidence that the punctate stimuli and the larger stimuli excite at least two different sets of receptors, and that the larger stimuli (when vibrated at frequencies above 60 Hz) excite an afferent system that may summate stimulus energy over time and space in a way that the other system seemed not to do. Much of Verrillo's theory (the duplex mechanoreceptor theory) found support in neurophysiology (Mountcastle, Talbot, Darian-Smith, & Kornhuber, 1967), and a highly specialized and much studied mechanoreceptor, the Pacinian corpuscle, was tagged as the one primarily responsive to large stimulators driven at high frequencies (Talbot, Darian-Smith, Kornhuber, & Mountcastle, 1968). Subsequent sensory physiological (Harrington & Merzenich, 1970; Johansson & Vallbo, 1979) and psychophysical (Bolanowski, Gescheider, Verrillo, & Checkosky, 1988) studies have indicated that no less than four types of mechanoreceptors mediate the sensitivity to pressure and vibration in humans.

The apparent existence of a "quadruplex" system for vibration and touch furnishes a prime example of the influence that more recent somesthetic research (particularly in neurophysiology) has had upon classical notions. Why dispute the possibility of separate receptors for vibration and pressure when vibration is itself served by more than one type of receptor? It has become increasingly apparent that touch stimuli are transduced by a variety of receptors that vary as to their optimal stimulus (e.g., velocity-sensitive vs. amplitude-sensitive), their structure (e.g., encapsulated or otherwise), their rate of adaptation to prolonged stimulation (quickly versus slowly adapting), and their location (in the skin versus in subdermal tissues) (Harrington & Merzenich, 1970; Iggo, 1968). The idea of differentiating separate skin senses of touch, warmth, cold, and pain in terms of only four separate receptors turns out to appear embarrassingly naive.

C. Adaptation

Like adaptation to thermal stimuli, adaptation to tactile stimuli has received less than deserved attention, perhaps in part because of a stronger motivation to explore the variables that initiate sensation than those that end it. In the early literature the persistence of sensation after removal of the nominal stimulus (after sensations) seemed to enjoy more attention than did adaptation (Hayes, 1912).

The first conceptions of adaptation placed its locus in the receptor. Zigler (1932) early voiced this idea and also called for a distinction between effects occurring during stimulation (stimulatory) and after stimulation (poststimulatory). He discovered that adaptation time varies (a) directly with force, and (b) when force is held constant, inversely with the area of the stimulus. These two findings may amount to the same thing because, as Zigler noted, the effective intensity of stimulation may diminish as force is distributed over a larger area (i.e., pressure decreases). On the assumption that stimulation constitutes disturbance of an equilibrium state in the receptor, Zigler concluded that adaptation results from a reestablishment of equilibrium during prolonged stimulation. (Brief stimulation allows insufficient time to regain equilibrium). He believed "the fact that adaptation time to more intense stimulation is longer [means that] adaptation obviously cannot consist in a fatigue-producing process in the receptor" (Zigler, 1932, p. 719).

A major contribution to our knowledge of tactile adaptation came a decade later, when Nafe and Wagoner concluded that adaptation was mechanical rather than neural, representing a failure of the stimulus to remain in motion.

The notion of stimulus failure established once and for all the importance of movement in tactile stimulation. The significance of movement has served to enhance the attractiveness of vibration as a tool for studying touch. It can be argued in this context that vibration is a more "natural" stimulus than pressure. It takes great care to produce a pure pressure stimulus on the skin. The impact imparted to the skin during active touching must generate complex spatiotemporal variations in pressure that more closely mimic the action of a vibrator than the local impression produced by a carefully placed von Frey hair.

D. Active Touch

Active touch (haptics), characterized by free palpation of objects, will be considered only briefly here; it is covered in more detail in another chapter of this volume. That research on active touch often receives separate treatment from tactile research, per se, may serve to indicate how fragmentary is our comprehension of the whole nature of somesthesis; for as Griffing (1895) wrote a century ago, "The great majority of so-called tactile sensations are in reality results of complex kinesthetic and haptic sensory elements" (p. 7). Mainstream research on touch has failed to reflect the preeminence of the touching process.

Why has active touch been ignored by most psychologists in their studies of cutaneous sensation? Probably the most important reason is, as Griffing pointed out, that active touch involves not only cutaneous sensations, but also kinesthetic ones. Even this oversimplifies, because visual and other cues can also play a role. The guiding light (albeit a flickering one) of tactile research—the classical theory—fails to address active versus passive touch, containing no provisions for movement either on the skin or by the skin (perceiver). As Gibson (1962) put it, sensory physiologists and psychologists have viewed the skin as a passive "mosaic of receptors, not an exploratory organ" (p. 477).

The most influential early contribution to research on active touch came from D. Katz in 1925 with publication of *Der Aufbau der Tastwelt* (reviewed by Zigler, 1926), in which Katz challenged the punctiform approach to tactile research and stressed the importance of movement in the perception of texture and form. This point he illustrated with the perception of roughness and smoothness: Discrimination was better when a surface was felt actively rather than passively. Gibson (1962) subsequently made the same point for perception of form, where active touching again proved superior to passive touching. Gibson (1966) also posed the engaging question (which he first raised in reference to vision) of perceptual invariance: Why does perception follow the object and not the stimulation? We feel an object outside of the skin, mysteriously abstracted from the complex spatiotemporal pattern of pressure sensations. Answers to these sorts of questions come hard. But, "in general, experimenters have not realized that to apply a stimulus to an observer is not the same as for an observer to *obtain* [italics in original] a stimulus" (Gibson, 1962, p. 490). There is evidence, however, that this parochial view is beginning to be replaced by a broader perspective that challenges some of the tenets of the classical theory.

VII. CONTEMPORARY DIRECTIONS

At the beginning of this chapter we defined touch as "sensations aroused through stimulation of receptors in the skin." We went on to say that we ordinarily fail to break down the experience into various attributes that simultaneously make up the total, unitary experience of touching. These statements contain two themes that have run like counterpoint through research on somesthesis since the time of Weber. Weber emphasized the unity of somesthetic perceptions, von Frey their punctiform nature. Spatial summation and synthetic heat illustrate the holistic view of Weber, whereas the ability to analyze many of our somesthetic percepts in terms of the four classical skin senses illustrates the reductionist view of von Frey.

The persistence of both of these views of somesthesis informs us that neither may be intrinsically superior to the other. Instead, we might conclude that the two approaches are both valid, but that their utility depends upon the goal of the investigator. Is the goal to understand the neural mechanisms underlying sensation, or is it to understand the nature of perceptual experience? Most of the research reported in this chapter has been directed toward the former question, utilizing psychophysics as a handmaiden to serve physiology rather than to help understand perception. However, recent trends in tactile research suggest the use of psychophysics as a probe of sensory physiology is no longer of primary interest to most researchers. As the zeitgeist of perception research has moved away from issues of sensation toward issues of cognition, and as advances in sensory physiology have led to a more detailed understanding of cutaneous innervation, psychophysicists have begun to choose topics of study that explore function and experience rather than peripheral physiology.

This trend is exemplified by the systematic studies by Klatzky and Lederman (e.g., Klatzky & Lederman, 1992; Lederman & Klatzky, 1987) of how information about the shape and surface features of objects is gathered during the touching and grasping process. But it is also evident in the recent, intriguing studies of perceptual plasticity that indicate the very structure of tactile perception can be profoundly altered by long-term changes in tactile stimulation (Craig, 1993; Halligan, Marshall, Wade, Davey, & Morrison, 1993). These studies, and the benchmark neurophysiological experiments that inspired them (Merzenich et al., 1983; Jenkins, Merzenich, Ochs, Allard, & Guic-Robles, 1990), challenge the notion, implicit in the classical theory, that perception is a purely passive process that plays out within a "hard-wired" sensory system. So far perceptual shifts in humans have been demonstrated only under extremely unnatural conditions of stimulation; capturing these effects psychophysically in free-living individuals whose daily experience with touch is so rich and varied is a difficult task. Yet recent psychophysical experiments in vision imply that plasticity may be an inherent attribute of perceptual processing in time frames spanning 1 sec or less (Kapadia, Gilbert, & Westheimer, 1994), suggesting that much more remains to be learned about tactile processing in the temporal domain.

Broadly consistent with the research on plasticity is the growing interest in how the aging process—from childhood through senescence—affects tactile perception (Kenshalo, 1986; Verrillo, 1980; J. C. Stevens, 1992). Motivated more by the realities of physiological changes during aging than by interest in neural plasticity per se, study of tactile perception throughout the life span nevertheless underscores the view that perception is far from invariant over time.

Finally, a subject that has spawned a good deal of research on vibrotactile perception over the last half decade—the attempt to use the skin as an alternate communication channel for visual- and hearing-impaired individuals (Geldard, 1957)—continues to exert its influence on tactile research (see Sherrick, 1982, 1991, for reviews). Indeed, no other area of study epitomizes more the struggle to reconcile the counterpoints of sensory specificity (discrimination) and perceptual integration. For example, the effort to substitute the skin for the ear as the receiver of speech information requires that the auditory signal be transformed and then transduced in ways that take maximum advantage of the powers of tactile discrimination, both spatial and temporal; yet the same transduction strategy must also enable users to synthesize the discriminable tactual elements into a coherent stream of speech information. Although uncertainty remains regarding the ultimate utility of vibrotactile speech perception, it may well be the case that diligence in this area will unearth principles of perceptual organization that bring us closer to understanding ordinary tactile perception. Or are such principles more likely to emerge from studies of haptics or perceptual plasticity? Very likely, no single approach holds the key to full understanding, and the most fruitful avenues of research may still be awaiting discovery. As always, diversity in research fueled by the tension between old views and unexpected discoveries holds the greatest promise for future progress.

Acknowledgments

Preparation of this chapter was supported in part by National Institutes of Health grants AG10295, DC00249, and ES04356, and by National Science Foundation grant BNS76-24341.

References

Alrutz, S. (1898). On the temperature senses. II. The sensation "hot". *Mind, 7,* 140–144.

Banks, W. (1976). Areal and temporal summation in the thermal reaction time. *Sensory Processes, 1,* 2–13.

Bentley, I. M. (1900). The synthetic experiment. *American Journal of Psychology, 11,* 405–425.

Bershansky, I. (1923). Thunberg's illusion. *American Journal of Psychology, 34,* 291–295.

Bishop, G. H. (1946). Neural mechanisms of cutaneous sense. *Physiological Review, 26,* 77–102.

Blix, M. (1884). Experimentelle Beiträge zur Lösung der Frage über die specifische Energie des Hautnerven [Experimental contributions towards the resolution of the question of the specific energies of the nerves supplying the integument]. *Zeitschrift für Biologie, 20,* 141–156. (Translated from an earlier Swedish publication).

Bolanowski, S. J., Jr., Gescheider, G. A., Verrillo, R. T., & Checkosky, C. M. (1988). Four channels mediate the mechanical aspects of touch. *Journal of the Acoustical Society of America, 84,* 1680–1694.

Boring, E. G. (1916). Cutaneous sensation after nerve-division. *Quarterly Journal of Physiology, 10,* 1–95.

Boring, E. G. (1942). *Sensation and perception in the history of experimental psychology.* New York: Appleton.

Craig, J. C. (1993). Anomalous sensations following prolonged tactile stimulation. *Neuropsychologia, 31,* 277–291.

Dallenbach, K. M. (1929). A bibliography of the attempts to identify the functional end-organs of cold and warmth. *American Journal of Psychology, 41,* 344.

Dallenbach, K. M. (1939). Pain: History and present status. *American Journal of Psychology, 52,* 331–347.

Darwin, E. (1796). *Zoonomia* (vol. 1). New York: T. & J. Swords.

Donaldson, H. H. (1885). On the temperature sense. *Mind, 10,* 399–416.

Eijkman, E., & Vendrik, A. J. H. (1961). Dynamic behavior of the warmth sense organ. *Journal of Experimental Psychology, 62,* 403–408.

Fechner, G. (1966). *Elements of psychophysics* (H. E. Adler, Trans.). New York: Holt. (Original work published 1860).

Friedline, C. L. (1918). Discrimination of cutaneous patterns below the two-point limen. *American Journal of Psychology, 29,* 400–419.

Geldard, F. A. (1940a). The perception of mechanical vibration: I. History of a controversy. *Journal of General Psychology, 22,* 243–269.

Geldard, F. A. (1940b). The perception of mechanical vibration: II. The response of pressure receptors. *Journal of General Psychology, 22,* 271–280.

Geldard, F. A. (1940c). The perception of mechanical vibration: III. The frequency function. *Journal of General Psychology, 22,* 281–308.

Geldard, F. A. (1957). Adventures in tactile literacy. *American Psychologist, 12,* 115–124.

Gibson, J. J. (1962). Observations on active touch. *Psychological Review, 69,* 477–490.

Gibson, J. J. (1966). *The senses considered as perceptual systems.* Boston: Houghton Mifflin.

Gilmer, B. von H. (1935). The measurement of the sensitivity of the skin to mechanical vibration. *Journal of General Psychology, 13,* 42–61.

Goff, G. D. (1967). Differential discrimination of frequency of cutaneous mechanical vibration. *Journal of Experimental Psychology, 74,* 294–299.

Goldscheider, A. (1884). Die spezifische Energie der Temperaturnerven. [The specific nerve energies of the temperature nerves]. *Monatshefte für Praktische Dermatologie, 3,* 198–208; 225–241.

Gordon, I. (1936). The sensation of vibration, with special reference to its clinical significance. *Journal of Neurology and Psychopathology, 17,* 107–134.

Green, B. G. (1977). Localization of thermal sensation: An illusion and synthetic heat. *Perception & Psychophysics, 22,* 331–337.

Green, B. G. (1979). Thermo-tactile interactions: Effects of touch on thermal localization. In D. R. Kenshalo (Ed.), *Sensory functions of the skin of humans* (pp. 223–240). New York: Plenum Press.

Griffing, H. (1895). On sensations from pressure and impact. *Psychological Review,* Monograph Supplement No. 1, 1–88.

Halligan, P. W., Marshall, J. C., Wade, D. T., Davey, J., & Morrison, D. (1993). Thumb in cheek? Sensory reorganization and perceptual plasticity after limb amputation. *Neuroreport, 4,* 233–236.

Hallin, R. G., Torebjörk, H. E., & Wiesenfeld, Z. (1981). Nociceptors and warm receptors innervated by c fibers in human skin. *Journal of Neurology, Neurosurgery and Psychiatry, 45,* 313–319.

Hardy, J. D., & Oppel, T. W. (1937). Studies in temperature sensation. III. The sensitivity of the body to heat and the spatial summation of the end organ responses. *Journal of Clinical Investigation, 16,* 533–540.

Hardy, J. D., & Oppel, T. W. (1938). Studies in temperature sensation. IV. The stimulation of cold sensation by radiation. *Journal of Clinical Investigation, 17,* 771–778.

Harrington, T., & Merzenich, M. M. (1970). Neural coding in the sense of touch. *Experimental Brain Research, 10,* 251–254.

Hayes, M. H. S. (1912). A study of cutaneous after-sensations. *Psychological Monographs, 14,* 1–66.

Head, H. (1920). *Studies in neurology.* London: Oxford Medical Publications.

Hendler, E., Hardy, J. D., & Murgatroyd, D. (1963). Skin heating and temperature sensation produced by infrared and microwave irradiation. In C. M. Herzfeld (Ed.), *Temperature: Its measurement and control in science and industry* (Vol. 3, Part 3, pp. 211–230). New York: Reinhold.

Hensel, H. (1950). Temperaturempfindung und intercutane Warmbewegung. [Sensation of temperature and intercutaneous heat transfer]. *Pflügers Archives, 252,* 165–215.

Hensel, H. (1973). Cutaneous thermoreceptors. In A. Iggo (ed.), *Handbook of sensory physiology: Somatosensory system* (Vol. 2, pp. 79–110). Berlin and New York: Springer-Verlag.

Herget, C. M., Granath, L. P., & Hardy, J. D. (1941). Warmth sense in relation to the area of skin stimulated. *American Journal of Physiology, 135,* 20–26.

Hering, E. (1880). Der Temperatursinn. [The temperature sense]. In L. Hermann (Ed.), *Handbuch der Physiologie* (vol. 3, Part 2), pp. 415–439.

Holway, A. H., & Crozier, W. J. (1937). The significance of area for differential sensitivity in somesthetic pressure. *Psychological Record, 1,* 178–184.

Iggo, A. (1968). Electrophysiological and histological studies of cutaneous mechanoreceptors. In D. R. Kenshalo (Ed.), *The skin senses* (pp. 84–111). Springfield, IL: Charles C. Thomas.

Jenkins, W. L. (1938). Studies in thermal sensitivity: Adaptation with a series of small annular stimulators. *Journal of Experimental Psychology, 22,* 164–177.

Jenkins, W. L. (1951). Somesthesis. In S. S. Stevens (Ed.), *Handbook of experimental psychology* (pp. 1172–1190). New York: Wiley.

Jenkins, W. M., Merzenich, M. M., Ochs, M. T., Allard, T., & Guic-Robles, E. (1990). Functional reorganization of primary somatosensory cortex in adult owl monkeys after behaviorally controlled tactile stimulation. *Journal of Neurophysiology, 63,* 82–104.

Johansson, R. S., & Vallbo, A. B. (1979). Tactile sensibility in the human hand: Relative and absolute densities of four types of mechanoreceptive units in glabrous skin. *Journal of Physiology (London), 286,* 283–300.

Johnson, K. O., & Phillips, J. R. (1981). Tactile spatial resolution. I. Two-point discrimination, gap detection, grating resolution, and letter recognition. *Journal of Neurophysiology, 46,* 1177–1191.

Jones, M. B., & Vierck, C. J. (1973). Length discrimination on the skin. *American Journal of Psychology, 86,* 49–60.

Kapadia, M. K., Gilbert, C. D., & Westheimer, G. (1994). A quantitative measure for short-term cortical plasticity in human vision. *The Journal of Neuroscience, 14,* 451–457.

Katz, D. (1925). *Der Aufbau der Tastwelt.* Leipzig: Barth, 1925. [For an English translation, see L. E. Krueger, Ed., *The world of touch.* Hillsdale, NJ: Lawrence Erlbaum Associates, 1989.]

Kenshalo, D. R. (1986). Somesthetic sensitivity in young and elderly humans. *Journal of Gerontology, 41,* 732–742.

Kenshalo, D. R., Decker, T., & Hamilton, A. (1967). Spatial summation on the forehead, forearm and back produced by radiant and conducted heat. *Journal of Comparative and Physiological Psychology, 63,* 510–515.

Kenshalo, D. R., Holmes, C. E., & Wood, P. B. (1968). Warm and cool thresholds as a function of rate of stimulus temperature change. *Perception & Psychophysics, 3,* 81–84.

Klatzky, R. L., & Lederman, S. J. (1992). Stages of manual exploration in haptic object identification. *Perception & Psychophysics, 52,* 661–670.

Knudsen, V. O. (1928). Hearing with the sense of touch. *Journal of General Psychology, 1,* 320–352.

Lederman, S. J., & Taylor, M. M. (1972). Fingertip force, surface geometry, and the perception of roughness by active touch. *Perception & Psychophysics, 12,* 401–408.

Lederman, S. J., & Klatzky, R. L. (1987). Hand movements: A window into haptic object recognition. *Cognitive Psychology, 19,* 342–368.

Loomis, J. M., & Collins, C. C. (1978). Sensitivity to shifts of a point stimulus: An instance of tactile hyperacuity. *Perception & Psychophysics, 24,* 487–492.

Marks, L. E. (1974). *Sensory processes: The new psychophysics.* New York: Academic Press.

Meenes, M., & Zigler, M. J. (1923). An experimental study of the perception of roughness and smoothness. *American Journal of Psychology, 34,* 542–549.

Melzack, R., Rose, G., & McGinty, D. (1962). Skin sensitivity to thermal stimuli. *Experimental Neurology, 6,* 300–313.

Melzack, R., & Wall, P. D. (1962). On the nature of cutaneous sensory mechanisms. *Brain, 85,* 331–356.

Merzenich, M. M., Kaas, J. H., Wall, J. T., Nelson, R. J., Sur, M., & Felleman, D. (1983). Topographic reorganization of somatosensory cortical areas 3b and 1 in adult monkeys following restricted deafferentation. *Neuroscience, 8,* 33–55.

Metcalf, J. T. (1928). Cutaneous and kinesthetic senses. *Psychological Bulletin, 25,* 569–581.

Mountcastle, V. B., Talbot, W. H., Darian-Smith, I., & Kornhuber, H. H. (1967). Neural basis of the sense of flutter-vibration. *Science, 155,* 597–600.

Nafe, J. P. (1929). A quantitative theory of feeling. *Journal of General Psychology, 2,* 199–211.

Nafe, J. P. (1942). Toward the quantification of psychology. *Psychological Review, 49,* 1–18.

Oppel, T. W., & Hardy, J. D. (1937a). Studies in temperature sensation: I. A comparison of the sensation produced by infra-red and visible radiation. *Journal of Clinical Investigation, 16,* 517–524.

Oppel, T. W., & Hardy, J. D. (1937b). Studies in temperature sensation: II. The temperature changes responsible for the stimulation of the heat end organs. *Journal of Clinical Investigation, 16,* 525–531.

Rein, H. (1925). Über die Topographie der Warmempfindung. [Concerning the topography of the sense of warmth]. *Zeitschrift für Biologie, 82,* 513–535.

Sherrick, C. E. (1953). Variables affecting sensitivity of the human skin to mechanical vibration. *Journal of Experimental Psychology, 45,* 273–282.

Sherrick, C. E. (1982). Cutaneous communication. In W. D. Neff (Ed.), *Contributions to sensory physiology* (Vol. 6, pp. 1–47). New York: Academic Press.

Sherrick, C. E. (1991). Vibrotactile pattern perception: Findings and applications. In M. A. Heller & W. Schiff (Eds.), *The psychology of touch* (pp. 189–217). Hillsdale, NJ: Lawrence Erlbaum Associates.

Stevens, J. C. (1979). Thermal intensification of touch sensation: Further extensions of the Weber phenomenon. *Sensory Processes, 3,* 240–248.

Stevens, J. C. (1983). Thermal sensation: Infrared and microwaves. In E. R. Adair (Ed.), *Microwave and thermal regulation* (pp. 191–201). New York: Academic Press.

Stevens, J. C. (1990). Perceived roughness as a function of body locus. *Perception & Psychophysics, 47,* 298–304.

Stevens, J. C. (1991). Thermal sensibility. In M. A. Heller & W. Schiff (Eds.), *Psychology of touch* (pp. 61–90). Hillsdale, NJ: Lawrence Erlbaum Associates.

Stevens, J. C. (1992). Aging and spatial acuity of touch. *Journal of Gerontology (Psychological Sciences), 47,* 35–40.

Stevens, J. C., & Green, B. G. (1978). Temperature–touch interaction: Weber's phenomenon revisited. *Sensory Processes, 2,* 206–219.

Stevens, J. C., Marks, L. E., & Simonson, D. C. (1974). Regional sensitivity and spatial summation in the warmth sense. *Physiology & Behavior, 13,* 825–836.

Stevens, S. S. (1951). Mathematics, measurement, and psychophysics. In S. S. Stevens (Ed.), *Handbook of experimental psychology* (pp. 1–49). New York: Wiley.

Stevens, S. S. (1957). On the psychophysical law. *Psychological Review, 64,* 153–181.

Strughold, H., & Porz, R. (1931). Die Dichte der Kaltpunkte auf der Haut des menschlichen Körpers. [Density of cold spots on the skin of the human body]. *Zeitschrift für Biologie, 91,* 563–571.

Sullivan, A. H. (1923). The perceptions of liquidity, semi-liquidity and solidity. *American Journal of Psychology, 34*, 531–541.

Talbot, W. H., Darian-Smith, I., Kornhuber, H. H., & Mountcastle, V. B. (1968). The sense of flutter-vibration: Comparison of the human capacity with response patterns of mechanoreceptive afferents from the monkey hand. *Journal of Neurophysiology, 31*, 301–334.

Trotter, W., & Davies, H. M. (1909). Experimental studies in the innervation of the skin. *Journal of Physiology (London), 38*, 134–246.

Verrillo, R. T. (1962). Investigation of some parameters of the cutaneous threshold for vibration. *Journal of the Acoustical Society of America, 34*, 1768–1773.

Verrillo, R. T. (1968). A duplex mechanism of mechanoreception. In D. R. Kenshalo (Ed.), *The skin senses* (pp. 139–159). Springfield, IL: Thomas.

Verrillo, R. T. (1980). Age related changes in the sensitivity to vibration. *Journal of Gerontology, 35*, 185–193.

Vierck, C. J., & Jones, M. B. (1969). Size discrimination on the skin. *Science, 163*, 488–489.

Vierordt, K. (1870). Die Abhängigkeit der Ausbildung des Raumsinnes der Haut von den Beweglichkeit der Körpertheile. [The dependence of the development of the spatial sense of the skin on movement of the body parts]. *Zeitschrift für Biologie, 6*, 53–72.

Weber, E. H. (1834). *De tactu*. Leipzig: Koehler.

Weber, E. H. (1846). Der Tastsinn und das Gemeingefühl. [Touch and Common Sensibility]. In R. Wagner (Ed.), *Handwörterbuch der Physiologie* (Vol. 3, pp. 481–588). Braunschweig.

Weber, E. H. (1978). Der Tastsinn. [Touch]. In H. E. Ross & D. J. Murray (Eds.), *E. H. Weber: The sense of touch* (pp. 139–264). New York: Academic Press.

Weddell, G. (1955). Somesthesis and the chemical senses. *Annual Review of Psychology, 6*, 119–136.

Weinstein, S. (1968). Intensive and extensive aspects of tactile sensitivity as a function of body part, sex and laterality. In D. R. Kenshalo (Ed.), *The skin senses* (pp. 195–222). Springfield, IL: Charles C. Thomas.

Wright, G. H. (1951). The latency of sensations of warmth due to radiation. *Journal of Physiology (London), 112*, 344–358.

Zigler, M. J. (1923). An experimental study of the perception of clamminess. *American Journal of Psychology, 34*, 550–561.

Zigler, M. J. (1926). Review of "Der Aufbau der Tastwelt" by D. Katz. *Psychological Bulletin, 23*, 326–336.

Zigler, M. J. (1932). Pressure adaptation-time: A function of intensity and extensity. *American Journal of Psychology, 44*, 709–720.

CHAPTER 2

The Psychophysics of Tactile Perception and Its Peripheral Physiological Basis

Joel D. Greenspan
Stanley J. Bolanowski

I. INTRODUCTION

This chapter examines the fundamental aspects of tactile psychophysics, particularly those research advances made since the previous edition of the *Handbook of Perception* (Kenshalo, 1978). A large component of this chapter is devoted to reviewing the role of mechanoreceptive afferent neurons in mediating the psychophysical phenomena discussed. This has proven to be a particularly fruitful avenue during the last two decades of research. Because this chapter is necessarily restricted in scope, the interested reader is referred to other relevant review articles and collected works on the topic of tactile somesthesis that have appeared since the last Handbook edition (Cholewiak & Collins, 1991; Darian-Smith, 1984; Granzén & Westman, 1991; Gescheider, Bolanowski, & Verrillo, 1992; Gordon, 1978; Greenspan & LaMotte, 1993; Heller & Schiff, 1991; Iggo & Andres, 1982; Johnson and Hsiao, 1992; Keidel, 1984; Kenshalo, 1979; Light & Perl, 1993; Loomis, 1981; Loomis & Lederman, 1986; Lynn, 1983; Mountcastle, 1984; Schiff & Foulke, 1982; Sherrick & Cholewiak, 1986; Vallbo, Hagbarth, Torebjörk and Wallin, 1979; Vallbo and Johansson, 1984; Verrillo & Bolanowski, 1993; von Euler, Franzén, Lindblom, & Ottoson, 1984; Willis & Coggeshall, 1991).

Pain and Touch

Tactile perceptions are derived from information provided by mechan- oreceptors innervating the skin, and in some cases, subcutaneous tissues. Cutaneous mechanoreceptors are defined functionally: those elements of the peripheral nervous system (PNS) that are selectively responsive to nonnox- ious mechanical stimulation of the skin. Strictly speaking, the mechan- oreceptor is the end organ that actually transduces mechanical energy into electrochemical energy, thus forming the neural signal. Often, however, both the transducing element and the associated nerve fiber is referred to as a mechanoreceptor. A large body of data exists concerning the neuro- physiological properties of cutaneous mechanoreceptors and their afferent fibers, only some of which we are able to review here. One of our tasks (both in this chapter and in sensory neuroscience in general) is to discern relationships between the properties of cutaneous structures and aspects of tactile perception. This is predominantly a correlative exercise, and the leap to causal relationship must be taken cautiously.

Several early theories attempted to ascribe physiological mechanisms to somesthetic perceptions, the two most prominent being *specificity theory* and *pattern theory*. Derived from the doctrine of specific nerve energies (Bell, 1869; Müller, 1838), and the pioneering "spot mapping" studies (Blix, 1882, 1884; Donaldson, 1885; Goldscheider, 1884), the theory of receptor speci- ficity proposed that there are distinguishable nerve endings selectively sensi- tive to one of the various forms of energy that impinge on the skin's surface. Furthermore, each specific nerve ending was associated with a specific sen- sation. Max von Frey (1895) elaborated upon this concept by proposing that there were specific receptors for heat (Ruffini endings), for cold (Krause end bulbs—mucocutaneous end organs), for touch (Meissner corpuscles and hair-follicle endings), and for pain (free nerve endings). In contrast to the idea of receptor specificity, pattern theory (Kenshalo & Nafe, 1962; Nafe, 1927; Sinclair, 1955; Weddell, 1955) proffered that there was no neural specificity either at the receptor level or elsewhere, and that the various somesthetic sensations arise from the overall pattern of neural activity reaching the central nervous system (CNS). It is now well accepted that the skin is innervated by sensory nerve fibers possessing distinct and specialized response properties. Furthermore, each fiber type is associated with a morphologically identified end organ. The specialized end organ structure contributes, at least in part, to the physiological properties of the fibers. These facts indicate that specificity of peripheral afferent fibers, rather than the pattern of activity across nonspecific fiber types, forms the basis of somesthetic perceptual qualities. It should be recognized, however, that most tactile perception involves activity across the various afferent fiber types.

Before describing the psychophysical aspects of the sense of touch and the physiological response profiles of mechanoreceptive afferents, we will consider the apparatus that houses the peripheral portion of the tactile system, and that interfaces with the external environment: the body organ that is called the skin.

II. SKIN ANATOMY AND THE PHYSICS OF CUTANEOUS STIMULATION

The skin is a highly complex living structure that incorporates not only the sensory receptors and nerve fibers, but blood vessels, sweat glands, and other specialized components such as hair, hoofs, claws, and nails. Skin is classified into three types, each having its own complement of mechanoreceptors and internal structure: glabrous or hairless skin typified by the skin of the palm; hairy skin, so called because of the presence of hairs; and mucocutaneous skin, which borders the entrances to the body's interior. The diverse structure of these three types of skin imparts differences in their mechanical properties. Furthermore, these mechanical properties (e.g., elasticity, resilience, thickness, attachment to subcutaneous tissues, etc.) can vary substantially as a function of body locus, age, gender, and species (Agache, Monneur, Lévêque, & De Rigal, 1980; Escoffier et al., 1989; Grahame, 1970; Marks, 1983). Because of this variability, stimulus specification and control at the receptor level is more difficult to achieve than for the senses of vision or hearing.

Many different stimulating techniques have been used in the study of taction: vibration stimuli applied perpendicularly or tangentially to the skin by pins or probes, periodic and aperiodic gratings moved across the skin, airpuffs, embossed letters, and even everyday items such as sandpaper, cloth, and steel wool. Each of these stimuli have provided knowledge about tactile perception and its mechanisms, although the precise relationship between stimuli applied to the skin surface and the actual stimulation of the tactile end organs is still an area requiring more investigation. To better understand the problem of stimulus specification and the interpretation of results to be presented later in this chapter, it is necessary to describe the general anatomical organization of skin and its mechanical properties.

The skin is composed of an outer layer (epidermis) and an inner layer (dermis; Figure 1). The dermis separates the epidermis from the underlying muscle, ligaments, and bone. Both the dermis and epidermis are formed by strata having specific characteristics. The epidermis consists of stratified squamous epithelial cells in various stages of differentiation. The outermost layer, stratum corneum, is composed of dead cell tissue of a tough, horny nature. Epidermal cells are made in the lowest layer, the stratum germinativum (or basal layer), and migrate outwardly, eventually reaching the stratum corneum where they die. The epidermis contains no blood supply. In the glabrous skin (Fig. 1A), the boundary between dermis and epidermis

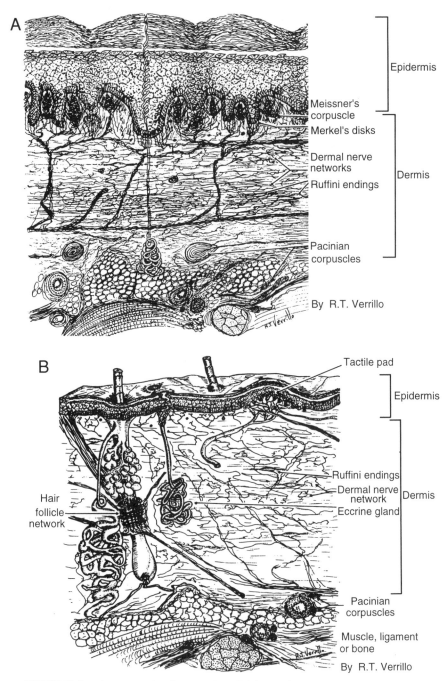

FIGURE 1   A cross-sectional perspective of glabrous skin (A) and hairy skin (B). (Adapted from T. S. Brown, 1975, Figs. 3.36 and 3.37, p. 102, with permission of the artist, R. T. Verrillo.)

is not a smooth plane; rather, there are regions called dermal pegs where the dermis and epidermis interdigitate. The physics of this arrangement appears important in determining mechanoreceptor activation.

The dermis, in contrast, is composed of connective tissue and elastic fibers floating in a semifluid, nonfibrillary, amorphous mixture called the ground substance. Also embedded in the ground substance are fat cells, blood vessels, the sensory receptors and the nerve fibers innervating them, smooth muscle, sweat glands, and a profuse blood and lympathic supply. The dermis is composed of two layers, the papillary layer (prominent in humans) and the reticular layer. The papillary layer is composed of collagenous elastin and reticular fibers in a meshwork of capillaries, all of which is surrounded by ground substance. The reticular layer is composed of coarse collagen fiber bundles arranged parallel to the skin surface and loose networks of elastin fibers between the collagen, blood vessels, and nerve fibers. It also contains connective tissue cells (fibroblasts and mast cells), pigment-bearing cells, melanocytes, macrophages, lymphocytes, and leukocytes.

From a physical standpoint, the skin can be thought of as an incompressible viscoelastic medium possessing the mechanical properties of viscous resistance (friction), stiffness (elastic restoring capabilities), and mass. Because of these properties, the skin has a measurable mechanical impedance, which varies across skin type and body location. As measured with vibratory stimuli, the skin's mechanical impedance decreases (is more easily moved) with increasing frequencies of stimulation to a minimum value, and then increases proportionally for higher frequencies. The minimum can be 80 Hz (at thenar eminence) or 200 Hz (at the fingertip; Lundström, 1984). The elasticity and viscosity of the tissue contribute to the impedance up to 100 Hz, but from 0.5–5 kHz, the viscosity component predominates.

Idealized models of the skin make simplifying assumptions about its properties, such as being a linear, homogeneous, isotropic, and incompressible viscoelastic medium (Phillips & Johnson, 1981b; Van Doren, 1989). Other models, for example the "waterbed" model, considers the skin of the fingertip as an elastic membrane enclosing an incompressible fluid. Such a model has provided better correspondence with in vivo data than other homogeneous models (Srinivasan, 1989; Srinivasan & Dandekar, 1996).

In considering the skin's physical response to mechanical stimuli, one needs to consider stresses (forces) and strains (deformations) occurring internal to the skin as well as those traveling across the skin. Within the skin, two kinds of internal waves are demonstrable: (a) the incompressible shear waves producing shear stresses and strains, and (b) irrotational compression waves producing normal stresses and strains (Oestreicher, 1950). Various types of neurophysiological and psychophysical studies have supported the theory that normal (compressive) strain, either perpendicular or parallel to the skin surface, is the adequate stimulus for transduction in tactile receptors (Greenspan, 1984; Phillips & Johnson, 1981a,b; Srinivasan & Dandekar, 1996; Van Doren, 1989). This is in contrast to the shear strain, or either shear or normal stresses, which are less well correlated with either mechanoreceptor response or tactile perception. The idea that strain is the adequate stimulus is also consistent with the earlier work of Loewenstein (1971) and Ilyinski (1965) who, using isolated Pacinian corpuscles, concluded that the adequate stimulus for mechanotransduction was deformation or strain. One should note that the actual strain transmitted through the skin tissue is strongly dependent on the spatial structure of the stimulus. In contrast, the transmission of strain through the skin is virtually independent of temporal frequency, at least in the range of 1–256 Hz (Van Doren, 1989). Assuming incompressibility of the skin, displacements at the surface of the skin produce deformations within the skin that follow a $1/d^3$ relationship (d = distance from the stimulus source). Thus, the strain is attenuated rapidly as the stimulus wave passes through the skin, and therefore is confined to the immediate vicinity of the stimulus.

   Traveling waves that propagate along the body surface from the point of stimulation are different from internal waves in several respects (see Franke, von Gierke, Oestreicher, & von Wittern, 1951; von Gierke, Oestreicher, Franke, Parrack, & von Wittern, 1952). The skin is a dispersive medium for surface waves, with increases in displacement frequency producing increases in the velocity of propagation of surface waves (Moore, 1970). Additionally, the amplitude of surface waves decay in a manner proportional to the inverse squared of the distance from the vibrating stimulus. This decay is similarly dependent upon stimulus frequency, in which higher frequencies result in greater decays for a given distance from the stimulus (Pubols, 1987). Given these rather complex effects at a distance from the stimulator, it is difficult to fully characterize such stimuli for all potentially activated mechanoreceptors. Some researchers have eliminated the confounding effects of surface waves by using a surround that confines the stimulus to the probe site (see Verrillo, 1963).

   At least one nonmechanical factor that should be taken into account in specifying the stimulus conditions is temperature of the skin. As described by Saxena (1983), the skin and subdermal tissue are greatly involved in homeostatic regulation of the body's core temperature. This is partially accomplished by modifying blood flow through the various skin tissues, and by perspiration. Factors involved in the amount of heat dissipated or absorbed are the specific heat of the tissue, its thermal conductivity, mass flow of the blood, and temperature of the blood. Indeed, it has been shown by Bolanowski and Verrillo (1982) that variation in skin surface temperature can significantly affect vibratory sensitivity, especially for the higher stimulus.

III. MECHANORECEPTOR ACTIVATION
AND TRANSDUCTION

Mechanoreceptors come in a variety of shapes and forms, and may or may not have nonneural elements attached or adjacent to the nerve terminations. The nonneural elements, referred to as accessory structures, are thought to participate in selective filtering of mechanical stimuli (Iggo, 1976), to allow for trophic influences (Loewenstein & Molins, 1958), and to regulate the ionic environment around the nerve ending (Ilyinski, Akeov, Krasnikova, & Elman, 1976). One major class of endings with accessory structures are the encapsulated endings (Pacinian corpuscles, Ruffini capsules and Meissner corpuscles). The capsules of these receptors are organized in specific ways. For example, the Pacinian corpuscle has an onion-like layered arrangement (Chouchkov, 1971) and the Ruffini capsule resembles a small muscle spindle (Halata, 1988). A second class of endings has accessory structures to which they are attached (Merkel cell–neurite complex, Iggo corpuscle, circumferential and palisade fibers on the shafts of hairs). Another major class of endings in the skin are the free nerve endings, which do not have accessory structures. This class of receptors is associated with the transduction of thermal, or noxious (strong mechanical, thermal, and/or chemical), or some types of tactile stimuli (i.e., C-mechanoreceptors).

The mechanoreceptor's neural component typically consists of an unmyelinated nerve terminal continuous with a myelinated nerve fiber, which has its soma in the dorsal root or trigeminal ganglion. The unmyelinated nerve terminal is the region where mechanotransduction takes place, whereas the myelinated nerve fiber is capable of transmitting action potentials. Tactile receptors are called primary receptors (Davis, 1961) because there is no synapse between the transducing portion of the mechanoreceptor and the site of action-potential generation.

The manner in which cutaneously applied mechanical stimuli are transduced into electrochemical energy is incompletely understood. One reason for this lack of knowledge is the receptors' inaccessibility for electrophysiological investigation while embedded in the skin. What is known regarding the transduction mechanisms of cutaneous mechanoreceptors has been derived primarily by studying Pacinian corpuscles isolated from the messentery of the cat (e.g., Bolanowski & Zwislocki, 1984a,b; Loewenstein, 1971). These corpuscles are anatomically (Chouchkov, 1971) and physiologically (Bolanowski & Zwislocki, 1984b) similar to those located within primate skin. It is likely that the mechanisms of transduction identified for the Pacinian corpuscle are relevant to the other tactile receptors (see Bell, Bolanowski, & Holmes, 1994, for a review of the structure and function of Pacinian corpuscles).

When a mechanical stimulus is applied to a Pacinian corpuscle, either through the skin or in vitro, the accessory capsule acts as a mechanical filter (Loewenstein & Skalak, 1966) transmitting the applied strain to the trans-ducing membrane. The strain induces hypothesized stretch-sensitive chan-nels to respond by increasing their conductance to certain ions, Na+ in particular (Diamond, Gray, & Inman, 1958), and possibly K+ (Akoev, Makovsky, & Volpe, 1980). This conductance increase produces the trans-membrane event known as the receptor potential at a very short latency (circa 0.2 msec from stimulus onset; Gray & Sato, 1953; Gottschaldt & Vahle-Hinz, 1981). Based on this short latency, the receptor potential can-not be mediated by neurotransmitters or other neuroactive substances po-tentially released by the surrounding accessory structure. If the receptor potential is of sufficient amplitude, an action potential will be generated and propagated along the peripheral axon to presynaptic endings within the CNS. The site of receptor-potential generation is likely to be different from the site of action-potential initiation (Gray & Sato, 1953). Specifically, the receptor potential arises in the unmyelinated portion of the neurite, and the action potential probably originates near the unmyelinated–myelinated juncture, perhaps at the most distal node of Ranvier (Loewenstein & Rath-kamp, 1958a,b). The specific location of the mechanoreceptive channels responsible for the receptor potential may be in the filopodia that extend from the unmyelinated nerve terminal (Bolanowski, Schyuler, & Slepecky, 1994; Bolanowski, Schyuler, Sulitka, & Pietras, 1996; Chouchkov, 1971; Nishi, Oura, & Pallie, 1969; Spencer & Schaumberg, 1973). Filopodia are present on many dermal and epidermal endings (Chouchkov, 1974) includ-ing Ruffini capsules (Gottschaldt, Fruhstorfer, Schmidt, & Kraft, 1982). However, other theories regarding the locus of transduction as it relates to the shape of the terminal neurite have been proposed (Ilyinski, Krylov, & Cherepnov, 1976).

In response to a ramp-and-hold stimulus, the Pacinian corpuscle's recep-tor potential first rises rapidly, and then decays (Loewenstein, 1959; Gray & Sato, 1953). This phenomenon, commonly called adaptation, is the basis for the rapidly adapting properties of the aciton-potential responses seen when recording from fibers innervating Pacinian corpuscles (see below). The rela-tionship between stimulus amplitude and the magnitude of the receptor potential is linear at low-stimulus amplitudes and saturates at higher inten-sities (Loewenstein & Altimirano-Orrego, 1958; Nishi & Sato, 1968). The greater the magnitude of the receptor potential, the greater the action-potential firing rate. The mechanism for increased receptor potential magni-tude with increased stimulus amplitude may involve a spatial sequestering of responses from discrete transduction sites operating in an all-or-none fashion (Loewenstein, 1959). These transduction sites may be the filopodia and/or the putative strain-sensitive channels on the nerve ending. The use of sinusoidal stimulation has revealed that the Pacinian corpuscle, at high stimulus amplitudes, can produce a receptor potential that shows frequency doubling (Bolanowski & Zwislocki, 1984b). This frequency doubling is the basis for the observation that, at sufficiently high-stimulus amplitudes, tactile mechanoreceptors can produce two action potentials for every cycle of vibratory stimulation (Bolanowski & Zwislocki, 1984a; Talbot, Darian-Smith, Kornhuber, & Mountcastle, 1968). Lastly, the frequency characteristics of the receptor potential of isolated Pacinian corpuscles is U-shaped, much like the frequency characteristics of action-potential firing rates recorded from the Pacinian's nerve fiber (Bolanowski & Zwislocki, 1984b). Thus the frequency selectivity of the Pacinian corpuscle's nerve fiber is largely determined at the level of transduction.

IV. MECHANORECEPTORS INNERVATING THE GLABROUS SKIN OF THE HAND

Electrophysiological studies have consistently identified four types of mechanoreceptive afferent fibers innervating the glabrous skin of the human hand (Fig. 2 and Table 1; Johansson, 1976; Johansson, 1978; Johansson & Vallbo, 1980; Johansson, Vallbo, & Westling, 1980; Knibestöl and Vallbo, 1970; Vallbo and Johansson, 1978; Vallbo & Johansson, 1984; Vallbo et al., 1979). There are two principle features that define these four types: adaptational properties and receptive field size. Two of these afferent types—SAI and SAII—are slowly adapting, whereas the other two—FAI (or RA) and FAII (or PC)—are fast (or rapidly) adapting.[1] The "adaptation" in these cases refers to how these afferent fibers respond to a sustained skin indentation. FA afferents respond while the skin is actively being indented, but do not generate action potentials when movement of the skin stops, even if the skin is still indented and under considerable force. SA afferents, in contrast, respond both while the skin is moving and during the period of sustained indentation. With a steadily maintained indentation, the discharge rate will slowly decrease in frequency over many seconds or minutes.[2]

The receptive field (RF) of a mechanoreceptor is the area of skin that, when

[1] Human neurophysiological reports generally use the FAI and FAII nomenclature, whereas animal neurophysiological reports use RA and PC. There is no doubt that FAI and RA refer to the same type of mechanoreceptive afferent; the same is true for FAII and PC.

[2] Despite the prevalent use of this terminology, the "fast adapting" mechanoreceptive afferents do not truly adapt. Rather, they are continuously responsive to dynamic stimuli, and exhibit little adaptation to suitable forms of stimulation, such as vibration. With this in mind, another classification scheme divided mechanoreceptive afferents into "detectors" of position, velocity, or transients (Burgess, 1973; Burgess & Perl, 1973). Although these terms are technically more accurate, the authors feel that it is overrestrictive and potentially misleading to characterize these sensory neurons as "feature detectors" per se.

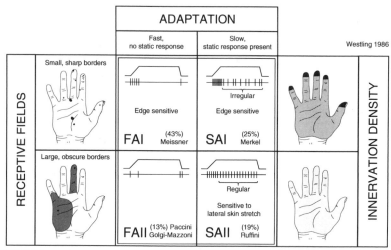

FIGURE 2   Characteristics of the four types of mechanoreceptive afferents that innervate the glabrous skin of the human hand. The center four panels schematically show the action potential discharges (lower traces) in response to a ramp-and-hold indentation of the skin (upper traces). The center boxes also show the relative frequencies of occurrence of the four afferent types, and the most likely morphological correlate. The black dots and shaded areas of the hand figures on the left represent typical receptive fields of type I (top) and type II (bottom) afferents. The hand drawings on the right depict the average density of type I (top) and type II (bottom) afferents, in which darker areas indicate higher densities. (Reproduced with permission from Westling, 1986, Fig. 1, p. 5.)

stimulated, will generate a response in that sensory neuron. The precise boundary of this area will depend upon the intensity of the stimulus used. But with any specific stimulus, the SAI and FAI afferents will be activated within a much smaller area of skin than the SAII and FAII afferents, which is to say that the type I fibers have smaller RFs than the type II fibers.

Although these two properties (adaptation and RF size) are the defining characteristics of these afferent types, there are other distinguishing features. Each of these four afferent types has been identified with a specialized structure within the skin: the FAI (RA) afferent with the Meissner corpuscle (Cauna, 1956; Jänig, 1971; Munger, Page, & Pubols, 1979), the FAII (PC) with the Pacinian corpuscle (Gray & Matthews, 1951; Gray & Sato, 1953; Loewenstein, 1958; Sato, 1961; for review, see Bell et al., 1994), the SAI with the Merkel cell–neurite complex (Iggo, 1962; Iggo & Muir, 1969; Munger et al., 1979), and the SAII with the Ruffini ending (Chambers, Andres, von Düring, & Iggo, 1972).[3] Another noteworthy aspect is the

---

[3]These structure–function relationships are primarily correlative in nature. Furthermore, there are various degrees of evidence supporting these correlations for each of the four mechanoreceptor types.

TABLE 1 Properties of Glabrous Skin Mechanoreceptive Afferents[a]

Physiological type	FAI (RA)	FAII (PC)	SAI	SAII
"Adaptational" property	Rapid	Rapid	Slow	Slow
Receptive field size	Small	Large	Small	Large
Putative receptor structure	Meissner	Pacinian	Merkel	Ruffini
Fiber innervation density (human; per cm^2)[b]				
Fingertip	140	21	70	49
Distal phalanx (other than tip)	77	15	42	28
Proximal and middle phalanx	37	10	30	14
Thenar eminence	25	9	8	16
Fiber innervation density (monkey; per cm^2)[c]				
Distal phalanx and fingertip	178	13	134	—
Middle phalanx	80	13	46	—
No. of axons per specialized structural ending	2–9[d]	1[e]	1–2[f]	1[g]
No. of specialized structural endings per axon	3–4 branches[d]	1–2[e]	1 or more[f]	> 1[g]
Conduction velocity (m/s)[h]	26–91	34–61	32–82	34–55
Sensations produced by INMS[i]	Tap, flutter	Vibration	Pressure	None

[a]FA, fast adapting; RA, rapidly adapting; PC, Pacinian corpuscle; SA, slowly adapting.

[b]Johansson & Vallbo, 1979.

[c]Darian-Smith & Kenins, 1980.

[d]Cauna, 1956; Halata, 1975.

[e]Quilliam & Sato, 1955; Cauna & Mannan, 1958.

[f]Iggo, 1962; Iggo & Muir, 1969; Jänig, 1971; Tapper, 1965.

[g]Goglia & Sklenska, 1969; Chambers, Andres, von Düring, & Iggo, 1972.

[h]Light & Perl, 1993.

[i]INMS, Intraneuronal microstimulation; Ochoa & Torebjörk, 1983; Vallbo, Olsson, Westburg, & Clark, 1984.

innervation density of these receptors within the hand. The Type I afferents (FAI and SAI) exhibit a large gradient of innervation: The greatest density is found in the fingertips, becoming progressively more sparse in the proximal direction. The Type II afferents show only a slight difference in density across the hand (Johansson & Vallbo, 1979). In the absence of mechanical cutaneous stimulation, the FAIs, FAIIs, and SAIs do not exhibit "spontaneous activity," whereas SAIIs usually do.

Based on these same criteria, investigators have consistently observed similar types of afferent fibers for other mammalian species, including rat, cat, opossum, raccoon, and monkey. Several studies have described three types of afferents in the glabrous skin of the monkey hand: RA, PC, and SA fibers. The SA type resembles the SAI found in humans. Sensory afferents resembling SAIIs have not been reliably identified in the monkey glabrous skin. Physiological data derived from such animal models have provided many useful points of reference for comparing human tactile perception to underlying physiological mechanisms. Several hypotheses regarding these psychophysiological relationships have been proposed, and they will be described in pertinent sections of this chapter.

Several interesting and important observations regarding the relationship between fiber type and tactile perception have been made by use of intra-neuronal microstimulation (INMS; Macefield, Gandevia, & Burke, 1990; Ochoa & Torebjörk, 1983; Schady, Torebjörk, & Ochoa, 1983; Torebjörk & Ochoa, 1980; Torebjörk, Vallbo, & Ochoa, 1987; Vallbo, 1981; Vallbo, Olsson, Westberg, & Clark, 1984; Table 1). In these studies, recording of single-unit activity (e.g., action potentials generated by a single axon) in human peripheral nerve is complemented with electrical stimulation through the same electrode, in order to selectively excite that individual axon. In most instances, the perceptions evoked from such stimulation (at near threshold intensity) are of a distinct quality, and are localized on the skin to an area often coincident with the receptive field of the recorded unit (e.g., the "perceptive field"). Several criteria (described most extensively in Ochoa & Torebjörk, 1983, and Torebjörk et al., 1987) provide confidence in inferring conditions of individual axon stimulation during these experiments.

When stimulating individual FAI fibers, people typically report an inter-mittent tapping sensation. For FAI fibers with RFs on the fingers, one electrical pulse (producing only one action potential in the fiber) is often sufficient to produce a sensation. Those FAI fibers with RFs on the palm require repetitive stimulus pulses (thus generating a train of action potentials) in order to evoke a sensation. At lower frequencies of continuous stimulation (<100 pulse/sec), people perceive repetitive taps matching the stimulus frequency. At higher frequencies of stimulation, people report no difference in perceived frequency, but rather an increase in perceived intensity of tactile sensation. Thus, individual FAIs appear to play a role in both frequency and intensity coding.

When stimulating an individual FAII fiber, a single impulse is insufficient to produce a perceptual experience. Rather, FAIIs require stimulation of 10–80 pulses/sec to evoke a sensation, which is uniformly described as a vibration. Increasing stimulus frequency up to 300 pulses/sec changes perceived frequency, but does not alter the quality of the sensation. This implies that any intensity coding in FAII fibers requires a population response.

Stimulation of an SAI fiber produces a sensation of sustained pressure, without a sense of contact per se, and without a sense of intermittence. Stimulus frequencies of 3–10 pulses/sec are necessary to elicit a sensation, and higher frequencies evoke more intense pressure sensations with no change in sensory quality.

Electrical stimulation of a SAII fiber rarely produces any sensation at all. When sensations are produced, they usually lack distinct, unitary qualities, suggesting that multiple axons are excited at that point. (But, see Macefield et al., 1990, for exceptions to this generalization.) SAII afferents are usually found to have ongoing activity without intentional cutaneous stimulation (5–20 spikes/sec; Ochoa & Torebjörk, 1983).

V. PSYCHOPHYSICAL CHANNELS AND VIBROTACTILE THRESHOLDS IN GLABROUS SKIN

Using vibratory (sinusoidal) stimuli in which the amplitude and frequency of displacements can be independently controlled and precisely specified, several investigators determined sensory threshold at various frequencies for the glabrous skin of the human hand (Hugony, 1935; Setzepfand, 1935; Sherrick, 1953; Verrillo, 1962; von Békésy, 1939; von Gilmer, 1935). Verrillo (1963, 1966a–c) noted that the threshold-frequency relationship contained two visibly distinct portions: a frequency-independent portion extending from 20–40 Hz and a higher frequency portion (40–700 Hz), which was strongly frequency dependent, having a maximum sensitivity around 250 Hz. This observation prompted him to propose the duplex theory of taction, in which two separate and independent channels combined to mediate the sense of touch. Verrillo (1966c) suggested that the high-frequency portion of the threshold-frequency characteristic was mediated by Pacinian corpuscles, based on Sato's (1961) physiological recordings from axons innervating Pacinian corpuscles. Verrillo defined this portion of the characteristic as the P (Pacinian) channel but left the low-frequency channel unspecified as to its physiological substrate (i.e., NP or non-Pacinian). Mountcastle and colleagues conducted psychophysical experiments with monkeys and human beings along with neurophysiological experiments on monkeys (LaMotte & Mountcastle, 1975; Mountcastle, 1967; Mountcastle, Talbot, & Kornhuber, 1966; Mountcastle, LaMotte, & Carli, 1972; Talbot et al., 1968). They showed that the low-frequency portion of the threshold-frequency characteristic matched the threshold responses of RA fibers, whereas the high-frequency portion paralleled PC afferent thresholds. They also described different sensory qualities associated with the two stimulus ranges: low frequencies produced a flutter sensation; high frequencies evoked the sensation of vibration.

Since that time many psychophysical, physiological, and anatomical experiments have been performed to determine additional relationships among the four physiologically defined fiber types in glabrous skin to the psychophysically defined threshold-frequency characteristic. Based on these studies, Bolanowski, Gescheider, Verrillo, and Checkosky (1988) proposed that there are four distinct psychophysical channels contributing to taction in the glabrous skin, each channel being mediated by a specific fiber type (see Figure 3). The threshold-frequency characteristics of the four channels were determined by means of manipulating stimulus parameters, such as the frequency and amplitude of the vibration (Gescheider, 1976; Verrillo, 1966a, 1968), probe or contactor size (Verrillo, 1963), stimulus duration (Checkosky & Bolanowski, 1992; Verillo, 1965), skin-surface temperature (Bolanowski & Verrillo, 1982; Verrillo & Bolanowski, 1986), and the application of various masking techniques (Capraro, Verrillo, Zwislocki, 1979; Gescheider, Frisina, & Verrillo, 1979; Gescheider, O'Malley, & Verrillo, 1983; Gescheider, Sklar, Van Doren, & Verrillo, 1985; Gescheider, Verrillo, & Van Doren, 1982; Hamer, Verrillo, & Zwislocki, 1983; Makous, Fried-

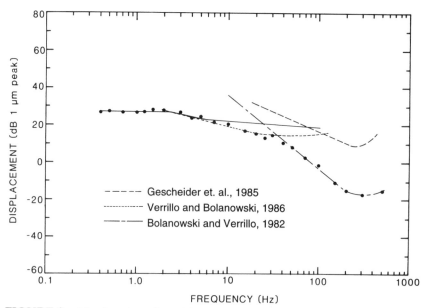

FIGURE 3   The four-channel model of vibrotaction showing the threshold-frequency characteristics of the various channels: — - —, PC; - - -, NPI; — — —, NPII; and ———, NPIII. Data points depict the psychophysically derived absolute threshold for detection, as determined for the thenar eminence using a 2.9-cm² stimulus area. (Reproduced with permission from Bolanowski et al., 1988, Fig. 8, p. 1691.)

man, & Vierck, 1995; Verrillo & Gescheider, 1977). The results of much of that work has been summarized in Table 2.

The model proposes that each psychophysical channel consists of specific end organs innervated by select groups of peripheral nerve fibers, the isolated activation of which can produce unique "unitary" sensations. Two major tenets of the model are that stimuli at threshold levels are signaled by the channel that is the most sensitive, and that suprathreshold sensations are the result of a combination of the neural activity transmitted by all activated channels. Not all of the four channels need be activated for a suprathreshold sensory experience. Indeed, for certain suprathreshold conditions only one or two channels will be activated (Bolanowski, Gescheider, & Verrillo, 1992a,b). In that instance a specific tactile sensation will result, albeit of a quality different from that experienced if all four channels are activated. Another important point is that the model does not intrinsically require the existence of four channels. The model simply proposes that the sense of touch utilizes information being transmitted by separate, independent channels. Of course, integration of the activity arriving over the different channels must ultimately be combined to provide a unified perception. The model proposes that the psychophysically defined channels, their end organ–nerve-fiber substrates, their respective unitary sensations, and the nominal frequency range over which they operate at threshold levels are (a) the channel, mediated by Pacinian corpuscles and PC fibers, producing the sensation of "vibration" in the frequency range of 40–500 Hz; (b) non-Pacinian I (NPI), mediated by Meissner corpuscles and RA fibers, producing the sensation of "flutter" in the frequency range of 2–40 Hz; (c) NPII, mediated by Ruffini end organs and SA Type II fibers (SAII), producing a buzz-like sensation in the frequency range of 100–500 Hz; and (d) NPIII, mediated by Merkel cell-neurite complexes and SA Type I fibers (SA I), producing the sensation of "pressure" in the frequency range of 0.4–2.0 Hz.

Generally, the results of mechanoreceptive neurophysiological studies have dovetailed nicely with the results of psychophysical studies as to the roles that specific mechanoreceptors play in tactile perception. There is a discrepancy, however, between the human microneurography results and vibrotactile psychophysics regarding the perceptual contribution of the SAII fibers. Although SAII activity has been correlated with the psychophysically determined NPII channel, electrical stimulation of individual SAII fibers often fails to evoke sensations. Bolanowski et al. (1988) have argued that unless the specific code for sensation is produced, the CNS may not be able to interpret activity in the artificially activated channel. It is possible that the activity produced by regular trains of electrical pulses may not provide for the adequate code for SAII fibers. Additionally, it is possible that more than one SAII fiber may need to be activated to elicit a perceptual experience.

TABLE 2   Four-Channel Model of Mechanoreception[a,b]

Psychophysical characteristics of channels	P channel[c,d,e,f]	NP I channel[c,e,f,h]	NP II channel[c,e,f,h]	NP III channel[c]
Sensation quality	"Vibration"	"Flutter"	Unknown	"Pressure"
Frequencies over which threshold defined				
Small contactor (0.008 cm²)	None	3–100 Hz	80–500 Hz	.4–3 Hz
Large contactor (2.9 cm²)	35–500 Hz	3–35 Hz	None	.4–3 Hz
Shape of frequency response function	U-shape	Flat (slight dip at 30 Hz)	U-shape	Flat
Slope of "tuning curve"	−12 dB/oct	−5.0 dB/oct	−5.5 dB/oct	Unknown
Best frequency (bf) varies with temperature	Yes	No	Yes	No
Low temperature (15°–25°C)	bf: 150–250 Hz	—	bf: 100–150 Hz	—
High temperature (30°–40°C)	bf: 250–400 Hz	—	bf: 225–275 Hz	—
Temperature sensitivity	Yes	Slight	Yes	Slight
Temporal summation	Yes	No	No[i]	Indeterminate
Spatial summation	Yes	No	Unknown	No
Putative mechanoreceptive afferent type	FAII(PC)	FAI(RA)	SAII	SAI
Psychophysical procedures				
To lower threshold				
Best frequency	250–300 Hz	25–40 Hz	150–400 Hz	.4–1 Hz
Surround presence	No	Yes	Unknown	Yes
Contactor size	Large (>2.9 cm²)	Any	Very small (<0.02 cm²)	Any

Skin temperature	>32°C	15–40°C	>32°C	15–40°C
Stimulus duration	>500 ms	> one cycle	>500 ms	> one cycle
To elevate threshold				
Frequency of masker/adapter	250–300 Hz	25–40 Hz	150–400 Hz	.4–10 Hz
Surround presence	Yes	No	Unknown	No
Contactor size	Very small (<0.02 cm^2)	Any	Unknown	Any
Skin temperature	Cold (15°C)	Any	Cold (15°C)	Any
Stimulus duration	<200 ms	Any	<200 ms	Any

[a]The information in this table is limited to the palm of the hand and is based on the studies referred to by letter. (Adapted from Table 2.2 in Cholewiak & Collins, 1991.)

[b]P, Pacinian; NP, non-Pacinian.

[c]Bolanowski et al., 1988.

[d]Bolanowski & Verrillo, 1982.

[e]Capraro, Verrillo, & Zwislocki, 1979.

[f]Gescheider et al., 1985.

[g]Verrillo, 1968.

[h]Verrillo & Bolanowski, 1986.

[i]Gescheider et al. (1985) described an effect of stimulus duration on threshold; however a subsequent study (Gescheider et al., 1994) found the effect due to the noise masker rather than temporal integration.

VI. MECHANORECEPTORS IN HAIRY SKIN

There are relatively few physiological studies of human hairy skin afferent fibers. Thus, much of the relevant neurophysiological data come from animal experiments, particularly from the feline. In general, one can find mechanoreceptive afferents in the hairy skin that are analogous to those in glabrous skin. Accordingly, sensory fibers with properties like SAI, SAII, RA, and PC afferents have been described in the hairy and facial skin of several mammalian species (including humans; Edin & Abbs, 1991; Edin et al., 1995; Essick & Edin, 1995). However, this scheme is complicated by two factors. The first is that some afferents do not neatly fit into one of these four categories. One example from human data involves the SA types, in which the distinction between SAI and SAII is not as clear as in glabrous skin, and some units fall into some "in-between" category (Edin & Abbs, 1991; Järvilehto, Hämäläinen, & Soininen, 1981). Another complicating factor is that RA-like afferents in hairy skin do not fit in a single, homogeneous category. Precisely how many subdivisions of RA-like afferents exist in hairy skin is not settled. The most elaborate categorization of (feline) cutaneous mechanoreceptive afferents was derived by Burgess, Horch, and colleagues (Burgess, Petit, & Warren, 1968; Horch, Tuckett, & Burgess, 1977). There is a distinction between hair follicle afferents and field afferents, the former being more sensitive to hair movement, and the latter being more sensitive to skin indentation. In the densely furred skin of most mammals, making this distinction requires delicate stimulation under microscopic viewing conditions. In humans, it is easier to identify rapidly adapting afferents that are excited independent of hair movement, thus defining field afferents (Järvilehto et al., 1981; Vallbo et al., 1995).

The hair follicle afferents have been subdivided in two ways. One consideration is with respect to the hair type associated with the afferent, which in cats and rabbits can be either tylotrich, guard, or down hairs (Brown & Iggo, 1967; Iggo, 1966). Tylotrich and guard hair afferents are very different from down hair afferents, in that down hair afferents are associated with slowly conducting myelinated (Aδ) fibers, whereas the others have rapidly conducting (Aβ) fibers. It is not apparent to what extent this subclassification of hair follicle afferents is relevant to human sensory innervation. Some human recording studies have identified low-threshold mechanoreceptive afferents in hairy skin with conduction velocities in the Aδ range (Adriaensen, Gybels, Handwerker, & Van Hees, 1983; Järvilehto et al., 1981; Konietzny & Hensel, 1977). Furthermore, when a compressive nerve block is sufficient to eliminate tactile perception (but not thermal perception) in glabrous skin, it is still possible to evoke tactile sensations in hairy skin, presumably mediated by Aδ or C-fiber afferents (Mackenzie, Burke, Skuse,

& Lethlean, 1975). It is not known what perceptual consequences are related to these conduction velocity differences (Kakuda, 1992).[4]

At another level, the guard hair and field afferents can be further subdivided. Originally, distinctions were made between guard hair type 1 (G_1) and guard hair type 2 (G_2) afferents, the former requiring a more rapid skin indentation to be activated (Burgess et al., 1968). Similarly, the field afferents were divided into F_1 and F_2. Subsequently, some afferents did not fit neatly into these divisions, and so an intermediate category was designated (G_{int} and F_{int}; Burgess & Perl, 1973; Horch et al., 1977). Some question exists as to whether these guard hair and field afferent subdivisions represent distinct categories, or just a continuum of functional properties. In either event, studies (in felines) using well-controlled tactile stimuli have shown significant variations in the stimulus–response profiles of both guard hair and field afferents, such that neither group can be considered a homogeneous functional class (Greenspan, 1992; Ray, Mallach, & Kruger, 1985; Tuckett, Horch, & Burgess, 1978). Recently, Essick and Edin (1995) reported an extensive and sophisticated analysis of human hairy skin and mechanoreceptor responses to well-controlled stimuli. Their results, which were very similar to results from feline experiments, support the idea that although any individual RA mechanoreceptor exhibits very consistent stimulus–response functions, any sample of such units exhibits considerable variation in the stimulus velocity response profiles.

The relationship between hairy skin afferent fibers and tactile perception is still largely unknown. One unexpected finding was that SAI afferents in (primate) hairy skin are responsive to levels of stimulation that are below (human) perceptual threshold (Harrington & Merzenich, 1970). Furthermore, considerable activity can be maintained in human hairy skin SAIs without evoking a perception (Järvilehto, Hämäläinen, & Laurinen, 1976). This is markedly different from the situation in glabrous skin, in which SAI activity is well correlated with perceived pressure. A recent study defining psychophysical channels in the hairy skin has distinguished three channels, tentatively associated with PC, RA, and SAII mechanoreceptor types (Bolanowski, Gescheider, & Verrillo, 1994).

[4]Recently, mechanoreceptive afferents associated with unmyelinated axons have been described in human nerve recordings (Nordin, 1990; Vallbo, Olausson, Wessberg, & Norrsell, 1993). While such C-mechanoreceptors have been known to exist in cats and monkeys for several decades (Bessou, Burgess, Perl, & Taylor, 1971; Douglas & Ritchie, 1957; Iggo, 1960; Iggo & Kornhuber, 1977; Kumazawa & Perl, 1977; Zotterman, 1939), their presence in humans has only recently been demonstrated. These afferents have been proposed to mediate the perception of tickle, because they are most responsive to slow, light tactile movement. However, one study reported that pressure block of the A-fibers of the radial nerve (while preserving C-fiber function) prevented the perception of tickle (Hallin & Torebjörk, 1976).

VII. INTENSIVE DIMENSIONS OF TACTILE PERCEPTION

The perceived magnitude of tactile sensations is related in an orderly way to the physical intensity of the stimulus applied to the skin. The precise relation between the perceived magnitude and the intensity of the stimulus, however, has been the object of study since the time of Ernst H. Weber (1795–1878).[5] Many of Weber's observations led to the idea, now called "Weber's Law," that there is a linear relationship between the intensity of a stimulus and the incremental increase of a stimulus from that level to obtain a "just noticeable difference" (JND; also referred to as the DL or *differenz limen*) in stimulus intensity. Stated another way, the change in stimulus intensity required to discern a difference in intensity is a constant proportion of the original intensity.

One obvious factor that determines the value of the Weber fraction is the manner in which stimulus intensity is measured. Classic work by von Frey and Kiesow (1899; as described in Boring, 1942) used controlled forces, and concluded that the appropriate metric for pressure sensation was "tension" (g/mm), which they defined as force divided by the perimeter of the contacting surface. In other words, there was an even trade-off between force applied and the linear dimension of the probe, resulting in the same (threshold) perception of pressure. This relationship was only found for rather small stimuli (stimulation of one or few touch spots), and does not hold for larger stimuli (Greenspan & McGillis, 1991). As mentioned earlier with reference to skin mechanics, recent studies have indicated that metrics related to strain (displacement) appear to be more relevant than those related to stress (force). One study compared psychophysical intensity functions using controlled forces versus controlled depth of skin indentation. Power functions derived with controlled skin indentation were consistently better fitting (e.g., less residual variance) than functions derived with controlled force (Greenspan, 1984).

Early work related to tactile intensity evaluated the difference thresholds for weights or pressures on the body. Very often, the stimuli were intense enough to necessarily activate subcutaneous mechanoreceptors, and thus did not specifically elicit cutaneous perceptions. Kiesow (1922) and Gatti and Dodge (1929) used von Frey hairs to evaluate cutaneous perception for small spots on the palm. They calculated Weber fractions for a large range of stimulus intensities (calculated as tension), and found remarkably consistent results (Figure 4). Their Weber fractions were not entirely constant across all values, but increased dramatically at the two intensity extremes. More recent studies assessing the DL to vibratory stimuli have found values rang-

---

[5]An English translation of his most influential work—originally published in 1834 and 1846—appears in Ross and Murray, 1978.

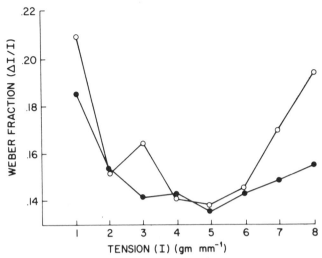

FIGURE 4   A comparison of Weber fractions for single skin indentations on the palm. Data from Kiesow, 1933 (open circles) and from Gatti and Dodge, 1929 (closed circles). (Drawn from data presented in Table 12 of Boring, 1942. Reproduced with permission from Kenshalo, 1978, Fig. 3, p. 38.)

ing from 0.05 (0.4 dB[6]; Knudsen, 1928) to 0.3 (2.3 dB; Sherrick, 1950), with many values in between (Craig, 1972, 1974; Fucci, Small, & Petrosino, 1982; Goble & Hollins, 1993; Schiller, 1953). Actually a "near miss" to Weber's Law was described for vibratory stimuli applied to the thenar eminence, the approximate DL equal to about 1.2 (1.5 dB; Gescheider, Bolanowski, Verrillo, Arpajian, & Ryan, 1990). Differences in methodology certainly contribute to the wide range of DLs obtained.

A. Perceived Intensity of Vibratory Stimulation

Georg von Békésy was one of the first to apply vibratory stimuli to the surface of the skin in attempting to answer a number of questions derived from the field of audition (von Békésy, 1955, 1957, 1958, 1959). He showed that the magnitude of tactile sensations increases more rapidly for those regions of the skin having a lower density of neural elements than for those regions with higher densities. It is now known that the rate of rise of sensation magnitude is inversely related to the level of sensitivity, regardless of how or why the sensitivity is manipulated (Verrillo, 1974; Verillo & Capraro, 1975; Verrillo & Chamberlain, 1972).

[6]Decibel (dB) is the most common amplitude measure for vibrotactile stimuli. It is defined as: $dB = 20 \log(D_1/D_{ref})$, in which D_1 is the skin displacement produced by the stimulus, and D_{ref} is a reference value, usually 1.0 μm peak.

S. S. Stevens (1968), using the procedure of magnitude estimation, measured the perceived magnitude of vibratory stimuli applied to the fingertip. He found that a power function with an average exponent of 0.56 described the stimulus–response relationship quite well. Other studies have further examined the psychophysical relationship for vibrotaction (Bolanowski, Zwislocki, & Gescheider, 1991; Franzén, 1966; Sherrick, 1960; Verrillo, 1974, 1979, 1991; Verrillo & Capraro, 1975; Verrillo & Chamberlain, 1972; Verrillo, Fraioli, & Smith, 1969). Many of these studies demonstrated that a single power-function exponent does not hold over the entire range to which the tactile system can respond. The relationship found for sinusoidal stimuli, and for ramp-and hold stimuli (Greenspan, 1984; Greenspan, Kenshalo, & Henderson, 1984), is more accurately described as a two-limbed function. With vibratory stimulation of the thenar eminence, the upper

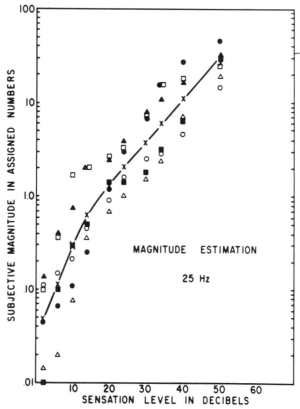

FIGURE 5   Numerical magnitude estimation of subjective vibration intensity as a function of sensation (stimulus) level. Symbols represent individual raw data. A solid line has been drawn through the geometric means. (From Verrillo et al., 1969, Fig. 2, p. 368. Reprinted with permission of the Psychonomic Society, Inc.)

limb of the magnitude estimation function has an exponent of approximately 0.5 in terms of stimulus energy (or 1.0 in terms of stimulus amplitude; Figure 5). The lower limb has an exponent of about 1.0 in terms of stimulus energy. That is, near threshold, the functions become directly proportional to stimulus energy (see Hellman & Zwislocki, 1961), thus affecting the analytic expression of the subjective magnitude function. Indeed, the presence of temporal summation requires that the functions become steeper (Zwislocki, 1960, 1965; Zwislocki, Adams, & Kletsky, 1970) for stimuli that at threshold preferentially activate the P channel (Checkosky & Bolanowski, 1994; Verrillo, 1965).

The exact form that the power functions take as obtained in ratio-scaling tasks, such as magnitude estimation, are somewhat different depending upon body locus (e.g., fingertip, thenar eminence vs. forearm; Verrillo & Capraro, 1975; Verrillo & Chamberlain, 1972), neural density and threshold level (Verrillo, 1974; Verrillo & Capraro, 1975; Verrillo & Chamberlain, 1972), and stimulus conditions such as contactor size (Verrillo, 1974; Verrillo & Capraro, 1975), stimulus frequency (Franzén, 1969; S. S. Stevens, 1968; Verrillo et al., 1969; Verrillo & Capraro, 1975; Verrillo & Chamberlain, 1972), and stimulus duration (Verrillo & Smith, 1976). For example, when the stimulus area is smaller, the sensation magnitude functions have steeper slopes (Verrillo, 1974). It was Fechner's hypothesis that the size of the intensity DL is inversely related to the slope of the sensation magnitude function. Surprisingly, Gescheider, Bolanowski, Zwislocki, Hall, and Mascia (1994) recently showed that the DL is not necessarily related to the sensation magnitude function slope for vibrotactile perception.

In other research, gender studies (Verrillo, 1979) have shown that although threshold sensitivities were the same between men and women, regardless of stimulus frequency, women perceive suprathreshold vibratory stimuli as more intense than do men. Lastly, the effect of age on sensation magnitude has been shown to result in an overall shift to lesser sensation levels for fixed stimulus intensities (Verrillo, 1982). This observation indicates that thresholds, too, are affected by aging, as shown for all four psychophysical channels by Gescheider, Bolanowski, Hall, Hoffman, and Verrillo (1994).

Another aspect of suprathreshold vibrotactile perception deals with equal-sensation magnitudes across vibratory frequencies. Goff (1967) and S. S. Stevens (1968) both performed experiments in which equal subjective-intensity curves were obtained for vibratory stimuli, analogous to the equal-loudness contours for hearing obtained by Fletcher and Munson (1933). Goff used the method of matching suprathreshold stimuli of different frequencies, whereas Stevens used a variant of magnitude estimation. A more thorough study used both magnitude estimation and matching techniques (Verrillo et al., 1969). The suprathreshold equisensation contours resembled

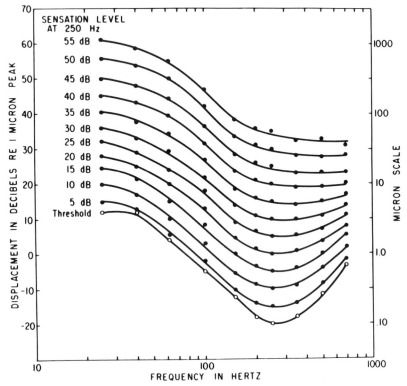

FIGURE 6   Contours of equal sensation magnitude for vibration derived from data obtained by the method of numerical magnitude balance. (From Verrillo et al., 1969, Fig. 8, p. 371. Reprinted with permission from the Psychonomic Society.)

the threshold frequency characteristic below 25 Hz, but as the sensation level increased, the high-frequency portion flattened (Figure 6). This is presumably due to the different receptor populations and channels contributing to sensation at the higher intensities.

B. Perceived Intensity of Nonvibratory Forms of Stimulation and Correlated Mechanoreceptive Afferent Response Properties

The most common forms of nonsinusoidal stimuli used to evaluate tactile sensory intensity have been trapezoidal ("ramp and hold") or triangular waveforms with a constant rate of change. The general result is that higher amplitude stimuli produce more intense tactile sensations. However, detailed examination of this relationship reveals a great deal of complexity.

Early magnitude-estimation studies using such stimuli indicated that the stimulus–response intensity relationship for the hand was a nearly linear function, being adequately described by a power function with an exponent close to 1.0 (F. N. Jones, 1960; Mountcastle, 1967). Another evaluation using controlled skin indentation reported power functions with exponents averaging 0.9 for glabrous skin (digit) and 0.4 for hairy skin (distal forearm; Harrington & Merzenich, 1970). These and related studies also examined the stimulus–response functions of slowly adapting mechanoreceptive afferents (of the monkey) to these same stimuli (Harrington & Merzenich, 1970; Mountcastle et al., 1966; Werner & Mountcastle, 1965). Considering either the total number of action potentials per stimulus or the response rate during the afferent's "steady state" period of response, the stimulus–response functions of these afferents were described by power functions with exponents close to 1.0 on the glabrous skin, and close to 0.5 on the hairy skin. Thus, there was a striking parallel between psychophysical and neurophysiological data, indicating that the relationship between the amplitude of skin indentation and perceived tactile intensity may be a direct reflection of the average response of SA mechanoreceptive afferents.

Subsequent work has revealed considerable intricacy associated with the question of tactile intensity perception and the role of cutaneous mechanoreceptors. At one level, there are questions concerning the consistency of results—both psychophysical and neurophysiological. At another level, there are various ways in which one can compare psychophysical and neurophysiological data related to response magnitude. The practice of comparing exponents of best-fit power functions is only one way, and it has its pitfalls (Kruger & Kenton, 1973). At a third level, stimulus dimensions other than intensity (e.g., temporal and spatial) have effects upon perceived tactile intensity, and therefore, these factors must also be taken into consideration when attempting to define the neural basis of intensity perception. Lastly, it is important to specify the energy spectrum of the applied displacements to identify what fiber populations are contributing to the response. For example, stimuli that have fast onsets, even for low-stimulus amplitudes, can activate PC units.

When magnitude estimates of tactile intensity were examined for individual subjects, there was large intersubject variation in power function exponents (Greenspan, 1984; Knibestöl & Vallbo, 1980). Additionally, Knibestöl and Vallbo (1980) compared magnitude estimates of tactile intensity with the response rates of human SA afferents innervating the glabrous skin of the hand. Although the normalized, average psychophysical power functions had an exponent close to 1.0, individual power–function exponents ranged from 0.36–2.09. This variability was largely attributable to intersubject variation, because individual subjects would consistently replicate their own results. At the same time, the stimulus–response functions of SA afferents from these same subjects had an average exponent of 0.7. When both neural and psychophysical data were compared within the same subject, there was no apparent correlation between the power function exponents for the two data sets (Figure 7). Thus, this study showed great disparity between perceived tactile intensity and SA afferent response rates.

As mentioned earlier, some studies indicated that two different power functions were necessary to adequately describe the stimulus–response relationship for tactile sensory intensity (Greenspan, 1984; Greenspan et al., 1984). With ramp and hold stimuli, the stimulus–response function was shallower for low-intensity stimuli than for higher intensity stimuli. Precisely where the break point between the two segments lay was dependent upon other stimulus variables, including the rate of stimulus change.

One possible reason for such variable psychophysical results (both within and across studies) is that the perceptual experience resulting from these tactile stimuli may not be simple or unitary. One series of studies demonstrated that there is a difference between perceived tactile intensity and the perceived depth of skin indentation (Burgess et al., 1983; Mei, Tuckett, Poulos, Horch, Wei, & Burgess, 1983; Poulos, Mei, Horch, Tuckett, Wei,

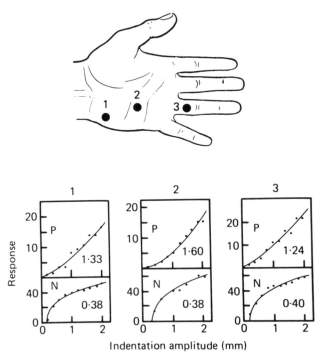

Indentation amplitude (mm)

FIGURE 7   Uniformity of discrepancy between stimulus–response functions of afferent units and psychophysical magnitude-estimation functions when tested at three different locations in the same subject. N refers to neural data, P to psychophysical. (Reproduced with permission from Knibestöl & Vallbo, 1980, Fig. 9, p. 262.)

Cornwall, & Burgess, 1984). In these experiments, stimuli consisted of controlled skin indentations at a constant rate, with either triangular or trapezoidal waveforms. When subjects were asked specifically to judge skin *indentation,* they could accurately report that information without being confounded by variations in the rate of indentation. In contrast, experiments in which stimulus *intensity* was judged showed influences by both depth and rate of indentation. The responses of primate cutaneous mechanoreceptive afferents to these same stimuli indicated that firing rate did not parallel either perceived intensity or perceived depth of indentation. Specifically, both RA and SA afferents showed much more of a change in response due to stimulus rate than either perceptual measure did. The authors concluded that neither of the two perceptual phenomenon (indentation depth or intensity) could be based strictly upon any mechanoreceptors' instantaneous firing rate, but they could be based on centrally integrated afferent activity (e.g., a "leaky integrator"). If this hypothesis is true, then different integrative time courses (and presumably neural processes) would be necessary in order to account separately for perceived intensity and perceived skin indentation depth.

A more recent study evaluated both the qualitative and intensive aspects of perceiving a ramp-and-hold force applied to the finger (Cohen & Vierck, 1993b). With rapidly applied forces (100 g/sec), there were qualitatively distinct perceptions associated with the onset ramp ("tap"), the plateau ("pressure"), and the offset ramp ("tap"). With slowly applied forces (1 g/sec), there was only a perception of gradually increasing and decreasing pressure, without accompanying tap sensations. Moderately applied forces (10 g/sec) typically evoked a tap sensation with stimulus onset, but not as often as with stimulus offset, and not as often as with the faster stimuli.

These same stimuli were used while recording activity from individual cutaneous mechanoreceptive afferents innervating the glabrous skin of the monkey hand (Cohen & Vierck, 1993a). Based on comparisons between afferent response rate and psychophysical intensity estimates, the authors concluded that (a) RA afferent responses were responsible for the tap sensations associated with stimulus onset and offset, and (b) SA afferent responses were responsible for the pressure sensations. These conclusions are consistent with Bolanowski et al. (1992b), who proposed that the intensity of tactile stimuli come about through a combination of activity across fiber types. At a quantitative level, however, there was not a simple relationship between afferent response and perceived magnitude. In general, simple summation of mechanoreceptive afferent activity largely overestimated the perceived magnitude. For instance, considering a 2-g stimulus applied at 100 g/sec, the relative perceptual magnitude of the stimulus onset versus plateau averaged 1.7:1.0. In contrast, the relative difference in response rate was 40:1 for all activated RAs and SAs, and 5.6:1 for just the SAs (see Figure

8). Another example of an important difference between afferent fiber response and perceptual magnitude relates to stimulus rate. The rate of the onset ramp greatly affected the perceived magnitude during the subsequent

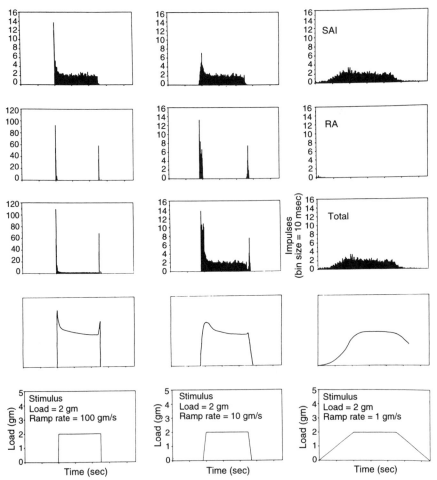

FIGURE 8   Mechanoreceptor response profiles and perceived intensity of a 2.0-g stimuli, presented at 1, 10, and 100 g/sec. The response of slowly adapting (SAI) (top row) and rapidly adapting (RA) (second row) receptors, and the total response (third row) to 100 g/sec (left column), 10 g/sec (center column), and 1.0 g/sec (right column). The response of each receptor type is weighed according to estimated densities of receptors innervating the fingertip region. Note the change in range of the axes that are needed to display the discharge of RAs and the total response at 100 g/sec. The bottom row shows the force-controlled stimuli presented at each rate. The fourth row shows the most frequently chosen line drawing that matches perceived tactile magnitude for each stimulus waveform. (Reprinted with permission from Cohen & Vierck, 1993b, Fig. 4, p. 127.)

plateau period: the relative perceived magnitudes during the plateau period following the 1, 10, and 100 g/sec ramps averaged 1.0:1.3:1.6. Although ramp rate did have an effect upon the SA response during stimulus onset, there was no effect of ramp rate upon the plateau period response. This result implies that some temporal integration must be occurring so that SAI responses during the onset ramp influence perceived magnitude during the plateau period. Thus, whatever the nature of the integrative processing, our perception of tactile intensity is not simply the instantaneous rate of mechanoreceptive afferent activity. Rather, the results are more consistent with the concept of a "leaky integrator," as suggested by Burgess and colleagues (Burgess et al., 1983; Mei et al., 1983; Poulos et al., 1984).

C. Spatial Aspects of Stimulation That Influence Perceived Intensity

The size of the stimulus area can have pronounced effects upon perceived tactile intensity. Specifically, increasing stimulus area can increase perceived intensity (or lower threshold)—the phenomenon of "spatial summation." Early studies of pressure perception thresholds described little or no effect of stimulus size (Nafe & Wagoner, 1941a,b; Zigler, 1932). However, with vibratory stimuli at threshold, high frequencies that engage the P channel show clear spatial summation, whereas low frequencies engaging the NPI channel do not (Verrillo, 1974; Verrillo & Capraro, 1975), although Muijser (1994) showed that at least one of the NP channels displays spatial summation when evaluated with an array of vibrating pins. With suprathreshold stimuli, there is evidence for spatial summation at all frequencies of stimulation (Gescheider, Verrillo, Capraro, & Hamer, 1977; Green & Craig, 1974; Verrillo, 1974; Verrillo & Capraro, 1975). These latter observations are consistent with the idea that suprathreshold stimuli can activate all of the psychophysically and physiologically defined channels (Bolanowski et al., 1988, 1992b).

Perceived tactile intensity can also be influenced by presenting multiple stimuli simultaneously. With appropriate stimulus parameters, multisite stimuli produce a more intense tactile perception localized to only one site of stimulation. This phenomenon, termed "funneling" by von Békésy (1958, 1959, 1960, 1967) can be observed even with stimuli widely separated on the body, and bilaterally on the fingers (Craig, 1968). A comparison of human psychophysics and feline mechanoreceptive afferent neurophysiology showed that funneling can occur with stimulus separations distinctly larger than a single afferent's RF (Gardner & Spencer, 1972). Thus, spatial summation of tactile intensity is not dependent upon summation of energy within an individual afferent's RF, but must instead depend upon CNS integration.

VIII. TEMPORAL DIMENSIONS OF TACTILE PERCEPTION

A. Difference Thresholds for Frequency Discrimination

One issue that has arisen in the psychophysical and physiological literature as a result of the use of vibratory stimuli is the ability of the human observer (or components of the nervous system) to discriminate different frequencies of vibration. Early interest in this capacity of the tactile system was driven by the attempts to use the tactile system as a surrogate channel for "hearing" speech (see Gault, 1924; Gault & Crane, 1928), although at the time little was known about the resolution with which frequencies of stimulation could be discriminated on the skin (see Geldard, 1957). One way to examine this capability is to measure DLs in the frequency domain. This is accomplished by measuring the JND between two suprathreshold vibratory stimuli of different frequencies. The ratio of the change in frequency required for this JND and the standard or reference stimulus defines the Weber fraction. Analogous experiments have been performed along the intensive dimension, as described above.

Several early studies attempted to measure frequency discrimination for vibratory stimuli (e.g., Dunlap, 1911, 1913; Joël, 1935; Knudsen, 1928; Périlhou, 1947; Roberts, 1932; Sherrick, 1954; von Békésy, 1962). As an example, Dunlap (1911, 1913) reported that at a reference of 440 Hz, the difference threshold was on the order of 26 Hz, the calculated Weber fraction equaling 0.06. Dunlap found that the DLs were greater for lower frequencies, but pointed out that intensive cues may have confounded the measurements. This was an important observation because in order to measure the exact difference threshold for frequencies of stimulation without the measurement being confounded with intensity discrimination, the stimuli must be matched for equal sensation level. For both threshold and suprathreshold stimuli it has been established that the tactile system acts as a U-shaped filter, the system being more sensitive in the 100–300-Hz frequency range than either lower of higher ranges (Goff, 1967; Verrillo et al., 1969). Thus as the frequency of stimulation is increased from, for example, 50 Hz to 100 Hz, the same intensity of stimulation will produce a greater magnitude of sensation. The intensity cues can be eliminated by first matching perceived intensity of all the comparison stimuli that will be used to the perceived magnitude of the standard or reference stimuli (Goff, 1967; LaMotte & Mountcastle, 1975; Mountcastle, Steinmetz, & Romo, 1990; Rothenberg, Verrillo, Zahorian, Brachman, & Bolanowski, 1977).

In a series of experiments where the subjective intensity of the stimuli were taken into account and where the stimuli were well defined, Goff (1967) determined frequency DLs on the finger at two sensation levels and five frequencies of stimulation. Goff found that the DLs were greater at larger subjective magnitude levels and that they also increased with stimulus frequency, the Weber fraction being approximately 0.2 at 25 Hz and 0.35 at 200 Hz. This latter observation, however, is at odds with that of Mountcastle et al. (1969, 1990) who found such DLs to be independent of stimulus frequency, at least over the range of 5–200 Hz. Regarding the absolute values of the DLs, Mountcastle and colleagues obtained DLs that were consistently lower than those reported by Goff (1967). For a 20-dB stimulus applied to the fingertip, the DL at 40 Hz was 0.07–0.11 (Mountcastle, Talbot, Sakata, Hyvärinen, 1969, 1990). This is roughly one-half the value obtained by Goff (1967) for a similar frequency and body site. Rothenberg et al. (1977) found that body site has an effect on the DLs obtained, being about 0.3 on the hairy skin of the forearm, 0.24 on the thenar eminence, and 0.19 on the fingertip for stimuli of the "warble-tone" variety.

More recent experiments dealing with frequency DLs have assessed the potential physiological basis for frequency discrimination. Mountcastle et al. (1990) conclude that serial-ordered processing of activity in RA units of primary somatosensory cortex may be responsible for frequency discrimination in the frequency range of 10–40 Hz. Other experiments have found that frequency discrimination can be improved by the presence of conditioning stimuli (Gobel & Hollins, 1994), implying an integrative role for higher order mechanisms. In fact, frequency discrimination has been used as a way of measuring plasticity effects that occur in the cortex in response to training. For example, Recanzone, Jenkins, Hradek, and Merzenich (1992) have shown that monkeys' DLs can be improved by a factor of 50% with training. These behavioral results parallel changes that occur in the somatosensory cortex as revealed by an increase in the cortical representation of the "exercised" skin (Recanzone, Merzenich, & Jenkins, 1992; Recanzone, Merzenich, Jenkins, Grajski, & Dinse, 1992). Thus it is possible that the neural code for frequency of a stimulus may incorporate a spatial component as well as the time-ordered serial code proposed by Mountcastle et al. (1990).

B. Perceived Velocity of Tactile Stimuli

Few studies have evaluated how well humans can perceive differences in the velocity of tactile stimuli. One problem with attempting to evaluate velocity perception per se is the fact that varying velocities can produce different qualities of sensation, and can also affect the perceived intensity of stimulation. (See section VII.) A second problem is that with a given stimulus amplitude, higher velocities are necessarily of shorter duration. Thus, one could judge the stimulus velocity by perceiving stimulus duration. In fact, subjects perform equivalently when they are asked to scale either the velocity or the duration of a brush stroked across the skin (Essick, Franzén, & Whitsel, 1988).

Using probes that indent the skin at different rates of movement, people can scale the apparent velocity of movement without being confounded by the stimulus amplitude (Pubols & Pubols, 1982). With magnitude estimation of perceived velocity, the stimulus–response function is fairly comparable for skin indentation of the index finger (between 0.1–10 cm/sec) and for surface parallel brushing of the index finger or the volar forearm (between 1.0–100 cm/sec; Franzén, Thompson, Whitsel, & Young, 1984). The fact that the psychophysical functions are equivalent for the index finger and the forearm suggests a central neural mechanism that is independent of absolute sensitivity or receptor density.

Feline hairy skin mechanoreceptive afferents were examined with the same surface parallel stimuli employed by Essick et al. (1988). Although all mechanoreceptive afferent types respond with greater response rates to faster moving stimuli, only the G_1 and F_1 types showed a difference in response rates that matched human perceptual discrimination of stimulus velocity (Greenspan, 1992). Thus, a simple rate code mediated by these afferents could account for perceptual velocity discrimination of surface-parallel movement on hairy skin.

More recently, the velocity sensitivity of human mechanoreceptors was evaluated using similar, well-controlled brushing stimuli (Edin et al., 1995; Essick & Edin, 1995). The results bore striking resemblance to the feline data. Furthermore, the authors examined the interaction of stimulus intensity and velocity upon mechanoreceptor responses. They pointed out that although any individual mechanoreceptor varied its response rate to both variables, thus making it impossible to code for both factors unequivocally, a population analysis of response rate and response duration could easily code for these two factors separately.

C. Adaptation to Sustained Stimuli

Tactile adaptation is defined as the reduction in perceived intensity in the presence of a constant cutaneous stimulus. Accordingly, this phenomenon could be considered under the general heading of temporal variations in perceived intensity. Several studies evaluated the time course of complete adaptation, that is, the time necessary for tactile perception to disappear altogether. In general, adaptation period increases with increasing stimulus intensity, decreases with increasing stimulus area, and varies inversely with the absolute sensitivity of different body sites (Crook & Crook, 1935; Zigler, 1932). A classic study by Nafe and Wagoner (1941a,b) measured the skin's continual movement with a constant, resting weight (force). Sensory adaptation was found to correlate with the rate of skin movement under this force, and complete adaptation occurred when the skin movement was below approximately 10 μm/sec. They argued that skin movement was the relevant stimulus, and tactile adaptation was actually the absence of sufficient skin movement. However, subsequent studies have shown that adaptation is not that simple. By using a stimulator that indents the skin a precise amount, one still experiences a prolonged tactile sensation that gradually adapts, even when a constant skin position is maintained (Horch, Clark, & Burgess, 1975).

One study, evaluated perceived intensity over an 18-sec period of constantly maintained skin indentation at the fingertip (Poulos et al., 1984). Additionally, these investigators evaluated the responses of primate SAI afferents to these same stimuli. Perceived intensity did not change very much during this period, overall. In contrast, the SAI response rate decreased dramatically over this time period. Thus, tactile adaptation is not a result of "stimulus failure" (e.g., the lack of skin movement), nor is it just a matter of mechanoreceptor adaptation. In fact, mechanoreceptor adaptation appears to precede perceptual adaptation by several seconds.

In a separate study, these same investigators evaluated the perceived depth of skin indentation using similar stimuli (Mei et al., 1983). During a steadily maintained indentation, the subjects' perceived a steadily *increasing* depth of indentation. In order to perceive a constant depth of indentation, the probe had to be retracted an average of 14% of the original depth over the 18-sec period. There are, of course, no mechanoreceptive afferents that increase their firing rate during a period of steadily maintained skin indentation. Thus, perceived depth of indentation (and to a lesser extent, perceived tactile intensity) is apparently based upon a temporally summating CNS network with a relatively long time constant.

IX. SPATIAL DIMENSIONS OF TACTILE PERCEPTION

There is a fundamental distinction between active and passive touch. It has long been recognized that certain aspects of tactile perception are enhanced by active exploration of an object (Gibson, 1962; Katz, 1925; Weber, 1834). In situations such as attempting to identify a relatively large object, this active exploration provides distinctly more information than could be acquired otherwise (Cronin, 1977; Heller & Myers, 1983; Magee & Kennedy, 1980), primarily by adding kinesthetic information. In other situations— such as distinguishing a microgeometric form or perceiving a texture— tangential movement is an important factor, but it is irrelevant whether the person moves his or her skin over a stationary surface or whether the surface moves over a stationary portion of skin (Cronin, 1977; Grunwald, 1966; Lamb, 1983b; Lederman, 1981; Schwartz, Perey, & Azulay, 1975; Vega-Bermudez, Johnson, & Hsiao, 1991).

This section will focus on those aspects of tactile spatial perception that are independent of active exploration. We will first consider the simpler situation in which stimuli are presented normal to the skin's surface, without lateral movement. Then we will consider more complex conditions involving tangential movement of the tactile stimulus.

A. Spatial Acuity with Surface-Normal Stimulation

1. Tactile Spatial Acuity: A Brief Historical Perspective[7]

There has been continuous interest in tactile spatial acuity since the pioneering scientific work of Ernst H. Weber. Weber devised two procedures for evaluating tactile acuity: two-point threshold and the error of localization. Conceptually, the two-point threshold is the minimal separation between two probes (placed simultaneously on the skin) that are perceived as two separate points. Weber's protocol for measuring the error of localization entailed applying a single probe to a skin site, and subsequently having the subject indicate the site where he felt the stimulus. The distance between the stimulus point and the subject's referred point is the localization error. Both of these procedures would be expected to measure the same phenomenon: the spatial resolving power of the tactile system. However, these two methods consistently produced different results, with localization being 3–4 times better than two-point discrimination (Kottenkamp & Ullrich, 1870; Weber, 1846, 1852; Weinstein, 1968; Zigler, 1935). This difference has been attributed to both stimulus protocol differences and task demands upon the subject (Zigler, 1935).

Although mechanoreceptor density and RF size are likely to have some relationship to tactile acuity, several observations indicate that more is involved. For instance, practice reduces two-point thresholds, both within and across sessions (Dresslar, 1894; Volkmann, 1858). Also, stretching the skin does not necessarily increase the two-point threshold in proportion to the degree of skin stretching (Hartmann, 1875). Another consideration is the perceptual experiences resulting from such stimuli. Subjects report a variety of sensory qualia with two-point stimulation at various separations. At one extreme, Gates (1915) identified nine perceptual categories, including a point, a circle, a line, a dumbbell, and two points. Thus, the results of a "simple" two-point discrimination task (reporting either "one" or "two") is dependent upon the criterion one chooses to use among these various sensory qualities (Titchener, 1916). Friedline (1918) argued that the practice effect in two-point discrimination studies is merely a change in the subject's criteria of what sensory experience he would call "two." In a more recent evaluation of tactile acuity, Johnson and Phillips (1981) reported that people

---

[7]Much of the 19th-century and early 20th-century work described in this section was adapted from Boring (1942) and Sherrington (1900). A more detailed historical review of this subject appears in Johnson, Van Boven, and Hsiao, 1994.

could distinguish between one and two 0.5-mm probes on the fingertip, even when there was no separation between the two probes. In this case, "two-point discrimination" could not be the result of distinguishing one from two points of stimulation, but rather distinguishing between two areas (or shapes) of stimulation.

Despite some similarities, there is an essential difference between localization error (identifying "where" on the skin), and two-point threshold (identifying "what" on the skin). Furthermore, there is a fundamental difference between stimulus discrimination ("Does A feel different from B?") and perceptual qualitative discrimination ("Does this feel like one or two points?"). Weber's two-point threshold is a test of perceptual quality, although in some experiments (i.e., Johnson and Phillips, 1981), it can be a stimulus-discrimination task. Stimulus-discrimination tasks are particularly useful for quantitative comparisons with neurophysiological data. On the other hand, perceptual discrimination tasks are necessary in order to learn anything about the quality of the perceptual experience.

2. Other Measures of Tactile Spatial Acuity

Since Weber, several other measures of tactile acuity have been devised, most of which demonstrate greater acuity than that found by two-point discrimination. The higher spatial resolution derived by point localization error has already been mentioned. When subjects were asked to discriminate size differences between two successively applied disks (Vierck & Jones, 1969) or edges (M. B. Jones & Vierck, 1973) on the forearm, they could resolve size differences 4–10 times smaller than classical two-point threshold. Loomis (1979) described several spatial-discrimination tasks applied to the fingertip, including tactile vernier resolution and line-separation discriminations. All of these tasks produced spatial acuity thresholds lower than the two-point threshold. The test providing the smallest threshold was a spatial interval discrimination, in which two bars separated by 5.0 mm was the standard stimulus, and two bars of slightly greater or lesser separation were the comparison stimuli. The threshold difference in separation for this task averaged 0.27 mm, which was approximately one-tenth of these subjects' two-point threshold. Even lower thresholds (between 0.1–0.15 mm) were derived when subjects were asked to discriminate the difference in spatial frequency of gratings (Morley, Goodwin, & Darian-Smith, 1983). These and other experiments clearly show that two-point threshold does not define absolute spatial acuity. It is also worth noting that despite the continued prevalence of two-point threshold measurements in clinical settings other tactile acuity protocols have been shown to be more sensitive and more reliable (Van Boven & Johnson, 1994a,b; Johnson et al., 1994; J. C. Stevens & Patterson, 1995).

One feature that many of these other acuity measures have in common is that the stimuli are more spatially extensive than just two points. Vierck and Cooper (1990) reported that the ability to distinguish between two gap sizes was much better when the stimulating surfaces consisted of a sequence of periodic gaps rather than just a single gap. There appeared to be a "critical length" for the stimulating surface (approximating the classic two-point threshold) before the more refined gap discrimination could be made. The authors argued, based on their findings, that the classic two-point threshold may represent a minimal extent of stimulation needed in order to make spatial discriminations. Once this critical extent is reached, finer discriminations can be made within that region.

3. Experimentally Assessed Factors That Influence Tactile Acuity

Body site is an important determinant of tactile acuity (Anstis & Loizos, 1967; Halar, Hammond, LaCava, Camann, & Ward, 1987; Kottenkamp & Ullrich, 1870; Vierordt, 1870; Weber, 1846; Weinstein, 1968). Despite the consistent difference in absolute value, both two-point threshold and localization error show the same pattern of body site variation (Fig. 9: note the similarity in pattern for A and B; also note the very different pattern for pressure threshold in C). Recognizing the prominent proximal–distal gradient of acuity along the upper extremity, Vierordt (1870) proposed the Law of Outward Mobility: acuity increases with increased mobility of body members. Although this seems generally true for the upper extremity, it is not as clearly applicable to other body areas.

At any given body site, the orientation of the stimulus also has an effect upon acuity measures. Weber (1846) was the first to report that two-point discrimination is better when points are separated transversely rather than longitudinally on the extremities. In another spatial-discrimination task—detecting the presence of a series of ridges on a metal disk—performance was significantly better when the ridges were oriented longitudinally at the fingertip (thus making the interridge gaps transversely oriented), as opposed to any other direction (Essock, Krebs, & Prather, 1992).

Tactile acuity has been found to be poorer in the elderly (Axelrod & Cohen, 1961; Bolton, Winkelmann, and Dyck, 1966; Gellis & Pool, 1977), but not always so (Potvin & Tourtellotte, 1985). Two studies employing well-controlled stimuli and forced-choice procedure to derive thresholds for perceiving two-point separation reported a significant increase in threshold with age for stimulation of the finger and the arm (J. C. Stevens, 1992; Woodward, 1993). In addition, Woodward (1993) measured skin compliance in her subjects (i.e., the amount of indentation resulting from a specified force). She found no correlation between compliance and threshold, suggesting that the aging effect is neural in nature, rather than due to

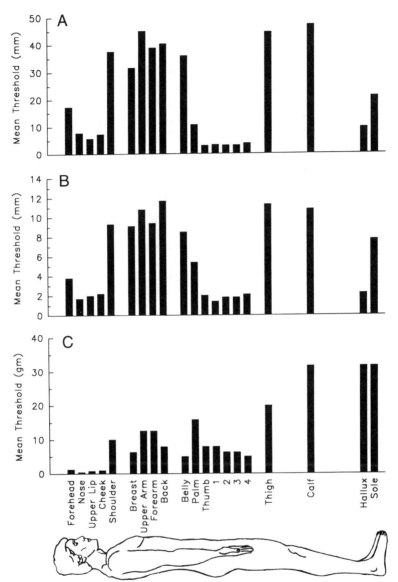

FIGURE 9 Three different measures of tactile perception threshold at various body sites, averaged over 24 male subjects. (A) two-point threshold; (B) point localization error; (C) threshold for the perception of pressure. (Adapted from data presented in Weinstein, 1968.)

changes in the biophysical properties of the skin. Alternatively, other aspects of skin biophysics do vary with age (Cook, 1989; Escoffier et al., 1989; Grahame, 1970) and may play a role here. An extensive evaluation of tactile spatial acuity was performed on a large sample of people aged 18–87 years (J. C. Stevens & Patterson, 1995). Four different perceptual phenomena were evaluated, including perceptions of length, locus, orientation, and discontinuity of tactile stimuli. All four of these spatial-acuity dimensions showed deterioration with age, exhibiting a rather constant threshold increase of ~1% per year between ages 20–80.

Most studies report no gender differences in tactile acuity (Nolan, 1985; J. C. Stevens, 1992; Weinstein, 1968; Woodward, 1993; J. C. Stevens & Patterson, 1995). Exceptions to this include reports in which many different body sites were tested, with one or two of them showing a significant difference between males and females (Weinstein, 1968; two sites out of 20; Nolan, 1985; one site out of 11). In marked contrast to his two-point threshold results, Weinstein (1968) observed that pressure thresholds were significantly higher for males at 19 of 20 body sites tested.

One might expect that the amount of force applied by the stimulus probes would affect spatial acuity. This is suggested by the fact that with higher forces, individual afferent firing rates increase, and a larger number of mechanoreceptors are excited due to the spatial spread of the stimulus. (Cohen & Vierck, 1993a; Johansson, 1978; Johnson, 1974; Pubols, 1987; Ray & Doetsch, 1990a,b). Despite these prominent intensity effects on mechanoreceptive afferent responses, however, there is no clear effect of stimulus intensity upon spatial acuity (Johnson & Phillips, 1981; M. B. Jones & Vierck, 1973). Perhaps there is an "informational trade-off" of intensity effects: the higher firing rate (a stronger signal to the CNS) versus a larger population of active mechanoreceptors (a spatially more diffuse signal). As further support for this perspective, the spatial gradient of mechanoreceptive afferent responses from the center of a stimulus is proportionally constant for a wide range of intensities (Pubols, 1987). Ray and Doetsch (1990a,b) evaluated the interrelationship between stimulus intensity and location within the RFs of glabrous skin mechanoreceptors in the rat. They showed that the firing rate of any individual mechanoreceptive afferent necessarily confounds stimulus intensity and stimulus location within the RF. Yet, they further demonstrated that a relatively simple integration of responses from a population of neighboring afferents can separately decode the intensive and spatial aspects of the stimulus.

J. C. Stevens (1982, 1989) demonstrated that two-point (and two-edge) separation thresholds were significantly better when the probes were either warmer or cooler than skin temperature. This effect was most pronounced on the forearm, but was also evident for the palm and forehead. Stevens argued that this thermal effect is not due to the additional thermal perception per se (which was not necessarily produced by the two-point stimulator), but rather by a thermal enhancement of mechanoreceptor activity at the point of stimulation. This effect would be different from simply increasing the force applied, because the enhanced activity would be confined to mechanoreceptors directly under the probes, without a lateral spread of additional excitation.

4. Mechanoreceptive Afferent Properties Related to Tactile Acuity

A proximal-distal gradient of spatial acuity is dramatically apparent across the glabrous skin of the hand. Accordingly, this site-dependent variation in perception allows for direct comparisons with mechanoreceptor variation along the hand. On the basis of the response properties of the glabrous skin mechanoreceptive afferents, one might expect that those with small RFs— FAI and SAI (see Fig. 2)—would be more likely to set the limits for spatial acuity than those with large RFs. Essentially, the logic is that individual SAI and FAI afferents represent a smaller area of skin than SAII or FAII afferent. Thus, activity from a given SAI or FAI would be more precisely referred to the skin area stimulated. If this argument has merit, one would expect to find an inverse relationship between tactile acuity and RF size of FAI and/or SAI units.

Johansson and Vallbo (1979, 1980; also see Vallbo & Johansson, 1978) compared the response properties of human cutaneous mechanoreceptive afferents with two-point threshold data for the hand. In close agreement with other studies (Greenspan, 1993; Vierordt, 1870; Weber, 1834), the average two-point values were 4.8 times higher on the palm versus the fingertip (7.7 vs. 1.6 mm). These results were compared to the receptive field sizes of mechanoreceptive afferents at different sites on the hand. The FAI afferents showed a proximal–distal gradient from fingertip to palm, but it was small compared to the difference in two-point threshold: the average RF diameter increased by a factor of 1.7 from the fingertip to the palm. Furthermore, the range of receptive field sizes was largely overlapping at these two sites. Thus, there was a poor relationship between FAI RF size and two-point threshold.[8] Additionally, Johansson and Vallbo (1980) re-ported that SAI RF size did not systematically vary across the hand. How-ever, Knibestöl (1975) reported that SAI RF areas averaged 4.4 times larger on the palm than on the distal phalanges, the difference being statistically significant. This discrepancy between studies must be resolved before de-ciding whether SAI RF size is a relevant factor for tactile spatial acuity.

An entity related to a mechanoreceptive afferent's RF is its "perceptive field," that is, the area of skin to which a person refers a sensory experience

---

[8]Knibestöl (1973) observed a comparable finger–palm difference in RF sizes of FAI afferents, and reported that the difference was not statistically significant. Using a computer-controlled, multiprobe stimulating device, Gardner and Palmer (1989) described similar results from the RA afferents of the monkey hand: average RF diameters were 1.34 times larger on the palm than the finger, and this difference was not statistically significant.

resulting from electrical stimulation of a mechanoreceptive fiber (INMS). Schady et al. (1983; also see Vallbo et al., 1984) asked subjects to estimate the location and size of perceptive fields resulting from INMS within the median nerve. The size of the perceptive fields associated with pressure sensations (putative SAIs) were 2–4 times larger than perceptive fields associated with tapping sensations (putative FAIs). Furthermore, the pressure perceptive fields were increasingly larger going from the fingertip to the palm, whereas the tapping perceptive fields did not show a proximal–distal gradient. It is not obvious how these perceptive field data are related to tactile acuity. On the one hand, the small perceptive fields for tapping sensations versus pressure sensations would imply that FAIs would set the lower limit for perceived spatial resolution. However, only the pressure perceptive fields demonstrated a proximal–distal gradient in size across the hand, which would parallel the regional trend in spatial acuity.

Johansson and Vallbo (1979) also considered the possible relationship between two-point threshold and mechanoreceptor density. FAI and SAI afferents are densely distributed at the fingertips, and progressively sparser at more proximal sites, paralleling the proximal–distal gradient of two-point thresholds. In contrast, the FAII and SAII afferents exhibited much less of a density change across the hand (Figure 10). This correlation suggests that receptor density of either FAI or SAI afferents (or both) is an important factor in determining tactile spatial acuity. The density of FAI afferents was estimated to be 5.7 times higher in the fingertip than in the palm, whereas that of the SAI afferents was 8.8 times higher.

Another feature of spatial acuity that can be related to mechanoreceptor properties is the effect of stimulus orientation. As mentioned earlier, spatial

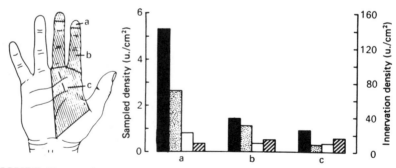

FIGURE 10   Mechanoreceptive afferent densities for the median nerve territory of (a) the fingertip, (b) the rest of the finger, and (c) the palm. Filled, stippled, hollow, and hatched bars represent FAI, SAI, FAII, and SAII afferents, respectively. Left ordinate indicates the number of units sampled per square centimeter. The right ordinate represents an estimate of the absolute number of units per square centimeter. (Reproduced with permission from Johansson & Vallbo, 1979, Fig. 4, p. 291.)

acuity measures on the hand or arm result in lower thresholds for transversely versus longitudinally separated stimuli. There is no reason to suspect that receptor distribution is anisotropic in order to account for this orientation effect (although relevant data are lacking). It has been proposed that RF orientation accounts for this phenomenon, given that mechanoreceptive afferent RFs are generally oval in shape, with the long axis frequently oriented longitudinally (Johansson & Vallbo, 1980; Johnson & Lamb, 1981). This same RF orientation is normally observed for tactually sensitive neurons of the primate thalamus (Kaas, Nelson, Sur, Dykes, & Merzenich, 1984; Loe, Whitsel, Dreyer, & Metz, 1977) and primary somatosensory cortex (Sur, Nelson, & Kaas, 1978, 1982). Accordingly, one would require a larger separation longitudinally versus transversely in order to excite the same differential pattern of mechanoreceptors with the two points. However, Fuchs and Brown (1984) demonstrated that the two-point threshold on the dorsal surface of the torso was significantly lower for longitudinally oriented than for transversely oriented probes, even though thalamic and cortical RFs tend to be oriented transversely (Loe et al., 1977; Kaas et al., 1984; Sur et al., 1978, 1982; comparable data on peripheral mechanoreceptive afferent RFs of the trunk are not available). Thus, factors other than RF orientation are necessary to explain this perceptual anisotropy. Fuchs and Brown pointed out that the consistent result (for both extremities and torso) is that the two-point threshold is lower when probes are oriented across dermatomic boundaries rather than along a dermatomic axis. Another noteworthy observation of Fuchs and Brown (also described by Weber, 1834) was that two-point thresholds were dramatically smaller when the two points straddled the torso midline, in contrast to when the two points were near, but on the same side of the midline. It is not likely that the peripheral mechanoreceptor density or the pattern of activation would be very different in these two situations. Rather, some differential aspect of the CNS processing of mechanoreceptive input from across the body midline must be responsible for this phenomenon.

If the SAI or FAI afferents are responsible for spatial acuity, then one would expect that stimulus parameters that selectively excite them would produce much better results than stimulus parameters that selectively exclude them. Two studies have attempted to selectively stimulate the P-channel and the NP-channels while evaluating spatial acuity. The results are not consistent with SAI or FAI afferents being particularly important for spatial acuity.

One study examined both two-point threshold and localization error on the forearm (Vierck & M. B. Jones, 1970). When the two-point threshold was determined either with steady force or with vibrating probes, the latter produced lower thresholds. There was little difference between 10-Hz or 300-Hz vibration, which should selectively excite NP and P channels, respectively.[9] Surprisingly, point localization error was worse with the vibrating probes than with steady indentation. Thus, these two measures showed opposite effects of applying vibratory versus sustained stimuli, but neither measure showed a significant frequency effect.

A more recent study measured localization accuracy with stimulation of the palm at either 25 or 250 Hz (Sherrick, Cholewiak, & Collins, 1990). Localization performance was better with the 25-Hz than with the 250-Hz stimulus, as one might expect, but only slightly better. Surprisingly, this frequency-dependent difference was only seen on the more proximal part of the palm. At a more distal part of the palm, localization performance improved for the 250-Hz stimulus, but not for the 25-Hz stimulus, resulting in comparable performances for the two frequencies at the more distal site. Thus, these results suggest that tactile localization can be nearly as well mediated by FAII afferents as by FAI or SAI afferents.

Most recently, Wheat, Goodwin, and Browning (1995) evaluated people's ability to localize pressure stimuli applied to the fingertip, and correlated that performance with monkey mechanoreceptors' response profiles. People were significantly better at localizing a stimulus site when the hemispheric probe had a greater curvature. SA and RA mechanoreceptors' response profiles were evaluated during stimulation at several different sites. SA mechanoreceptors exhibited significant differences in response profiles with stimulation at different sites on the fingertip. Akin to the psychophysical results, these differences were greater when stimulating with the more curved probe. A population response model developed by Wheat et al. (1995) showed a very close correspondence between localizing ability of human subjects and the changes in response profiles of SA mechanoreceptors. Thus, these units have the ability to provide information that can be used to localize tactile stimuli. Only about half of the RA mechanoreceptors responded reliably to these same stimuli. Those that did respond, however, showed similar trends as the SA mechanoreceptors, and thus could provide some information relevant to stimulus localization.

Another measure of spatial acuity is the threshold for detecting a gap in an otherwise smooth surface. Johnson and Phillips (1981) evaluated gap detection with stimuli applied to the fingertip without lateral movement (Figure 11, II). The absolute threshold for gap detection averaged 0.87 mm. In another experiment, a fixed grating pattern (e.g., a series of parallel ridges) was applied to the skin in one of two orthogonal orientations (Figure 11, III). The subject then reported the orientation of the ridges in a two-alternative, forced-choice paradigm. The minimal ridge separation necessary to perform this task on the fingertip averaged 0.85 mm. This result

---

[9]In this study, stimuli were applied to the hairy skin of the forearm, for which channel specificity is not well documented (however, see Bolanowski, Gescheider, & Verillo, 1994).

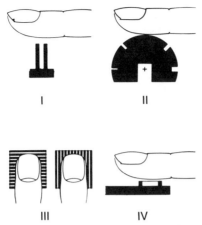

FIGURE 11   Diagrams showing the relationship between the subjects' right index finger and the stimuli employed to measure tactile spatial resolution by Johnson and Phillips, 1981. I, two-point discrimination; II, gap detection; III, grating resolution; IV, letter recognition (edge of letter shown). (Reproduced with permission from Johnson & Phillips, 1981, Fig. 1, p. 1179.)

closely matched the threshold size of a perceptible gap, suggesting that both discriminations were based on the same neural mechanisms.[10]

In parallel neuorophysiological experiments, Phillips and Johnson (1981a) applied similar stimuli to the receptive fields of mechanoreceptors innervating the monkey fingerpad. Of the three mechanoreceptor types evaluated with these stimuli, only the SAs showed differential responses to gaps in the range of 1.0 mm. The RAs required gaps of approximately 3 mm before they exhibited responses noticeably different from responses to flat surfaces. One PC was examined, and it did not differentiate any gap separation. This would have been predicted based on this afferent's large RF (See Figure 2). Thus, only the SAs exhibited the sensitivity in response to gaps or gratings that corresponded to human perceptual threshold for gap detection and orientation discrimination.

B. Shape Discrimination with Surface-Normal Stimulation

A few studies have evaluated how well people can discriminate the shape of an object pressed against the skin. Goodwin and colleagues (Goodwin,

[10]Loomis (1979) assessed people's ability to determine the orientation of two parallel bars when touched to the fingerpad. He found the threshold separation averaged 1.6 mm, twice that of Johnson and Phillips (1981). Despite these being basically the same task—two-alternative, forced-choice orientation discrimination—there were some stimulus differences, including the size and number of bars constituting the stimulus. This result is reminiscent of the previously cited report of Vierck and Cooper (1990), describing higher resolution gap-size discrimination with a periodic sequence of gaps than with just a single gap.

John, & Marceglia, 1991; Goodwin & Wheat, 1992) evaluated human discriminability of surface curvature by pressing hemispheric probes against the fingertips. In the best cases, when the standard stimulus had a high degree of curvature, subjects' difference thresholds were 10–12% (in terms of the radius of a hemispheric probe). When the standard stimulus had a low degree of curvature, (a) DLs tended to be higher, but (b) discriminability improved if probe size was allowed to covary with curvature. A recent study by the same group demonstrated that the response properties of monkey SA mechanoreceptors (but not RA or PC mechanoreceptors) could account for this psychophysical performance, and also provide for unambiguous coding of stimulus force (Goodwin, Browning, & Wheat, 1995).

Another study examined the neural basis for sensory discrimination of differences in the shape of a step (Srinivasan & LaMotte, 1987). Each step (or edge) had a different "steepness," but each was 0.5 mm in height, in the shape of a half-sinusoid. When these sinusoidal shapes were pressed against the monkey fingerpad, the SA afferents responded with greater activity to the curved part of the step than to the upper or lower sections of the step. Furthermore, they responded at a higher rate to the steeper steps. The RA afferents, on the other hand, showed little difference in response to the presence or the steepness of the steps. The extent to which people could discriminate among the different steps paralleled the differential responses of SAs, but were dissimilar to the less discriminating RAs. Thus, extrapolating the primate electrophysiological results to humans, the ability to recognize small differences in spatial features of objects pressed against the skin is most likely determined by the response properties of SAI afferents.

C. Suprathreshold Perception of Spatially Distinct Stimuli

There is relatively little data concerning the perception of tactile stimuli that are suprathreshold for perceived separation. Weber (1834, 1846) reported that people's perception of suprathreshold separation was not necessarily veridical. Furthermore, the perceived separation for a given stimulus pair would systematically change across body sites, more or less paralleling the two-point thresholds. These observations were replicated by Goudge (1918) and extended by Green (1982). Green showed that perceived distance between two simultaneously applied probes were mostly underestimates of true distance. On some body sites (most notably, arm and thigh), longitudinally oriented stimuli were perceived much closer together than transversely oriented stimuli. When subjects were asked to localize one of the two stimulation sites (and ignore the other), they still showed a tendency to localize that one site closer to its partner than it actually was. Thus, for both somatic localization and distance estimation, simultaneous stimulation produces a perceptual compression, that is, a "spatial illusion" in the direction of proximity.

D. Temporal Aspects of Stimulation Influencing Spatial Perception

Many aspects of tactile perception improve by allowing for temporal variation in the stimulus. For instance, it has long been recognized that two-point threshold is lower when the points are applied successively, or even "rocked" in place, than when applied simultaneously (Weber, 1834). The same type of temporal manipulation will improve acuity when attempting to distinguish the orientation of two points, or the orientation of an edge on the forearm (Gould, Vierck, & Luck, 1979).

As alluded to earlier, the lower value for localization error versus two-point threshold is partially attributable to the difference between successively and simultaneously applied stimuli (Zigler, 1935). F. N. Jones (1956) systematically varied the spacing and the time interval between two electrocutaneous stimuli, and determined the threshold spatial separation necessary for people to recognize the two stimuli as spatially separate (in essence, a successive two-point threshold). With intervals ranging from 2–1000 msec, the threshold separation progressively reduced with longer intervals.

With even more spatially separated stimuli, stimulus timing has a demonstrably greater effect upon spatial perception. Helson and King (1931) expanded upon Gelb's (1914) early work demonstrating the "Tau effect" with touch: the illusory spatial localization of a stimulus due to its timing relative to a sequence of stimuli. This phenomenon can be observed when three equidistant points on the skin are stimulated sequentially. If the time interval between stimulation of points 1 and 2 is less than the interval between stimulation of points 2 and 3, then the distance between points 1 and 2 will be perceived as closer than the distance between points 2 and 3. Of course, the extent to which one observes this Tau effect depends upon choosing appropriate distances and time intervals.

Related to the Tau effect is the illusion of the "cutaneous rabbit" (or cutaneous saltation; Geldard, 1975; Geldard & Sherrick, 1972). An early description of this phenomenon used three tactile-stimulating probes placed 10 cm apart in a line on the forearm. Each probe applied five rapid taps in succession. Instead of perceiving five taps at each of three sites, the 15 taps appear to be distributed uniformly between the site of the first and the third probe. "The impression is that of a tiny rabbit hopping up the arm" (Geldard, 1975, p. 30). In the experimental configuration known as the "reduced rabbit," only two stimulus sites are used. At the first site, two pulses are applied, followed by a third pulse at the other site. Depending upon the temporal parameters, the second pulse at the first site can be perceived as being anywhere between the two actual probe sites. Variations in the intensity of this second pulse will also affect its perceived location.[11] This phenomenon can be produced at any body site. Interestingly, it is vividly described when the stimulus sites are within a single dermatome, but it often fails when attempting to have "the rabbit" cross dermatomes. This is reminiscent of the dermatomic significance of two-point threshold, and may indicate an important difference between integration of tactile spatial information within versus across dermatomes. Another noteworthy observation is that the skin surface area for which cutaneous saltation can be demonstrated is progressively larger as one goes from the distal finger pad (2.3 cm^2) to the palm (31 cm^2) and to the ventral forearm (146 cm^2; Geldard & Sherrick, 1983).

Recently, Kilgard and Merzenich (1995) demonstrated two components to this perceptual illusion: (mis)localization on the body and the (mis)perceived distance separating two stimuli. The former was shown to be easily manipulated by directing the attention of the subjects, whereas the latter was very consistent for a given set of stimulus parameters. These results further support the idea of separate neural mechanisms for stimulus localization ("where on the body") versus stimulus spatial discrimination ("what on the body").

E. Spatial Perception of Tangentially Moving Stimuli

1. Acuity of Tangential Movement

When the skin is in contact with an object, and that object moves parallel to the skin's surface, one can easily recognize that movement has occurred. This is true even when the movement is small enough so that the skin is only stretched, and there is no "slip" of the object—that is, the object and skin move together. With such skin stretch stimuli, one can detect 0.1–0.2 mm of movement, and one can accurately perceive the direction of movement (or "pull" of the skin) with less than 1.0 mm of movement (Gould et al., 1979). Based on both human psychophysics and primate neurophysiological recordings, information from such stimuli is probably conveyed by SA afferents, but not by RAs or PCs (Srinivasan, Whitehouse, & LaMotte, 1990). Even though human SAII afferents are usually described as more sensitive to lateral skin stretch than SAIs, there is insufficient data available to determine which SA type sets the limit for resolution of skin stretch. Furthermore, there may be important differences in

---

[11] A similar phenomenon was described by von Békésy (1959) using two vibrators that applied different amplitude stimuli.

the stretch sensitivity of SA afferents in glabrous versus hairy skin (Edin & Abbs, 1991). If an object is completely smooth, one cannot distinguish the difference between movement that only produces skin stretch from movement that also causes slippage of the surface over the skin. In order to distinguish actual movement of an object relative to the skin from skin stretch alone, there needs to be some surface feature present (Srinivasan et al., 1990).

Morley et al. (1983) determined the DL for discriminating a change in the spatial frequency of a grating, which the subject would scan horizontally with his or her fingertip. With a standard spatial period of either 1.00 or 0.77 mm, DL values were approximately 5%, meaning that people could distinguish spatial periods that varied as little as 40–50 μm. DL values rose to 12% if subjects were not allowed to move their fingers tangentially.

It has been suggested, not unreasonably, that the additional acuity afforded by tangential movement of a stimulus is due to the engagement of RA afferents. Indeed, the development of the "moving two-point threshold" procedure was based on this assumption (Dellon, 1981). However, other factors must be considered as well. Johnson and Lamb (1981) showed that the response rate of primate SA afferents increased tenfold when a textured stimulus was stroked across the skin versus when that same stimulus was only pressed against the skin. Also, the addition of tangential movement provides for a sequential activation of a larger sample of mechanoreceptors of all types. Therefore, the "perceptual advantage" afforded by tangential scanning of tactile stimuli is not readily ascribable to any particular mechanoreceptor or mechanism.

One series of studies used plates of etched glass, which featured a raised surface in the shape of a dot. For both human psychophysical and monkey mechanoreceptor electrophysiological experiments, the digits were held in place, and a glass plate was stroked across the surface of the skin. The ability to detect the presence of the dot was dependent upon both the height and the radius of the dot (Johansson & LaMotte, 1983; Fig. 12). With a standard stimulus consisting of a 550-μm diameter dot, the height required for threshold detection was 1–3 μm. RA afferents had a comparable range of response thresholds, whereas the most sensitive SA afferents had thresholds of 8 μm and PCs had thresholds over 20 μm. Thus, the ability to detect a small elevation on the glass surface stroked across the fingertip was attributable to the activation of RA afferents (LaMotte & Whitehouse, 1986; Srinivasan et al., 1990). In comparison, the lateral movement of a texture made up of a set of smaller dots (1-μm high, 50-μm diameter, spaced 0.1-mm apart) could be perceived in psychophysical experiments, but not as individual dots. Rather, the movement of this surface produced a vibratory sensation, and only evoked consistent activity in PCs (Srinivasan et al., 1990).

Another set of studies compared human psychophysical discriminations

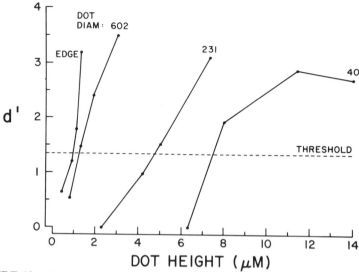

FIGURE 12   Psychometric functions representing the capacities of one subject to detect the presence of sharp edges or raised dots of different diameters. Sensitivity (d′) is plotted as a function of the height of the edge and the heights of dots with mean diameters of 602, 231, and 40 μm. A d′ of 1.35 was defined as detection threshold (dashed line). (Reproduced with permission from Johansson & LaMotte, 1983, Fig. 2, p. 26.)

with monkey mechanoreceptive afferent responses using surfaces with periodically spaced dots (Lamb, 1983a,b). The major stimulus variable was the spacing between the dots. In the psychophysical experiments, people were asked to distinguish between a standard stimulus (with dots spaced either 1.0 or 2.0 mm apart) and stimuli that differed from the standard by slightly greater spacing of the dots (from 1–8% increases in spacing). Subjects could discriminate 2% differences in dot spacing when the spacing difference was parallel to the direction of movement across the finger. (Subjects required 4–5% differences in dot spacing when the spacing difference was perpendicular to the direction of movement.) Furthermore, stimulus discriminability increasingly improved with larger differences in dot spacing. When the standard dot spacing was 2.0 mm, the RA afferents exhibited increases in mean firing rate with increased dot spacing. In contrast, the SAs did not show increased responses, and the PCs showed variable responses. Alternatively, when the standard dot spacing was 1.0 mm, only the PCs responded in a manner consistent with psychophysical discrimination (e.g., increased firing rate with increased dot spacing). Thus, in this situation, the discriminability of small changes in dot spacing could be mediated by differences in the firing rate of either RAs (with a 2-mm standard) or PCs (with a 1-mm standard). However, as we will see below, aspects of mechanoreceptive afferents' responses other than firing rate may be relevant to tactile perception of textures.

2. Perceiving Direction and Distance of Movement across the Skin

As described earlier, one can discriminate the direction of skin stretch on the forearm with as little as 1.0 mm of lateral displacement (Gould et al., 1979). In contrast, with little or no skin stretch, more than 10 mm of movement is required to correctly distinguish the direction of movement of a stimulus along the forearm (Dreyer, Hollins, & Whitsel, 1978; Gould et al., 1979; Norrsell & Olausson, 1992; Olausson & Norrsell, 1993; Whitsel, Dreyer, Hollins, & Young, 1979). Just how much tangential movement is necessary for correct directional discrimination (the "critical length") depends upon the body site and the speed of movement (Fig. 13; Dreyer et al., 1978; Essick, Bredehoeft, McLaughlin, and Szaniszlo, 1991; Essick, Franzén, McMillian, and Whitsel, 1991; Whitsel et al., 1979). For a given length and speed of stimulation, directional discrimination is best on the digits, nearly as good on the face, and shows a proximal–distal gradient in sensitivity along the upper extremity (Fig. 14, A and C). Furthermore, as one goes

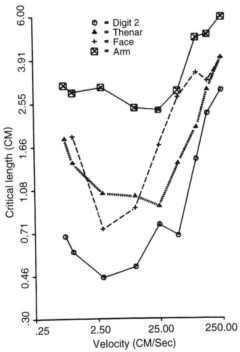

FIGURE 13   The critical length (*y*-axis) for making directional discriminations at various speeds of movement across the skin (*x*-axis) for four body sites. Critical length is calculated as the length of skin that needs to be traversed in order to correctly identify the direction of movement 75% of the time. In this way, it can be conceived as a threshold distance for directional discriminability of movement on the skin. (Reproduced with permission from Whitsel et al., 1979, Fig. 14, p. 99.)

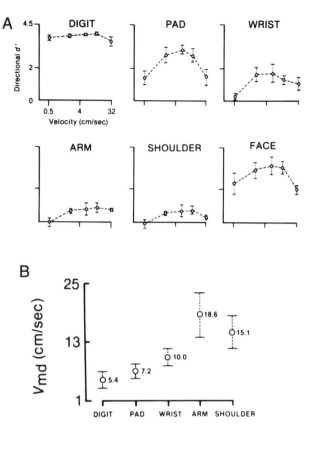

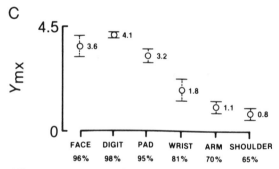

FIGURE 14   Three representations of directional discriminability of movement along the skin. (A) Mean directional sensitivity in d' units (*y*-axis; ±1 *SE*; *n* = 4) versus velocity of brush sweep for each of six test sites. The length of skin traversed was held constant at 0.75 cm. (B) Mean of the optimal velocity (*y*-axis; ±1 *SE*) for directional discrimination on each site of the upper limb. (C) Mean maximum sensitivity in d' units (*y-axis; ±1 SE; n* = 4) for each of six skin sites. The length of skin traversed was held constant at 0.75 cm. The performance, in percent correct, expected of an unbiased observer is shown under the abscissa labels for each site. (Reproduced with permission from Essick, Bredehoft, McLaughlin, & Szaniszlo, 1991, Figs., 3, 5, and 7; pp. 16–18.)

from fingers to shoulders, the optimal speed of movement for directional discrimination increases (Fig. 14, B).

Directional discriminability of a probe drawn across the forearm is improved with increasing loads on the skin, up to at least 8 g (Norrsell & Olausson, 1992; Olausson & Norrsell, 1993). Interestingly, the directional discriminability of an air-jet swept across the forearm (which produces no lateral skin stretch) is independent of the force applied (Norrsell & Olausson, 1994). Thus, different means of producing tactile movement across the skin may invoke different neural mechanisms for directional perception.

Gardner and Sklar (1994) used an OPTACON[12] stimulator to evoke apparent motion along the skin of the fingertip by applying a successive series of tactile pulses. By independently manipulating the number of pulses, the pulse spacing, pulse timing, and overall stimulus length, they evaluated the importance of these variables on perceived direction of movement. As one might expect, directional discrimination improved as the number of successive pulses increased from two to eight. Most interestingly, the distance between successive pulses of stimulation was irrelevant to the subjects' directional discrimination. Thus, the number of successive pulses, but not the overall length of stimulation, was the important variable for directional discrimination. In other words, there is not a "critical length" of stimulation necessary for directional discrimination, but rather a critical number of activated mechanoreceptors. Subjects could not determine the direction of apparent movement when only two pulses were applied, regardless of whether the sites were separated by 1.2 or 4.8 mm. One should recall that 4.8 mm is much larger than any measure of threshold for tactile spatial acuity on the fingertip. (See section IX.A.) Thus, the authors argue that perception of the direction of movement along the skin does not involve a processing of successive spatial localizations. Furthermore, this processing apparently relies on central nervous mechanisms that are independent of those used for other spatial-discrimination tasks.[13]

Not only is directional discrimination poorer with slower velocities, but

[12]The OPTACON (Optical-to-Tactile Conversion) is a device that allows for independent control of a 6-×-24 array of small probes (each probe being 0.25 mm in diameter). Adjacent rows of probes are separated by 1.2 mm, and adjacent columns are separated by 2.5 mm. By successively activating rows of probes, one can evoke the perception of continuous movement along the skin. For further information about this device, see Bliss, Katcher, Rogers, and Shepard (1970); Craig and Sherrick (1982); and Gardner and Palmer (1989).

[13]One should note that the OPTACON only activates RA and PC mechanoreceptors (Gardner & Palmer, 1989), whereas most objects moving over the skin will excite all types of mechanoreceptors. Despite this difference, the discriminability indices (d' values) reported by Gardner and Sklar (1994) are close to those reported in studies using a moving brush to evaluate directional discrimination (Essick et al., 1991).

subjects describe illusory paths when tested with such stimuli, as depicted in Figure 15 (Langford, Hall, & Monty, 1973). A straight, continuous movement along the forearm is perceived as "meandering" with stimulus velocities <10 cm/sec. This effect is more pronounced as velocity decreases.

Another rate-related phenomenon is the perceived distance traveled by a tangentially moving stimulus. In general, faster rates of movement are perceived as traversing smaller lengths of skin (Essick, Franzén, McMillan, & Whitsel, 1991; Langford et al., 1973; Whitsel et al., 1986). There is a range of velocities, 5–20 cm/sec at which perceived traverse length is relatively constant and the most veridical. Interestingly, this range largely coincides with the velocities that are optimal for directional discrimination of the forearm (Whitsel et al., 1986).

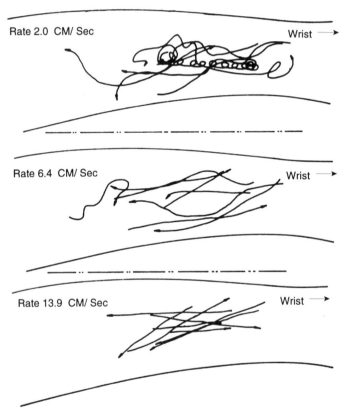

FIGURE 15   Characteristic drawings made by one subject on the outline of a forearm as a point moved across the skin in a straight line. Each segment of the drawing shows lines going both toward and away from the wrist, as indicated by the direction of the arrows. (Reprinted with permission from Langford et al., 1973, Fig. 2, p. 60. Copyright © 1973 by the American Psychological Association.)

X. COMPLEX PERCEPTUAL QUALITIES AND ASSOCIATED MECHANORECEPTOR PROPERTIES

As one moves beyond the simple intensive, temporal, and spatial aspects of touch, one encounters an expanse of largely uncharted territory. There are indeed some areas that have received considerable attention, but much awaits scientific scrutiny. One study sought to determine if there are fundamental perceptual dimensions of touch, and what those dimensions might be (Hollins, Faldowski, Rao, & Young, 1993). Subjects tactually evaluated several ordinary objects, and by using multidimensional scaling procedures, a model of tactual perceptual space was created. The results indicated that three dimensions adequately accounted for the various tactile perceptions. These three dimensions were best described as (a) roughness–smoothness, (b) hardness–softness, and (c) sponginess. Of these dimensions, roughness has been given the most attention in terms of identifying its neural substrate, and is reviewed below.

A. Roughness Perception

By moving the fingers over a surface, one experiences a sense of roughness that is related to surface geometry. Several studies examined roughness perception using commercially available substances such as sandpaper (Ekman, Hosman, & Lindström, 1965; S. S. Stevens & Harris, 1962; Stone, 1967). However, what was needed to understand the physical and neural bases of roughness perception was a set of precise stimuli that allowed for systematic and independent manipulation of relevant physical features. Two types of stimuli that were created for this purpose were (a) surfaces with regularly spaced ridges (gratings), and (b) surfaces with raised dots. A series of studies by Lederman and colleagues used surfaces with regularly spaced ridges to identify the variables relevant to roughness perception. Two major factors were found: (a) the spacing between neighboring ridges and (b) the amount of force applied. Both had a strong, positive correlation with perceived roughness as measured by magnitude estimation. A minor factor was the size of the ridge, which had a weak, negative correlation with perceived roughness (Lederman, 1983; Lederman & Taylor, 1972). One factor that had little effect upon perceived roughness was the speed at which the finger scanned the surface (Lederman, 1974, 1983). Another important observation was that perceived roughness magnitude for a given surface was comparable whether the finger moved over the surface or the surface moved over a stationary finger, provided that the forces were the same (Lederman, 1981). This result implied that only the cutaneous mechanoreceptors provided signals relevant to roughness perception, independent of kinesthetic input.

Taylor and Lederman (1975) explored several candidate models relating physical properties of stimulation to perceived roughness. Based on their psychophysical results, they concluded that the most relevant features are compressive stresses or strains, as opposed to lateral (shear) forces. This conclusion is consistent with the previously cited work suggesting that normal strains are best related to mechanoreceptor activation. (See section II).

How do mechanoreceptive afferents respond to these types of stimuli? Darian-Smith and Oke (1980) were the first to describe the responses of primate cutaneous mechanoreceptive afferents to systematically applied grating stimuli. A description of that apparatus is presented in Figure 16. They

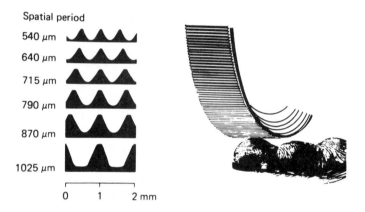

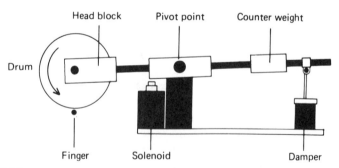

FIGURE 16 Details of stimulator used for presenting gratings to finger pad skin in Darian-Smith and Oke, 1980. The grating was mounted on a rotating drum 100 mm in diameter (upper right). The profile of each of the six gratings used is shown (upper left), along with its spatial period. The lower diagram illustrates the mechanisms for controlling the period of contact of the grating moving across the finger pad skin. The drum was mounted at one end of a counter-balanced level and rotated at a preset velocity. This drum was positioned 1.0 mm above the skin surface: an actuated solenoid held the drum off the skin except for the required contact period. The perpendicular force at which the moving grating was applied to the skin during this contact period was determined by the counterweight: This could be set in the range 20–100 g. Similar devices are used to present dot patterns and raised letters in other studies. (Reprinted with permission from Darian-Smith & Oke, 1980, Fig. 1, p. 119.)

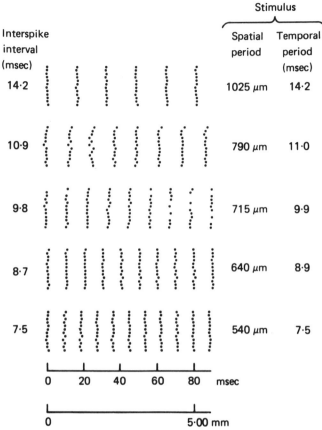

FIGURE 17   Responses of a rapidly adapting afferent to different gratings moving across its receptive field on the terminal phalanx of the thumb. The tangential velocity was 72 mm/sec in a direction at right angles to the long axis of the finger, and the applied force was 60 g for all records; successive stimuli were presented every 3 sec. Each row of dots indicates the occurrence of action potentials in response to a single passage of the grating across the skin; twelve successive responses are illustrated for each grating. The 80-msec response segment illustrated had its onset at approximately 500 msec after the beginning of stimulation. With these records there was both precise alignment of the time of occurrence of action potentials after the onset of stimulation, and also alignment relative to the instantaneous position of the grating on the skin. The stimulus spatial and temporal periods are indicated for each data block on right side of figure. The mean interspike interval is to the left of each data block. (Reprinted with permission from Darian-Smith & Oke, 1980, Fig. 3, p. 122.)

demonstrated that the responses of all types of mechanoreceptive afferents to a moving grating were primarily related to the stimulus temporal frequency (cycles/sec), defined as the scanning velocity divided by the spatial period of the grating (Fig. 17). For example, the response to a grating with a 1.0-mm spatial period moving at 10 cm/sec would be comparable to the response to a grating with a 0.5-mm spatial period moving at 5 cm/sec. An important consequence of this observation is that the spatial period of a stimulus (or its inverse, spatial frequency) cannot be unambiguously represented in the firing rate of individual mechanoreceptive afferents. Thus, information about the spatial features of the stimulus (independent of the temporal features produced by movement) must be extracted from a population of mechanoreceptors by the CNS. Another important observation was that each mechanoreceptor type had a particular range of temporal frequencies for which it was best "tuned": SA, 20–60 Hz; RA, 60–200 Hz; and PC, 100–300 Hz. This is reminiscent of the vibratory frequency tuning described earlier, although not strictly comparable.

A series of papers by Goodwin and colleagues examined in great detail the mechanoreceptive afferents' responses to gratings stroked across their receptive fields (Goodwin & Morley, 1987a,b; Goodwin, John, Sathian, & Darian-Smith, 1989; Morley & Goodwin, 1987; Sathian, Goodwin, John, & Darian-Smith, 1989). In these experiments, the gratings were square edged rather than sinusoidal, thus allowing for independent manipulation of the size of the groove and the ridge, rather than just spatial period per se. With most stimulus parameters, the afferents responded with a pattern coincident with an individual ridge contacting the receptive field. There may be only one action potential generated per ridge, or there may be more, depending upon stimulus parameters and mechanoreceptor type. In some instances, a ridge would fail to produce an action potential. The responses of mechanoreceptive fibers could be calculated as either the frequency of action potentials generated per ridge (ridge response rate) or the frequency of action potentials generated for each sweep of the stimulus across the RF (mean cycle response rate). All the mechanoreceptive types exhibited an increase in response rate (however calculated) as groove width increased from 0.18–2.0 mm. As ridge width increased (from 0.16–2.75 mm), PC afferents showed a small reduction in cycle response rate, whereas the RA and SA afferents showed no consistent change. All mechanoreceptor types showed a small increase in ridge response rate with increasing ridge width. (This "ridge effect" was attributed to the inevitable decrease in temporal frequency of the stimulus, when it was scanned at the same velocity.) As the speed of movement increased, RA and PC afferents showed an increase in cycle response rate, whereas SA afferents showed no consistent change. All mechanoreceptor types showed a decrease in ridge response rate with increased velocity.

These results are summarized in Figure 18, along with relevant psychophysical results reported by Lederman (1974, 1983; Lederman & Taylor, 1972) and Sathian et al. (1989). Neither response measure from any single mechanoreceptor type matches all of the psychophysical aspects of roughness perception. Thus, it appears that perceived roughness is not simply a function of any mechanoreceptor's response rate.

Stimulus Factor	Neurophysiological Results						Psychophysical Results
	RA		PC		SA		Perceived Roughness
	Ridge rate	Cycle rate	Ridge rate	Cycle rate	Ridge rate	Cycle rate	
Groove width: Increasing width:	↗	↗	↗	↗	↗	↗	↗
Ridge width: Increasing width:	↗	→	↗	↘	↗	→	↘
Velocity of Scan: Increasing velocity:	↘	↗	↘	↗	↘	→	→

FIGURE 18 A visual summary of neurophysiological and psychophysical experiments related to roughness perception of moving gratings (based on results of Goodwin and Morley, 1987b; Goodwin et al., 1989; Lederman, 1974; Lederman & Taylor, 1972; Morley & Goodwin, 1987; Sathian et al., 1989). An upward arrow indicates a positive relationship between a specified stimulus factor and response measure. A downward arrow indicates a negative relationship. A horizontal arrow indicates no functional relationship. The angle of the arrow approximates the strength of the effect. RA, rapidly adapting; PC, Pacinian corpuscle; SA, slowly adapting.

Another means of studying roughness perception is the use of systematically varied dot patterns. Connor. Hsiao, Phillips, and Johnson (1990) scanned dot patterns across the RFs of primate cutaneous mechanoreceptors innervating the fingertip, and had human subjects laterally scan these same surfaces with their index fingertip. Stimulus surfaces differed as to the diameter of the dots (0.5–1.2 mm) and in interdot spacing (1.5–6.0 mm). The perceived magnitude of roughness was affected by both variables. For a given dot spacing, the smaller dots felt rougher. For all dot sizes, the largest roughness estimates were at 3.0-mm spacing, and estimates decreased at both shorter and longer spacings. The mechanoreceptors typically exhibited their highest mean firing rates at a dot spacing of 2.0 mm, often with a remarkably lower response rate at a dot spacing of 3.0 mm. For all three types of mechanoreceptor, there was a poor correlation between mean response rate and the magnitude estimates of roughness. However, perceived roughness was better correlated with the *variability* in mechanoreceptor response rates, particularly of SA afferents. This was true whether considering temporal variation exhibited by a single afferent, or spatial variation at one moment in time.

In a more revealing study from the same laboratory, perceived roughness and mechanoreceptive afferent responses were measured with dot spacings independently varied in orthogonal directions (Connor & Johnson, 1992). Roughness magnitude increased with larger dot separations in both directions, but more so for separations in the horizontal (scanned) direction (see Figure 19, A). The mean response rate of RA or SA afferents did not parallel the psychophysical data (Fig. 19, B). The temporal variation in response of individual afferents showed a somewhat better correlation, but still showed

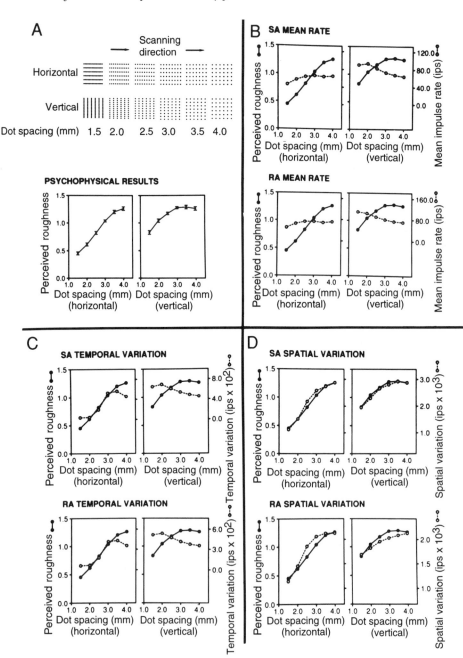

FIGURE 19   (A) Stimulus patterns. The stimuli were patterns of dots embossed on plastic surfaces. Each dot had the shape of a truncated cone 0.5 mm in height with a flat surface 0.5 mm in diameter. For the horizontal-row patterns, center-to-center spacing between dots varied from 1.5–4.0 mm in the horizontal direction, and remained constant at 4.0 mm in the

clear differences from the psychophysical results (Fig. 19, C). Finally, the between-fiber spatial variation in firing rates of SA afferents (and to a slightly lesser extent, RA afferents), provided a very good match with the shape of the psychophysical function (Fig. 19, D). In considering different spatial ranges, perceived roughness was best related to differences in discharge rate over distances of 1–2 mm on the skin. On the fingertip, this distance would correspond to differences in rates between neighboring SA or RA afferents.

Even though the effects of varying dot spacings were similar for both the SA and RA afferents, the effects of varying dot size were quite different. A recent review of the aforementioned work showed that increasing the dot sizes (independent of dot spacing) had opposite effects upon the spatial variation responses of SA vs. RA mechanoreceptors. Furthermore, perceived roughness followed the trend of the SA mechanoreceptors, not the RAs (Johnson & Hsiao, 1994).

The hypothesis that roughness is due to spatial variations in responses between mechanoreceptive nerve fibers was further supported by a study in which the ulnar nerve was cooled unilaterally at the elbow during roughness-discrimination tasks by human subjects (Phillips & Matthews, 1993). Subjects compared the roughness of raised dot patterns scanned by the fifth digit of each hand. A given surface felt smoother on the digit of the cooled nerve. This nerve cooling was associated with a lengthening of the absolute refractory period of the large myelinated (mechanoreceptive) fibers. Interestingly, reduction of the fibers' conduction velocity per se was not associated with alteration in roughness perception. The authors conclude that the increase in refractory period, which essentially lowers the fibers' maximum firing rate, was the major basis of reduced roughness magnitude. With the mechanoreceptors' response range reduced, they could not exhibit the same degree of response variation as they normally would. Based on the electrophysiologically recorded responses of mechanoreceptors to the same dot patterns stroked over the human finger pad (Phillips, Johansson, & Johnson,

---

vertical direction. For the vertical-column patterns, spacing varied from 1.5–4.0 mm in the vertical direction, and remained constant at 4.0 mm in the horizontal direction. The direction of motion relative to the skin surface is indicated. (A, lower) Psychophysical results. Roughness magnitude was averaged across 36 subjects after normalization within subjects to correct for differences in scale. The relationship of roughness to horizontal dot spacing is shown in the left panel, and to vertical dot spacing in the right panel. Error bars indicate *SE*. These same data are reproduced in the other graphs of this figure. (B) Mean firing rate (and roughness magnitude) versus dot spacing. The left-side ordinates represent normalized roughness magnitude, which is plotted with solid symbols. The right-side ordinates represent mean firing rate in impulses per second, plotted with open symbols. The two ordinates are scaled and translated relative to one another according to a linear regression between roughness and mean rate. (C) Temporal variation in mechanoreceptive afferent response (and roughness magnitude) versus dot spacing. Same format as for (A). (D) Spatial variation in mechanoreceptive afferent response (and roughness magnitude) versus dot spacing. (Reprinted with permission from Connor & Johnson, 1992, Figs. 2, 4, 8, 9, and 10; pp. 3415–3422.)

1992), Phillips and Matthews modeled the effects of the increased refractory period. Spatial variation in response rate decreased with simulated cooling in SAI, FAI, and FAII afferents (but not SAIIs), thereby paralleling the decrease in perceived roughness, and providing more correlative support for the Connor and Johnson model.

B. Complex Form (Character) Perception

Several investigators have examined people's ability to identify complex tactile forms, including Braille and letters of the alphabet. This topic cannot be adequately surveyed within this chapter. However, we will describe a few research results that address the question of mechanoreceptive afferent coding for form perception.

Cutaneous mechanoreceptive afferents' responses to dot patterns revealed the extent to which such patterns could be isomorphically represented by "spatial event plots" (SEPs; Darian-Smith, Davidson, & Johnson, 1980; Johnson & Lamb, 1981).[14] Based on primate neurophysiological recordings, the spatial event plots of SA and RA afferents more closely resembled the stimulus pattern than that of the PCs. More recently, these same observations have been made on human mechanoreceptive afferents (as shown in Figure 20; Phillips, Johansson, & Johnson, 1990; Phillips et al., 1992). Only the SAI and FAI afferents produced SEPs that spatially resembled the dot patterns. Additionally, the SAI afferents made visibly more precise SEPs than the FAI afferents. In both species, the SA (SAI) SEPs start to lose their resemblance to the stimulus form at a dot spacing of 1.0–1.5 mm. Thus, the SAs (SAIs) appear best able to represent spatial information, and this ability is limited to spatial features that are approximately 1.0 mm.

Similar results have been reported with respect to the responses of (monkey) afferents tested with raised letters of the alphabet (Bankman, Hsiao, & Johnson, 1990). The SAs provided responses that more precisely matched the pattern of the letters (see Fig. 21). Furthermore, the types of errors that people often made in recognizing letters of the alphabet by touch were matched by comparable ambiguity of the responses of SAs to these same letters (Loomis, 1982; Vega-Bermudez et al., 1991). For instance, the stimulus letter B is often incorrectly identified as D, but the stimulus D is seldom misidentified as B. The SEPs of SA afferents show a great similarity for the letters B and D. This similarity is largely due to the fact that the B's SEP only faintly represents the middle horizontal bar, thus making it more

[14]It should be pointed out that SEPs are a very useful way of representing neural responses to two-dimensional stimuli. However, it is not necessarily the case that CNS neural mechanisms interpreting this input use such an isomorphic representation. Some reports address this issue (Johnson, Phillips, Hsiao, & Bankman, 1991; Phillips, Johnson, & Hsiao, 1988; Warren, Hämäläinen, Palmer, & Gardner, 1991), but more research is required before settling this question.

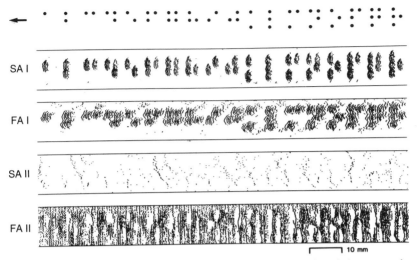

SA I

FA I

SA II

FA II

10 mm

FIGURE 20   Spatial event plots reconstructed from responses of single human mechanoreceptors to scanned Braille characters. Top panel shows representations of characters (A through R) that were repeatedly scanned across the receptive fields at 60 mm/sec in the direction of the arrow at top left. Application force = 60 *g*. Between each scan the characters were shifted by 200 microns normal to the scanning direction. Each successive panel shows the responses of a single afferent (SAI, FAI, SAII, and FAII) to repeated scanning of the character set across the receptive field. Each small dot represents the occurrence of one action potential; each horizontal row of small dots represents the response pattern resulting from a single scan. All four afferents shown had receptive fields on the distal phalanx of the index or middle finger. Direction of scanning was proximal to distal across the skin in each case. (Reprinted with permission from Phillips et al., 1990, Fig. 1, p. 591.)

Peripheral SA

Peripheral RA

FIGURE 21   Spatial event plots (SEPs) reconstructed from one slowly adapting (SA) and one rapidly adapting (RA) afferent fiber in response to raised letters of the alphabet. (Reprinted with permission from Bankman et al., 1990, Fig. 3, p. 614.)

closely resemble the shape of the letter D. Vega-Bermudez et al. (1991) have argued that this type of asymmetric error indicate that the CNS processes the spatial pattern of the SEPs per se, and attempts to match them with idealized (remembered) representations of the letters.

XI. CONCLUDING REMARKS

This review encompassed many of the scientific advances over the last two decades related to tactile perception and its peripheral physiological basis. Two features of the progress represented in this review stand out as particularly noteworthy. First, we have found the comparison between psychophysical and neurophysiological data to be particularly valuable in those experiments employing well-controlled and reproducible stimuli. The use of single-unit recordings in awake human subjects has enhanced this correlative endeavor considerably, and has largely validated the decades of research employing mammalian neurophysiological recordings of mechanoreceptive afferents as a model for human beings. Second, the development of more sophisticated models of somatosensation has provided greater insight into the neural basis of tactile perception, and has allowed for a wider range of experimental questions to be considered. This applies to global models, such as the four-channel model of touch for the glabrous skin of the hand, as well as to models of more restricted phenomenon, such as the waterbed model of skin mechanics or the SAI mechanoreceptor spatial integration model of roughness perception. Indeed, future developments in this field will depend not only on continued and more elaborate data collection, but also in improved analysis of data based on progressively developing models of somesthetic function.

Acknowledgments

Preparation of this chapter was assisted by National Institutes of the Health grants R01-NS28559 (JDG), P01-DC00380 and R01-DC00098 (SJB), and by National Science Foundation grants IBN-9211561 (SJB) and BNS-8808337 (JDG). We greatly appreciate the editorial assistance of Sandra L. B. McGillis.

References

Adriaensen, H., Gybels, J., Handwerker, H. O., & Van Hees, J. (1983). Response properties of thin myelinated fibers in human skin nerves. *Journal of Neurophysiology, 49*, 111–122.
Agache, P. G., Monneur, C., Lévêque, J.-L., & De Rigal, J. (1980). Mechanical properties and Young's modulus of human skin in vivo. *Archives of Dermatological Research, 269*, 221–232.
Akoev, G. N., Makovsky, V. S., & Volpe, N. O. (1980). Effects of tetrathylammonium on mechano- and electrosensitive channels of Pacinian corpuscles. *Neuroscience Letters, 19*, 61–66.

Anstis, S. M., & Loizos, C. M. (1967). Cross-modal judgments of small holes. *American Journal of Psychology, 80,* 51–58.

Axelrod, S., & Cohen, L. D. (1961). Senescence and embedded-figure performance in vision and touch. *Perceptual and Motor Skills, 12,* 283–288.

Bankman, I. N., Hsiao, S. S., and Johnson, K. O. (1990). Neural image transformation in the somatosensory system of the monkey: Comparison of neurophysiological observations with responses in a neural network model. *Cold Spring Harbor Symposium on Quantitative Biology, 55,* 611–620.

Bell, C. (1869). Idea of a new anatomy of the brain. *Journal of Anatomy and Physiology, 3,* 154–157.

Bell, J., Bolanowski, S. J., Jr., & Holmes, M. H. (1994). The structure and function of Pacinian corpuscles: A review. *Progress in Neurobiology, 42,* 79–128.

Bessou, P., Burgess, P. R., Perl, E. R., & Taylor, C. B. (1971). Dynamic properties of mechanoreceptors with unmyelinated fibers. *Journal of Neurophysiology, 34,* 116–131.

Bliss, J. C., Katcher, M. H., Rogers, C. H., & Shepard, R. P. (1970). Optical-to-tactile image conversion for the blind. *IEEE Transactions on Man-Machine Systems, MMS-11,* 58–64.

Blix, M. (1882). Experimentela bidrag till lösning af fragan om hudnerveras specifika energies. [Experimental research on the solution of the question on the specific energies of the skin nerves]. *Upsala Läkareförening Förhandlingar, 18,* 87–102.

Blix, M. (1884). Experimentelle Beiträge zur Lösung der Frage über die specifische Energie des Hautnerven. [Experimental research on the solution of the question on the specific energies of skin nerves]. *Zeitschrift für Biologie, 20,* 141–156.

Bolanowski, S. J., Jr., Gescheider, G. A., & Verrillo, R. T. (1992a). Individual magnitude estimates correlate with neural intensity characteristics. *Society for Neuroscience Abstracts, 18,* 1544.

Bolanowski, S. J., Jr., Gescheider, G. A., & Verrillo, R. T. (1992b). Individual magnitude estimation functions and their relation to the Pacinian (P) and non-Pacinian (NP) channels. *Journal of the Acoustical Society of America, 92,* 2436.

Bolanowski, S. J., Jr., Gescheider, G. A., & Verrillo, R. T. (1994). Hairy skin: Psychophysical channels and their physiological substrates. *Somatosensory Motor Research, 11,* 279–290.

Bolanowski, S. J., Jr., Gescheider, G. A., Verrillo, R. T., & Checkosky, C. M. (1988). Four channels mediate the mechanical aspects of touch. *Journal of the Acoustical Society of America, 84,* 1680–1694.

Bolanowski, S. J., Jr., Schuyler, J. E., & Slepecky, N. B. (1994). Semi-serial electron micrographic reconstruction of presumed transducer sites in Pacinian corpuscles. *Somatosensory Motor Research, 11,* 205–218.

Bolanowski, S. J., Jr., Schuyler, J. D., Sulitka, D., & Pietras, B. (1996). Mitochondrial distribution within the terminal neurite of the Pacinian corpuscle. *Somatosensory Motor Research, 13,* 49–58.

Bolanowski, S. J., Jr., & Verrillo, R. T. (1982). Temperature and criterion effects in a somatosensory subsystem: A neurophysiological and psycho-physical study. *Journal of Neurophysiology, 48,* 836–855.

Bolanowski, S. J., Jr., & Zwislocki, J. J. (1984a). Intensity and frequency characteristics of Pacinian corpuscles. I. Action potentials. *Journal of Neurophysiology, 51,* 793–811.

Bolanowski, S. J., Jr., & Zwislocki, J. J. (1984b). Intensity and frequency characteristics of Pacinian corpuscle. II. Receptor potentials. *Journal of Neurophysiology, 51,* 812–830.

Bolanowski, S. J., Jr., Zwislocki, J. J., & Gescheider, G. A. (1991). Intersensory generality of psychological units. In S. J. Bolanowski, Jr., & G. A. Gescheider (Eds.), *Ratio scaling of psychological magnitude* (pp. 260–276). Hillsdale, NJ: Lawrence Erlbaum.

Bolton, C. F., Winkelmann, M. D., & Dyck, P. J. (1966). A quantitative study of Meissner's corpuscles in man. *Neurology, 16,* 1–9.

Boring, E. G. (1942). *Sensation and perception in the history of experimental psychology.* New York: Appleton-Century-Crofts.

Brown, A. G., & Iggo, A. (1967). A quantitative study of cutaneous receptors and afferent fibres in the cat and rabbit. *Journal of Physiology, 193,* 707–733.

Brown, T. S. (1975). General biology of sensory systems. In B. Scharf (Ed.), *Experimental sensory psychology* (pp. 69–111). Glenview, IL: Scott, Foresman and Co.

Burgess, P. R. (1973). Cutaneous mechanoreceptors. In E. C. Carterette & M. P. Friedman (Eds.), *Handbook of perception. Vol. 3: Biology of perceptual systems* (pp. 219–249). New York: Academic Press.

Burgess, P. R., & Perl, E. R. (1973). Cutaneous mechanoreceptors and nociceptors. In A. Iggo (Ed.), *Handbook of sensory physiology, Vol. 2: Somatosensory system* (pp. 29–78). Berlin: Springer-Verlag.

Burgess, P. R., Mei, J., Tuckett, R. P., Horch, K. W., Ballinger, C. M., & Poulos, D. A. (1983). The neural signal for skin indentation depth. I. Changing indentations. *Journal of Neuroscience, 3,* 1572–1585.

Burgess, P. R., Petit, D., & Warren, R. M. (1968). Receptor types in cat hairy skin supplied by myelinated fibers. *Journal of Neurophysiology, 31,* 833–848.

Capraro, A. J., Verrillo, R. T., & Zwislocki, J. J. (1979). Psychophysical evidence for a triplex system of mechanoreception. *Sensory Processes, 3,* 334–352.

Cauna, N. (1956). Nerve supply and nerve endings in Meissner's corpuscles. *American Journal of Anatomy, 99,* 315–350.

Cauna, N., & Mannan, G. (1958). The structure of human digital Pacinian corpuscles (Corpuscula lamellosa) and its functional significance. *Journal of Anatomy, 92,* 1–20.

Chambers, M. R., Andres, K. H., von Düring, M., & Iggo, A. (1972). The structure and function of the slowly adapting Type II mechanoreceptor in hairy skin. *Quarterly Journal of Experimental Physiology, 57,* 417–445.

Checkosky, C. M., & Bolanowski, S. J., Jr. (1992). Effects of stimulus duration on the response properties of Pacinian corpuscles: Implications for the neural code. *Journal of the Acoustical Society of America, 91,* 3372–3380.

Checkosky, C. M., & Bolanowski, S. J., Jr. (1994). The effect of stimulus duration on frequency-response functions in the Pacinian (P) channel. *Somatosensory Motor Research, 11,* 47–56.

Cholewiak, R. W., & Collins, A. A. (1991). Sensory and physiological bases of touch. In M. A. Heller & W. Schiff (Eds.), *The psychology of touch* (pp. 23–60). Hillsdale, NJ: Lawrence Erlbaum.

Chouchkov, H. N. (1971). The ultrastructure of Pacinian corpuscles in men and cats. *Zeitschrift für Mikroskopich-Anatomische Forschung, 83,* 17–32.

Chouchkov, H. N. (1974). An electron microscopic study of the intraepidermal innervation of human glabrous skin. *Acta Anatomica, 88,* 33–46.

Cohen, R. H., & Vierck, C. J., Jr. (1993a). Population estimates for responses of cutaneous mechanoreceptors to a vertically indenting probe on the glabrous skin of monkeys. *Experimental Brain Research, 94,* 105–119.

Cohen, R. H., & Vierck, C. J., Jr. (1993b). Relationships between touch sensations and estimated population responses of peripheral afferent mechanoreceptors. *Experimental Brain Research, 94,* 120–130.

Connor, C. E., & Johnson, K. O. (1992). Neural coding of tactile texture: Comparison of spatial and temporal mechanisms for roughness perception. *Journal of Neuroscience, 12,* 3414–3426.

Connor, C. E., Hsiao, S. S., Phillips, J. R., & Johnson, K. O. (1990). Tactile roughness: Neural codes that account for psychophysical magnitude estimates. *Journal of Neuroscience, 10,* 3823–3836.

Cook, T. H. (1989). Mechanical properties of human skin with aging. In A. K. Balin & A. M. Kligman (Eds.), *Aging and the skin* (pp. 205–225). New York: Raven Press.

Craig, J. (1968). Vibrotactile spatial summation. *Perception and Psychophysics, 4,* 351–354.

Craig, J. C. (1972). Difference threshold for intensity of tactile stimuli. *Perception and Psychophysics, 11*, 150–152.

Craig, J. C. (1974). Vibrotactile difference thresholds for intensity and the effect of a masking noise. *Perception and Psychophysics, 15*, 123–127.

Craig, J. C., & Sherrick, C. E. (1982). Dynamic tactile displays. In W. Schiff & E. Foulke (Eds.), *Tactual perception: A sourcebook*, (pp. 209–233). Cambridge: Cambridge University Press.

Cronin, V. (1977). Active and passive touch at four age levels. *Developmental Psychology, 13*, 253–256.

Crook, M. N., & Crook, H. (1935). Adaptation to cutaneous pressure. *American Journal of Psychology, 47*, 301–308.

Darian-Smith, I. (1984). The sense of touch: Performance and peripheral neural processes. In J. M. Brookhart & V. B. Mountcastle (Eds.), *Handbook of physiology. Section 1: The nervous system. Vol. 3: Sensory processes* (pp. 739–788). Bethesda, MD: American Physiological Society.

Darian-Smith, I., & Kenins, P. (1980). Innervation density of mechanoreceptive fibres supplying glabrous skin of the monkey's index finger. *Journal of Physiology, 309*, 147–155.

Darian-Smith, I., & Oke, L. (1980). Peripheral neural representation of the spatial frequency of a grating moving across the monkey's finger pad. *Journal of Physiology, 309*, 117–133.

Darian-Smith, I., Davidson, I., & Johnson, K. O. (1980). Peripheral neural representation of spatial dimensions of a textured surface moving across the monkey's finger pad. *Journal of Physiology, 309*, 135–146.

Davis, H. (1961). Some principles of somatosensory action. *Physiological Reviews, 41*, 319–416.

Dellon, A. L. (1981). *Evaluation of sensibility and re-education of sensation in the hand*. Baltimore: Williams and Wilkins.

Diamond, J., Gray, J. A. B., & Inman, D. R. (1958). The relation between receptor potentials and the concentration of sodium ions. *Journal of Physiology, 142*, 382–394.

Donaldson, H. H. (1885). On the temperature sense. *Mind, 10*, 399–416.

Douglas, W. W., & Ritchie, J. M. (1957). Non-medullated fibres in the saphenous nerve which signal touch. *Journal of Physiology, 139*, 385–399.

Dresslar, F. B. (1894). Studies in the psychology of touch. *American Journal of Psychology, 6*, 313–368.

Dreyer, D. A., Hollins, M., & Whitsel, B. L. (1978). Factors influencing cutaneous directional sensitivity. *Sensory Processes, 2*, 71–79.

Dunlap, K. (1911). Palmesthetic difference sensibility for rate. *American Journal of Physiology, 29*, 108–114.

Dunlap, K. (1913). Palmesthetic beats and difference tones. *Science, 37*, 532–535.

Edin, B. B., & Abbs, J. H. (1991). Finger movement responses of cutaneous mechanoreceptors in the dorsal skin of the human hand. *Journal of Neurophysiology, 65*, 657–670.

Edin, B. B., Essick, G. K., Trulsson, M., & Olsson, K. Å. (1995). Receptor encoding of moving tactile stimuli in humans. I. Temporal pattern of discharge of individual low-threshold mechanoreceptors. *Journal of Neuroscience, 15*, 830–847.

Ekman, G., Hosman, J., & Lindström, B. (1965). Roughness, smoothness, and preference of quantitative relations in individual subjects. *Journal of Experimental Psychology, 70*, 18–26.

Escoffier, C., De Rigal, J., Rochefort, A., Vasselet, R., Lévêque, J.-L., & Agache, P. G. (1989). Age-related mechanical properties of human skin: An in vivo study. *Journal of Investigative Dermatology, 93*, 353–357.

Essick, G. K., Bredehoeft, K. R., McLaughlin, D. F., & Szaniszlo, J. A. (1991). Directional sensitivity along the upper limb in humans. *Somatosensory Motor Research, 8*, 13–22.

Essick, G. K., & Edin, B. B. (1995). Receptor encoding of moving tactile stimuli in humans. II. The mean response of individual low-threshold mechanoreceptors to motion across the receptive field. *Journal of Neuroscience, 15*, 848–864.

Essick, G. K., Franzén, O., McMillian, A., & Whitsel, B. (1991). Utilization of temporal and spatial cues to judge the velocity and traverse length of a moving tactile stimulus. In O. Franzén & J. Westman (Eds.), *Information processing in the somatosensory system* (pp. 341–352). New York: Stockton.

Essick, G. K., Franzén, O., & Whitsel, B. L. (1988). Discrimination and scaling of velocity of stimulus motion across the skin. *Somatosensory Motor Research, 6*, 21–40.

Essock, E. A., Krebs, W. K., & Prather, J. R. (1992). An anisotropy of human tactile sensitivity and its relation to the visual oblique effect. *Experimental Brain Research, 91*, 520–524.

Fletcher, H., & Munson, W. A. (1933). Loudness, its definition, measurement and calculation. *Journal of the Acoustical Society of America, 5*, 82–108.

Franke, E. K., von Gierke, H. E., Oestreicher, H. L., & von Wittern, W. (1951). *The propagation of surface waves over the human body.* (Technical Report No. 6464). Dayton, OH: USAF WADC.

Franzén, O. (1966). *On summation: A psychophysical study of the tactual sense.* (Speech Transmission Laboratory Quarterly Progress and Status Report). Stockholm Sweden: Royal Institute of Technology.

Franzén, O. (1969). The dependence of vibrotactile threshold and magnitude functions on stimulation frequency and signal level. *Scandinavian Journal of Psychology, 10*, 289–298.

Franzén, O., & Westman, J. (1991). *Information processing in the somatosensory system.* New York: Stockton.

Franzén, O., Thompson, F., Whitsel, B., & Young, M. (1984). Peripheral coding mechanisms of touch velocity. In C. Von Euler, O. Franzén, U. Lindblom, & D. Ottoson (Eds.), *Somatosensory mechanisms* (pp. 213–226). New York: Plenum.

Friedline, C. L. (1918). The discrimination of cutaneous patterns below the two-point limen. *American Journal of Psychology, 29*, 400–419.

Fucci, D., Small, L. H., & Petrosino, L. (1982). Intensity difference limens for lingual vibrotactile stimuli. *Bulletin of the Psychonomic Society, 1*, 54–56.

Fuchs, J. L., & Brown, P. B. (1984). Two-point discriminability: Relation to properties of the somatosensory system. *Somatosensory Research, 2*, 163–169.

Gardner, E. P., & Palmer, C. I. (1989). Stimulation of motion on the skin. I. Receptive fields and temporal frequency coding by cutaneous mechanoreceptors of OPTACON pulses delivered to the hand. *Journal of Neurophysiology, 62*, 1410–1434.

Gardner, E. P., & Sklar, B. F. (1994). Discrimination of the direction of motion on the human hand: A psychophysical study of stimulation parameters. *Journal of Neurophysiology, 71*, 2414–2429.

Gardner, E. P., & Spencer, W. (1972). Sensory funneling. I. Psychophysical observations of human subjects and responses of cutaneous mechanoreceptive afferents in the cat to patterned skin stimuli. *Journal of Neurophysiology, 35*, 925–953.

Gates, E. J. (1915). The determination of the limens of single and dual impression by the method of constant stimuli. *American Journal of Psychology, 26*, 152–157.

Gatti, H., & Dodge, R. (1929). Über die Unterschiedsempfindlichkeit bei der Reizung eines einzelnen isolierten Tastorgans. [On the differences of sensitivity in the stimulation of a single isolated organ of touch]. *Archiv für die gesamte Psychologie, 69*, 405–426.

Gault, R. H. (1924). Progress in experiments on tactual interpretation of oral speech. *Journal of Abnormal and Social Psychology, 19*, 155–159.

Gault, R. H., & Crane, G. W. (1928). Tactile patterns from certain vowel qualities instrumentally communicated from a speaker to a subject's fingers. *Journal of General Psychology, 1*, 353–359.

Gelb, T. (1914). Versuche im Gebiete der Raum- und Zeitanschauung. [Experiments in the field of the idea of space and time]. In *Berichte über den VI Kongress für experimentelle Psychologie in Göttingen*, (pp. 36–42). [Reports on the VI Congress for Experimental Psychology at Göttingen]. Leipzig.

Geldard, F. A. (1957). Adventures in tactile literacy. *American Psychologist, 12,* 115–124.

Geldard, F. A. (1975). *Sensory saltation. Mestastability in the perceptual world.* Hillsdale, NJ: Lawrence Erlbaum.

Geldard, F. A., & Sherrick, C. E. (1972). The cutaneous "rabbit": A perceptual illusion. *Science, 178,* 178–179.

Geldard, F. A., & Sherrick, C. E. (1983). The cutaneous saltatory area and its presumed neural basis. *Perception and Psychophysics 33,* 299–304.

Gellis, M., & Pool, R. (1977). Two-point discrimination distances in the normal hand and forearm. *Plastic and Reconstructive Surgery, 59,* 57–63.

Gescheider, G. A. (1976). Evidence in support of the duplex theory of mechanoreception. *Sensory Processes, 1,* 68–76.

Gescheider, G. A., Bolanowski, S. J., Jr., Hall, K. L., Hoffman, K. E., & Verrillo, R. T. (1994a). The effects of aging on information-processing channels in the sense of touch: I. Absolute sensitivity. *Somatosensory Motor Research, 11,* 345–357.

Gescheider, G. A., Bolanowski, S. J., Jr., & Verrillo, R. T. (1992). Sensory, cognitive and response factors in the judgement of sensory magnitude. In D. Algom (Ed.), *Psychophysical approaches to cognition* (pp. 575–621). Amsterdam: Elsevier.

Gescheider, G. A., Bolanowski, S. J., Jr., Verrillo, R. T., Arpajian, D. J., & Ryan, T. F. (1990). Vibrotactile intensity discrimination measured by three methods. *Journal of the Acoustical Society of America, 87,* 330–338.

Gescheider, G. A., Bolanowski, S. J., Jr., Zwislocki, J. J., Hall, K. L., & Mascia, C. (1994). The effects of masking on the growth of vibrotactile sensation magnitude and on the amplitude difference limen: A test of the equal senstion magnitude—equal difference limen hypothesis. *Journal of the Acoustical Society of America, 96,* 1479–1488.

Gescheider, G. A., Frisina, R. D., & Verrillo, R. T. (1979). Selective adaptation of vibrotactile thresholds. *Sensory Processes, 3,* 37–48.

Gescheider, G. A., Hoffman, K. E., Harrison, M. A., Travis, M. L., & Bolanowski, S. J., Jr. (1994). The effects of masking on vibrotactile temporal summation in the detection of sinusoidal and noise signals. *Journal of the Acoustical Society of America, 95,* 1006–1016.

Gescheider, G. A., O'Malley, M. J., & Verrillo, R. T. (1983). Vibrotactile forward masking: Evidence for channel independence. *Journal of the Acoustical Society of America, 74,* 474–485.

Gescheider, G. A., Sklar, B. F., Van Doren, C. L., & Verrillo, R. T. (1985). Vibrotactile forward masking: Psychophysical evidence for a triplex theory of cutaneous mechanoreception. *Journal of the Acoustical Society of America, 74,* 534–543.

Gescheider, G. A., Verrillo, R. T., Caparo, A. J., & Hamer, R. D. (1977). Enhancement of vibrotactile sensation magnitude and predictions from the duplex model of mechanoreception. *Sensory Processes, 1,* 187–203.

Gescheider, G. A., Verrillo, R. T., & Van Doren, C. L. (1982). Prediction of vibrotactile masking functions. *Journal of the Acoustical Society of America, 72,* 1421–1426.

Gibson, J. J. (1962). Observations on active touch. *Psychological Review, 69,* 477–491.

Gobel, A. K., & Hollins, M. (1993). Vibrotactile adaptation enhances amplitude discrimination. *Journal of the Acoustical Society of America, 93,* 418–424.

Gobel, A. K., & Hollins, M. (1994). Vibrotactile adaptation enhances frequency discrimination. *Journal of the Acoustical Society of America, 96,* 771–780.

Goff, G. D. (1967). Differential discrimination of frequency of cutaneous mechanical vibration. *Journal of Experimental Psychology, 74,* 294–299.

Goglia, G., & Sklenska, A. (1969). Richerche ultrastrutturali sopra i corpuscoli di Ruffini delle capsule articolari nel coniglio. [Research on the ultrastructure of the Ruffini corpuscle from the articular capsule of the rabbit]. *Quaderni Anatomia Pratia, 25,* 14–27.

Goldscheider, A. (1884). Die specifishe energie der gefühlsnerven der haut. [The specific energy of the sensory nerves of the skin]. *Monatshefte für practische Dermatologie, 3,* 49–67.

92 Joel D. Greenspan and Stanley J. Bolanowski

Goodwin, A. W., Browning, A. S., & Wheat, H. E. (1995). Representation of curved surfaces in responses of mechanoreceptive afferent fibers innervating the monkey's fingerpad. *Journal of Neuroscience, 15,* 798–810.

Goodwin, A. W., John, K. T., & Marceglia, A. H. (1991). Tactile discrimination of curvature by humans using only cutaneous information from the fingerpads. *Experimental Brain Research, 86,* 663–672.

Goodwin, A. W., John, K. T., Sathian, K., & Darian-Smith, I. (1989). Spatial and temporal factors determining afferent fiber responses to a grating moving sinusoidally over the monkey's fingerpad. *Journal of Neuroscience, 9,* 1280–1293.

Goodwin, A. W., & Morley, J. W. (1987a). Sinusoidal movement of a grating across the monkey's fingerpad: Effect of contact angle and force of the grating on afferent fiber responses. *Journal of Neuroscience, 7,* 2192–2202.

Goodwin, A. W., & Morley, J. W. (1987b). Sinusoidal movement of a grating across the monkey's fingerpad: Representation of grating and movement features in afferent fiber responses. *Journal of Neuroscience, 7,* 2168–2180.

Goodwin, A. W., & Wheat, H. E. (1992). Human tactile discrimination of curvature when contact area with the skin remains constant. *Experimental Brain Research, 88,* 447–450.

Gordon, G. (1978). *Active touch: The mechanism of recognition of objects by manipulation: A multi-disciplinary approach.* Oxford: Pergamon Press.

Gottschaldt, K.-M. Fruhstorfer, H., Schmidt, W., & Kraft, I. (1982). Thermosensitivity and its possible fine-structural basis in mechanoreceptors in the beak skin of geese. *Journal of Comparative Neurology, 205,* 219–245.

Gottschaldt, K.-M., & Vahle-Hinz, C. (1981). Merkel cell receptors: structure and transducer function. *Science, 214,* 183–186.

Goudge, M. E. (1918). A qualitative and quantitative study of Weber's illusion. *American Journal of Psychology, 29,* 81–119.

Gould, W. R., Vierck, C. J., Jr., & Luck, M. M. (1979). Cues supporting recognition of the orientation or direction of movement of tactile stimuli. In D. R. Kenshalo (Ed.), *Sensory functions of the skin of humans* (pp. 63–78). New York: Plenum Press.

Grahame, R. (1970). A method for measuring human skin elasticity in vivo with observations on the effects of age, sex, and pregnancy. *Clinical Science, 39,* 223–238.

Gray, J. A. B., & Matthews, P. B. C. (1951). A comparison of the adaptation of the Pacinian corpuscle with accommodation of its own axon. *Journal of Physiology, 144,* 454–464.

Gray, J. A. B., & Sato, M. (1953). Properties of the receptor potential in Pacinian corpuscles. *Journal of Physiology, 122,* 610–636.

Green, B. G. (1982). The perception of distance and location for dual tactile pressures. *Perception and Psychophysics, 31,* 315–323.

Green, B. G., & Craig, J. C. (1974). The role of vibration amplitude and static force in vibrotactile spatial summation. *Perception and Psychophysics, 16,* 503–507.

Greenspan, J. D. (1984). A comparison of force and depth of skin indentation upon psychophysical functions of tactile intensity. *Somatosensory Research, 2,* 33–48.

Greenspan, J. D. (1992). Influence of velocity and direction of surface-parallel cutaneous stimuli on responses of mechanoreceptors in feline hairy skin. *Journal of Neurophysiology, 68,* 876–889.

Greenspan, J. D. (1993). A laboratory exercise in somesthesis that is expeditious, inexpensive, and suitable for large classes. *Advances in Physiology Education, 10,* S2–S9.

Greenspan, J. D., Kenshalo, D. R., Sr., & Henderson, R. (1984). The influence of rate of skin indentation on threshold and suprathreshold tactile sensations. *Somatosensory Research, 1,* 379–393.

Greenspan, J. D., & LaMotte, R. H. (1993). Cutaneous mechanoreceptors of the hand: Experimental studies and their implications for clinical testing of tactile sensation. *Journal of Hand Therapy, 6,* 75–82.

Greenspan, J. D., & McGillis, S. L. B. (1991). Stimulus features relevant to the perception of sharpness and mechanically evoked cutaneous pain. *Somatosensory and Motor Research, 8,* 137–147.

Grunwald, A. P. (1966). A braille-reading machine. *Science, 154,* 144–146.

Halar, E. M., Hammond, M. C., LaCava, E. C., Camann, C., & Ward, J. (1987). Sensory perception threshold measurement: An evaluation of semiobjective testing devices. *Archives of Physical Medicine and Rehabilitation, 68,* 499–507.

Halata, Z. (1975). The mechanoreceptors of mammalian skin. Ultrastructure and morphological classification. *Advances in Anatomy, Embryology and Cell Biology, 50,* 1–77.

Halata, Z. (1988). Ruffini corpuscle—a stretch receptor in the connective tissue of the skin and locomotor apparatus. In W. Hamann & A. Iggo (Eds.), *Progress in brain research, Vol. 74: Transduction and cellular mechanisms in sensory receptors* (pp. 221–229). Amsterdam: Elsevier.

Hallin, R. G., & Torebjörk, H. E. (1976). Studies on cutaneous A and C fiber afferents, skin nerve blocks and perception. In Y. Zotterman (Ed.), *Sensory function of the skin* (pp. 137–148). Oxford: Pergamon Press.

Hamer, R. D., Verrillo, R. T., & Zwislocki, J. J. (1983). Vibrotactile masking of Pacinian and non-Pacinian channels. *Journal of the Acoustical Society of America, 73,* 1293–1303.

Harrington, T., & Merzenich, M. (1970). Neural coding in the sense of touch: Human sensations of skin indentation compared with the responses of slowly adapting mechanoreceptive afferents innervating the hairy skin of monkeys. *Experimental Brain Research, 10,* 251–264.

Hartmann, G. (1875). Der Raumsinn der Haut des Rumpfes und des Halses. [Spatial sense of the skin of the trunk and neck]. *Zeitschrift für Biologie, 11,* 79–101.

Heller, M. A., & Myers, D. S. (1983). Active and passive tactual recognition of form. *Journal of General Psychology, 108,* 225–229.

Heller, M. A., & Schiff, W. (1991). *The psychology of touch,* Hillsdale, NJ: Lawrence Erlbaum.

Hellmann, R. P., & Zwislocki, J. J. (1961). Some factors affecting the estimation of loudness. *Journal of the Acoustical Society of America, 33,* 687–694.

Helson, H., & King, S. M. (1931). The tau effect: An example of psychological relativity. *Journal of Experimental Psychology, 14,* 202–217.

Hollins, M., Faldowski, R., Rao, S., & Young, F. (1993). Perceptual dimensions of tactile surface texture: A multidimensional scaling analysis. *Perception and Psychophysics, 54,* 697–705.

Horch, K. W., Clark, F. J., & Burgess, P. R. (1975). Awareness of knee joint angle under static conditions. *Journal of Neurophysiology, 38,* 1436–1477.

Horch, K. W., Tuckett, R. P., & Burgess, P. R. (1977). A key to the classification of cutaneous mechanoreceptors. *Journal of Investigative Dermatology, 69,* 75–82.

Hugony, A. (1935). Uber die Empfindung von schwingunzen mittels des Tastsinnes. [On the perception of vibrations by means of the sense of touch]. *Zeitschrift für Biologie, 96,* 548–553.

Iggo, A. (1960). Cutaneous mechanoreceptors with afferent C fibres. *Journal of Physiology, 152,* 337–353.

Iggo, A. (1962). New specific sensory structures in hairy skin. *Acta Neurovegetativa, 24,* 175–180.

Iggo, A. (1966). Relation of single receptor activity to parameters of stimuli. Cutaneous receptors with a high sensitivity to mechanical displacement. In A. V. S. De Reuck and J. Knight (Eds.), *Ciba Foundation Symposium on Touch, Heat, and Pain* (pp. 237–256). Boston: Little, Brown, And Company.

Iggo, A. (1976). Is the physiology of cutaneous receptors determined by morphology? In A. Iggo & O. B. Ilyinski (Eds.), *Progress in brain research, Vol. 43: Somatosensory and visceral receptor mechanisms* (pp. 15–31). Amsterdam: Elsevier.

Iggo, A., & Andres, K. H. (1982). Morphology of cutaneous receptors. In W. M. Cowan,

Z. W. Hall, & E. R. Kandel (Eds.), *Annual review of neuroscience* (Vol. 5, pp. 1–32), Palo Alto, CA: Annual Reviews.

Iggo, A., & Kornhuber, H. H. (1977). A quantitative study of C-mechanoreceptors in hairy skin of the cat. *Journal of physiology, 271,* 549–565.

Iggo, A., & Muir, A. R. (1969). The structure and function of a slowly adapting touch corpuscle in hairy skin. *Journal of Physiology, 200,* 763–796.

Ilyinski, O. B. (1965). Process of excitation and inhibition in single mechanoreceptors (Pacinian corpuscles). *Nature, 208.* 351–353.

Ilyinski, O. B., Akoev, G. N., Krasnikova, T. L., & Elman, S. I. (1976). K and Na ion content in the Pacinian corpuscle fluid and its role in the activity of receptors. *Pflügers Archiv, 361,* 279–285.

Ilyinski, O. B., Krylov, B. V., & Cherepnov, V. L. (1976). Study of the optimal form of nerve ending of encapsulated tissue mechanoreceptor (Pacinian corpuscle). *Neurophysiologia, 9,* 423–428.

Jänig, W. (1971). Morphology of rapidly and slowly adapting mechanoreceptors in the hairless skin of the cat's hind foot. *Brain Research, 28,* 217–231.

Järvilehto, T., Hämäläinen, H. A., & Laurinen, P. (1976). Characteristics of single mechanoreceptive fibres innervating hairy skin of the human hand. *Experimental Brain Research, 25,* 45–61.

Järvilehto, T., Hämäläinen, H. A., & Soininen, K. (1981). Peripheral neural basis of tactile sensations in man: 2. Characteristics of human mechanoreceptors in the hairy skin and correlations of their activity with tactile sensations. *Brain Research, 219,* 13–27.

Joël, W. (1935). On the tactile perception of vibration frequencies. *Psychological Review, 42,* 267–273.

Johansson, R. S. (1976). Receptive field sensitivity profile of mechanosensitive units innervating the glabrous skin of the human hand. *Brain Research, 104,* 330–334.

Johansson, R. S. (1978). Tactile sensibility in the human hand: Receptive field characteristics of mechanoreceptive units in the glabrous skin area. *Journal of Physiology, 281,* 101–123.

Johansson, R. S., & LaMotte, R. H. (1983). Tactile detection thresholds for a single asperity on an otherwise smooth surface. *Somatosensory Research, 1,* 21–32.

Johansson, R. S., & Vallbo, A. B. (1979). Tactile sensibility in the human hand: Relative and absolute densities of four types of mechanoreceptive units in glabrous skin. *Journal of Physiology, 286,* 283–300.

Johansson, R. S., & Vallbo, A. B. (1980). Spatial properties of the population of mechanoreceptive units in the glabrous skin of the human hand. *Brain Research, 184,* 353–366.

Johansson, R. S., Vallbo, A. B., & Westling, G. (1980). Thresholds of mechanosensitive afferents in the human hand as measured with von Frey hairs. *Brain Research, 184,* 343–351.

Johnson, K. O. (1974). Reconstruction of population response to a vibratory stimulus in quickly adapting mechanoreceptive afferent fiber population innervating glabrous skin of the monkey. *Journal of Neurophysiology, 37,* 48–72.

Johnson, K. O., & Hsiao, S. S. (1992). Neural mechanisms of tactual form and texture perception. *Annual Review of Neuroscience, 15,* 227–250.

Johnson, K. O., & Hsiao, S. S. (1994). Evaluation of the relative roles of slowly and rapidly adapting afferent fibers in roughness perception. *Canadian Journal of Physiology and Pharmacology, 72,* 488–497.

Johnson, K. O., & Lamb, G. D. (1981). Neural mechanisms of spatial tactile discrimination: Neural patterns evoked by Braille-like dot patterns in the monkey. *Journal of Physiology, 310,* 117–144.

Johnson, K. O., & Phillips, J. R. (1981). Tactile spatial resolution. I. Two-point discrimination, gap detection, grating resolution, and letter recognition. *Journal of Neurophysiology, 46,* 1177–1191.

Johnson, K. O., Phillips, J. R., Hsiao, S. S., & Bankman, I. N. (1991). Tactile pattern recognition. In O. Franzén & J. Westman (Eds.), *Information processing in the somatosensory system* (pp. 305–318). New York: Stockton.

Johnson, K. O., Van Boven, R. W., & Hsiao, S. S. (1994). The perception of two points is not the spatial resolution threshold. In J. Boivie, P. Hansson, and U. Lindblom (Eds.), *Touch, Temperature, and Pain in Health and Disease*, (pp. 389–404), Seattle: IASP Press.

Jones, F. N. (1956). Space–time relationships in somesthetic localization. *Science, 124,* 484.

Jones, F. N. (1960). Subjective intensity functions in somesthesis. In G. R. Hawkes (Ed.), *Symposium on cutaneous sensitivity* (pp. 63–72), Fort Knox: U.S. Army Medical Research Laboratory.

Jones, M. B., & Vierck, C. J., Jr. (1973). Length discrimination on the skin. *American Journal of Psychology, 86,* 49–60.

Kaas, J. H., Nelson, R. J., Sur, M., Dykes, R. W., & Merzenich, M. M. (1984). The somatotopic organization of the ventroposterior thalamus of the squirrel monkey, Saimiri sciureus. *Journal of Comparative Neurology, 226,* 111–140.

Kakuda, N. (1992). Conduction velocity of low-threshold mechanoreceptive afferent fibers in the glabrous and hairy skin of human hands measured with microneurography and spike-triggered averaging. *Neuroscience Research, 15,* 179–188.

Katz, D. (1925). *Der Aufbau der Tastwelt. [The world of touch].* Barth: Leipzig.

Keidel, W. D. (1984). The sensory detection of vibrations. In W. W. Dawson & J. M. Enoch (Eds.), *Foundations of sensory science* (pp. 465–512). Berlin: Springer-Verlag.

Kenshalo, D. R., Sr. (1978). Biophysics and psychophysics of feeling. In E. C. Carterette & M. P. Friedman (Eds.), *Handbook of perception, Vol. VIB: Feeling and hurting* (pp. 29–74). New York: Academic Press.

Kenshalo, D. R., Sr. (1979). *Sensory functions of the skin of humans,* New York: Plenum Press.

Kenshalo, D. R., & Nafe, J. P. (1962). A quantitative theory of feeling: 1960. *Psychological Review, 69,* 17–33.

Kiesow, F. (1922). Über die taktile Unterschiedsempfindlichkeit bei suczessiver Reizung einzelner Empfindungsorgane. [On the tactile differences in sensitivity in successive stimulations of a single sense organ]. *Archiv für die gesamte Psychologie, 43,* 11–23.

Kilgard, M. P., & Merzenich, M. M. (1995). Anticipated stimuli across skin. *Nature, 373,* 663.

Knibestöl, M. (1973). Stimulus response functions of rapidly adapting mechanoreceptors in the human glabrous skin area. *Journal of Physiology, 232,* 427–452.

Knibestöl, M. (1975). Stimulus–response functions of slowly adapting mechanoreceptors in the human glabrous skin area. *Journal of Physiology, 245,* 63–80.

Knibestöl, M., & Vallbo, A. B. (1970). Single unit analysis of mechanoreceptor activity from the human glabrous skin. *Acta Physiologica Scandinavica, 80,* 178–195.

Knibestöl, M., & Vallbo, A. B. (1980). Intensity of sensation related to activity of slowly adapting mechanoreceptive units in the human hand. *Journal of Physiology, 300,* 251–267.

Knudsen, V. O. (1928). Hearing with the sense of touch. *Journal of General Psychology, 1,* 320–352.

Konietzny, F., & Hensel, H. (1977). Response of rapidly and slowly adapting mechanoreceptors and vibratory sensitivity in human hairy skin. *Pflügers Archiv, 368,* 39–44.

Kottenkamp, R., & Ullrich, H. (1870). Versuche über den raumsinn der haut der oberen extremität. [Experiments on the spatial sense of the skin of the upper extremity]. *Zeitschrift für Biologie, 6,* 37–52.

Kruger, L., & Kenton, B. (1973). Quantitative neural and psychophysical data for cutaneous mechanoreceptor function. *Brain Research, 49,* 1–24.

Kumazawa, T., & Perl, E. R. (1977). Primate cutaneous sensory units with unmyelinated (C) afferent fibers. *Journal of Neurophysiology, 40,* 1325–1338.

Lamb, G. D. (1983a). Tactile discrimination of textured surfaces: Psychophysical performance measurements in humans. *Journal of Physiology, 338,* 551–565.

Lamb, G. D. (1983b). Tactile discrimination of textured surfaces: Peripheral neural coding in the monkey. *Journal of Physiology, 338,* 567–587.

LaMotte, R. H., & Mountcastle, V. B. (1975). Capacities of humans and monkeys to discriminate between vibratory stimuli of different frequency and amplitude: A correlation between neural events and psychophysical measurements. *Journal of Neurophysiology, 38,* 539–559.

LaMotte, R. H., & Whitehouse, J. M. (1986). Tactile detection of a dot on a smooth surface: Peripheral neural events. *Journal of Neurophysiology, 56,* 1109–1128.

Langford, N., Hall, R. J., & Monty, R. A. (1973). Cutaneous perception of a track produced by a moving point across the skin. *Journal of Experimental Psychology, 97,* 59–63.

Lederman, S. J. (1974). Tactile roughness of grooved surfaces: The touching process and effects of macro- and microsurface structure. *Perception and Psychophysics, 16,* 385–395.

Lederman, S. J. (1981). The perception of surface roughness by active and passive touch. *Bulletin of the Psychonomic Society, 18,* 253–255.

Lederman, S. J. (1983). Tactual roughness perception: Spatial and temporal determinants. *Canadian Journal of Psychology, 37,* 498–511.

Lederman, S. J., & Taylor, M. M. (1972). Fingertip force, surface geometry, and the perception of roughness by active touch. *Perception and Psychophysics, 12,* 401–408.

Light, A. R., & Perl, E. R. (1993). Peripheral sensory systems. In P. J. Dyck, P. K. Thomas, J. W. Griffin, P. A. Low, & J. F. Poduslo (Eds.), *Peripheral neuropathy* (Vol. 1, pp. 149–165). Philadelphia: W. B. Saunders.

Loe, P. R., Whitsel, B. L., Dreyer, D. A., & Metz, C. B. (1977). Body representation in ventrobasal thalamus of macaque: A single-unit analysis. *Journal of Neurophysiology, 40,* 1339–1355.

Loewenstein, W. R. (1958). Generator processes of repetitive activity in a Pacinian corpuscle. *Journal of General Physiology, 41,* 825–845.

Loewenstein, W. R. (1959). The generation of electrical activity in a nerve ending. *Annals of the New York Academy of Sciences, 81,* 367–387.

Loewenstein, W. R. (1971). Mechano-electric transduction in the Pacinian corpuscle. Initiation of sensory impulses in mechanoreceptors. In W. R. Loewenstein (Ed.), *Handbook of sensory physiology* (pp. 269–290). Berlin: Springer-Verlag.

Loewenstein, W. R., & Altimirano-Orrega, R. (1958). Generation and propagation of impulses during refractoriness in a Pacinian corpuscle. *Nature, 181,* 124.

Loewenstein, W. R., & Molins, D. (1958). Cholinesterase in a receptor. *Science, 128,* 1284.

Loewenstein, W. R., & Rathkamp, R. (1958a). Localization of generator structures of electrical activity in a Pacinian corpuscle. *Science, 123,* 341–342.

Loewenstein, W. R., & Rathkamp, R. (1958b). The sites for mechano-electric conversion in a Pacinian corpuscle. *Journal of General Physiology, 411,* 1245–1265.

Loewenstein, W. R., & Skalak, R. (1966). Mechanical transmission in a Pacinian corpuscle. An analysis and a theory. *Journal of Physiology, 182,* 346–378.

Loomis, J. M. (1979). An investigation of tactile hyperacuity. *Sensory Processes, 3,* 289–302.

Loomis, J. M. (1981). Tactile pattern perception. *Perception, 10,* 5–27.

Loomis, J. M. (1982). Analysis of tactile and visual confusion matrices. *Perception and Psychophysics, 31,* 41–52.

Loomis, J. M., & Lederman, S. J. (1986). Tactual perception. In K. R. Boff, L. Kaufman, & J. P. Thomas (Eds.), *Handbook of perception and human performance. Vol. II. Cognitive processes and performance* (pp. 1–41). New York: John Wiley and Sons.

Lundström, R. (1984). Local vibrations—Mechanical impedance of the human hand's glabrous skin. *Journal of Biomechanics, 17,* 137–144.

Lynn, B. (1983). Cutaneous sensation. In L. A. Goldsmith (Ed.), *Biochemistry and physiology of the skin* (pp. 654–684). Oxford: Oxford University Press.

Macefield, G., Gandevia, S. C., & Burke, D. (1990). Perceptual responses to microstimulation of single afferents innervating joints, muscles and skin of the human hand. *Journal of Physiology, 429,* 113–129.

Mackenzie, R. A., Burke, D., Skuse, N. F., & Lethlean, A. K. (1975). Fibre function and perception during cutaneous nerve block. *Journal of Neurology, Neurosurgery, and Psychiatry, 38,* 865–873.

Magee, L. E., & Kennedy, J. M. (1980). Exploring pictures tactually. *Nature, 283,* 287–288.

Makous, J. C., Friedman, R. M., & Vierck, C. J., Jr. (1995). A critical band filter in touch. *Journal of Neuroscience, 15,* 2808–2818.

Marks, R. (1983). Mechanical properties of the skin. In L. A. Goldsmith (Ed.), *Biochemistry and physiology of the skin* (pp. 1237–1254). Oxford: Oxford University Press.

Mei, J., Tuckett, R. P., Poulos, D. A., Horch, K. W., Wei, J. Y., & Burgess, P. R. (1983). The neural signal for skin indentation depth II. Steady indentations. *Journal of Neuroscience, 12,* 2652–2659.

Moore, T. J. (1970). A survey of the mechanical characteristics of skin and tissue in response to vibratory stimulation. *IEEE Transactions on Man-Machine Systems, 11,* 79–84.

Morley, J. W., & Goodwin, A. W. (1987). Sinusoidal movement of a grating across the monkey's fingerpad: Temporal patterns of afferent fiber responses. *Journal of Neuroscience, 7,* 2181–2191.

Morley, J. W., Goodwin, A. W., & Darian-Smith, I. (1983). Tactile discrimination of gratings. *Experimental Brain Research, 49,* 291–299.

Mountcastle, V. B. (1967). The problem of sensing and the neural coding of sensory events. In G. C. Quarton, T. Melnechuk, & F. O. Schmitt (Eds.), *The neurosciences* (pp. 393–408), New York: Rockefeller University Press.

Mountcastle, V. B. (1984). Central nervous mechanisms in mechanoreceptive sensibility. In J. M. Brookhart, V. B. Mountcastle, I. Darian-Smith, & S. R. Geiger (Eds.), *Handbook of physiology. The nervous system, Volume III. Sensory processes* (pp. 789–878). Baltimore: Williams & Wilkins.

Mountcastle, V. B., LaMotte, R. H., & Carli, G. (1972). Detection thresholds for vibratory stimuli in humans and monkeys: Comparison with threshold events in mechanoreceptive afferent nerve fibers innervating the monkey hand. *Journal of Neurophysiology, 35,* 122–136.

Mountcastle, V. B., Steinmetz, M. A., & Romo, R. (1990). Frequency discrimination in the sense of flutter: Psychophysical measurements correlated with postcentral events in behaving monkeys. *Journal of Neuroscience, 10,* 3032–3044.

Mountcastle, V. B., Talbot, W. H., & Kornhuber, H. H. (1966). The neural transformation of mechanical stimuli delivered to the monkey's hand. In A. V. S. De Reuck & J. Knight (Eds.), *Touch, heat, and pain* (pp. 325–345). Boston: Little, Brown and Company.

Mountcastle, V. B., Talbot, W. H., Sakata, H., & Hyvärinen, J. (1969). Cortical neuronal mechanisms in flutter-vibration studied in unanesthetized monkeys. Neuronal periodicity and frequency discrimination. *Journal of Neurophysiology, 32,* 452–484.

Muijser, H. (1994). An indication for spatial integration in a non-Pacinian mechanoreceptor system? *Journal of the Acoustical Society of America, 96,* 781–785.

Munger, B. L., Page, R. B., & Pubols, B. H., Jr. (1979). Identification of specific mechanosensory receptors in glabrous skin of dorsal root ganglionectomized primates. *Anatomical Record, 193,* 630–631.

Müller, J. (1838). *Handbuch der Physiologie des Menschen, II, Bk. V.* [Handbook of human physiology]. Koblenz, Germany: J. Hölscher.

Nafe, J. P. (1927). The psychology of felt experience. *American Journal of Psychology, 39,* 367–389.

Nafe, J. P.,, & Wagoner, K. S. (1941a). The nature of pressure adaptation. *Journal of General Psychology, 25,* 323–351.

Nafe, J. P., & Wagoner, K. S. (1941b). The nature of sensory adaptation. *Journal of General Psychology, 25,* 295–321.

Nishi, K., & Sato, M. (1968). Depolarizing and hyperpolarizing receptor potentials in the non-myelinated nerve terminal in Pacinian corpuscles. *Journal of Physiology, 199,* 383–396.

Nishi, K., Oura, C., & Pallie, W. (1969). Fine structure of the Pacinian corpuscles in the mesentery of the cat. *Journal of Cell Biology, 43,* 539–553.

Nolan, M. F. (1985). Quantitative measure of cutaneous sensation. Two-point discrimination values for the face and trunk. *Physical Therapy, 65,* 181–185.

Nordin, M. (1990). Low-threshold mechanoreceptive and nociceptive units with unmyelinated (C) fibres in the human supraorbital nerve. *Journal of Physiology, 426,* 229–240.

Norrsell, U., & Olausson, H. (1992). Human, tactile, directional sensibility and its peripheral origins. *Acta Physiologica Scandinavica, 144,* 155–161.

Norrsell, U., & Olausson, H. (1994). Spatial cues serving the tactile directional sensibility of the human forearm. *Journal of Physiology, 478,* 533–540.

Ochoa, J. L., & Torebjörk, H. E. (1983). Sensations evoked by intraneural microstimulation of single mechanoreceptor units innervating the human hand. *Journal of Physiology, 342,* 633–654.

Oestreicher, H. L. (1950). *On the theory of the propagation of mechanical vibrations in human and animal tissue.* (Technical Report No. 6244). Dayton, OH: USAF WADC.

Olausson, H., & Norrsell, U. (1993). Observations on human tactile directional sensibility. *Journal of Physiology, 464,* 545–559.

Périlhou, P. (1947). The vibratory sense. *Journal of General Psychology, 36,* 23–28.

Phillips, J. R., & Matthews, P. B. C. (1993). Texture perception and afferent coding distorted by cooling the human ulnar nerve. *Journal of Neuroscience, 13,* 2332–2341.

Phillips, J. R. Johansson, R. S., & Johnson, K. O. (1990). Representation of braille characters in human nerve fibres. *Experimental Brain Research, 81,* 589–592.

Phillips, J. R. Johansson, R. S., & Johnson, K. O. (1992). Responses of human mechanoreceptive afferent to embossed dot arrays scanned across fingerpad skin. *Journal of Neuroscience, 12,* 827–839.

Phillips, J. R., & Johnson, K. O. (1981a). Tactile spatial resolution. II. Neural representation of bars, edges, and gratings in monkey primary afferents. *Journal of Neurophysiology, 46,* 1192–1203.

Phillips, J. R., & Johnson, K. O. (1981b). Tactile spatial resolution. III. A continuum mechanics model of skin predicting mechanoreceptor responses to bars, edges, and gratings. *Journal of Neurophysiology, 46,* 1204–1224.

Phillips, J. R., Johnson, K. O., & Hsiao, S. S. (1988). Spatial pattern representation and transformation in monkey somatosensory cortex. *Proceedings of the National Academy of Sciences USA, 85,* 1317–1321.

Potvin, A. R., & Tourtellotte, W. W. (1985). *Quantitative examination of neurologic functions. Vol. II: Methodology for test and patient assessments and design of a computer-automated system.* Boca Raton, FL: CRC Press.

Poulos, D. A., Mei, J., Horch, K. W., Tuckett, R. P., Wei, J. Y., Cornwall, M. C., & Burgess, P. R. (1984). The neural signal for the intensity of a tactile stimulus. *Journal of Neuroscience, 4,* 2016–2024.

Pubols, B. H., Jr. (1987). Effect of mechanical stimulus spread across glabrous skin of raccoon and squirrel monkey hand on tactile primary afferent fiber discharge. *Somatosensory Research, 4,* 273–308.

Pubols, B. H., & Pubols, L. M. (1982). Magnitude scaling of displacement and velocity of tactile stimuli applied to the human hand. *Brain Research, 233,* 409–413.

Quilliam, T. A., & Sato, M. (1955). The distribution of myelin on the nerve fibres from Pacinian corpuscles. *Journal of Physiology, 129,* 167–176.

Ray, R. H., & Doetsch, G. S. (1990a). Coding of stimulus location and intensity in populations of mechanosensitive nerve fibers of the raccoon: I. Single fiber response properties. *Brain Research Bulletin, 25,* 517–532.

Ray, R. H., & Doetsch, G. S. (1990b). Coding of stimulus location and intensity in populations of mechanosensitive nerve fibers of the raccoon: II. Across-fiber response patterns. *Brain Research Bulletin, 25*, 533–550.

Ray, R. H., Mallach, L. E., & Kruger, L. (1985). The response of single guard and down hair mechanoreceptors to moving air-jet stimulation. *Brain Research, 346*, 333–347.

Recanzone, G. H., Jenkins, W. M., Hradek, G. T., & Merzenich, M. M. (1992). Progressive improvement in discriminative abilities in adult owl monkeys performing a tactile frequency discrimination task. *Journal of Neurophysiology, 67*, 1015–1030.

Recanzone, G. H., Merzenich, M. M., & Jenkins, W. M. (1992). Frequency discrimination training engaging a restricted skin surface results in an emergence of a cutaneous response zone in cortical area 3a. *Journal of Neurophysiology, 67*, 1057–1070.

Recanzone, G. H., Merzenich, M. M., Jenkins, W. M., Grajski, K. A., & Dinse, H. R. (1992). Topographic reorganization of the hand representation in cortical area 3b of owl monkeys trained in a frequency-discrimination task. *Journal of Neurophysiology, 67*, 1031–1056.

Roberts, W. H. (1932). A two-dimensional analysis of the discrimination of differences in the frequency of vibration by means of the sense of touch. *Journal of the Franklin Institute, 213*, 283–311.

Ross, H. E., & Murray, D. J. (1978). *E. H. Weber: The sense of touch.* New York: Academic Press.

Rothenberg, M., Verrillo, R. T., Zahorian, S. A., Brachman, M. L., & Bolanowski, S. J., Jr. (1977). Vibrotactile frequency for encoding a speech parameter. *Journal of the Acoustical Society of America, 62*, 1003–1012.

Sathian, K., Goodwin, A. W., John, K. T., & Darian-smith, I. (1989). Perceived roughness of a grating: Correlation with responses of mechanoreceptive afferents innervating the monkey's fingerpad. *Journal of Neuroscience, 9*, 1273–1279.

Sato, M. (1961). Response of Pacinian corpuscles to sinusoidal vibration. *Journal of Physiology, 159*, 391–409.

Saxena, V. P. (1983). Temperature distribution in human skin and subdermal tissues. *Journal of Theoretical Biology, 102*, 277–286.

Schady, W. J. L., Torebjörk, H. E., Ochoa, J. L. (1983). Peripheral projections of nerve fibres in the human median nerve. *Brain Research, 277*, 249–261.

Schiff, W., & Foulke, E. (1982). *Tactual perception.* Cambridge: Cambridge University Press.

Schiller, H. (1953). *Über die Amplituden Unterschiedsschwellen des Vibrationssines beim Menschen.* [On the amplitudes of difference thresholds of the vibration sense in humans]. Unpublished doctoral dissertation, University of Erlanger, Germany.

Schwartz, A. S., Perey, A. J., & Azulay, A. (1975). Further analysis of active and passive touch in pattern discrimination. *Bulletin of the Psychonomic Society, 6*, 7–9.

Setzepfand, W. Z. (1935). Zur Frequenzabhängigkeit der vibrationsempfindung des Menschen. [On the frequency dependence of the vibration sense in humans]. *Zeitschrift für Biologie, 96*, 236–240.

Sherrick, C. E. (1950). *Measurement of the differential sensitivity of the human skin to mechanical vibration.* Unpublished Master's thesis, University of Virginia.

Sherrick, C. E. (1953). Variables affecting sensitivity of the human skin to mechanical vibration. *Journal of Experimental Psychology, 45*, 273–282.

Sherrick, C. E. (1954). *Measurement of delta f.* (Technical Report 24). University of Virginia, Project NR140-598, Office of Naval Research.

Sherrick, C. E. (1960). Observations relating to some common psychophysical functions applied to skin. In G. R. Hawkes (Ed.), *Symposium on Cutaneous Sensitivity* (pp. 147–158). Ft. Knox, KY: U.S. Army Research Laboratory.

Sherrick, C. E., & Cholewiak, R. W. (1986). Cutaneous sensitivity. In K. R. Boff, L. Kaufman, & J. P. Thomas (Eds.), *Handbook of perception and human performance. Vol. I. Sensory processes and perception* (pp. 1–58). New York: John Wiley and Sons.

Sherrick, C. E., Cholewiak, R. W., & Collins, A. A. (1990). The localization of low- and high-frequency vibrotactile stimuli. *Journal of the Acoustical Society of America, 88,* 169–179.

Sherrington, C. S. (1900). Cutaneous sensations. In E. A. Schäfer (Ed.), *Textbook of physiology* (pp. 920–1001). Edinburgh: Young J. Pentland.

Sinclair, D. C. (1955). Cutaneous sensation and the doctrine of specific energy. *Brain, 78,* 584–614.

Spencer, P. S., & Schaumberg, H. H. (1973). An ultrastructural study of the inner core of the Pacinian corpuscle. *Journal of Neurocytology, 2,* 217–235.

Srinivasan, M. A. (1989). Surface deflection of primate fingertip under line load. *Journal of Biomechanics, 22,* 343–349.

Srinivasan, M. A., & Dandekar, K. (1996). An investigation of the mechanics of tactile sense using two dimensional models of the primate fingertip. *Journal of Biomechanical Engineering, 118,* 48–55.

Srinivasan, M. A., & LaMotte, R. H. (1987). Tactile discrimination of shape: Responses of slowly and rapidly adapting mechanoreceptive afferents to a step indented into the monkey fingerpad. *Journal of Neuroscience, 7,* 1682–1697.

Srinivasan, M. A., Whitehouse, J. M., & LaMotte, R. H. (1990). Tactile detection of slip: Surface microgeometry and peripheral neural codes. *Journal of Neurophysiology, 63,* 1323–1332.

Stevens, J. C. (1982). Temperature can sharpen tactile acuity. *Perception and Psychophysics, 31,* 577–580.

Stevens, J. C. (1989). Temperature and the two-point threshold. *Somatosensory Motor Research, 6,* 275–284.

Stevens, J. C. (1992). Aging and spatial acuity of touch. *Journal of Gerontology, 47,* P35–40.

Stevens, J. C., & Patterson, M. Q. (1995). Dimensions of spatial acuity in the touch sense: Changes over the life span. *Somatosensory and Motor Research, 12,* 29–47.

Stevens, S. S. (1968). Tactile vibration: Changes of exponent with frequency. *Perception and Psychophysics, 3,* 223–228.

Stevens, S. S., & Harris, J. R. (1962). The scaling of subjective roughness and smoothness. *Journal of Experimental Psychology, 64,* 489–494.

Stone, L. A. (1967). Subjective roughness and smoothness for individual judges. *Psychonomic Science 9,* 347–348.

Sur, M., Nelson, R. J., & Kaas, J. H. (1978). The representation of the body surface in somatosensory area I of the grey squirrel. *Journal of Comparative Neurology, 179,* 425–449.

Sur, M., Nelson, R. J., & Kaas, J. H. (1982). Representation of the body surface in cortical areas 3b and 1 of squirrel monkeys: Comparisons with other primates. *Journal of Comparative Neurology, 211,* 177–192.

Talbot, W. H., Darian-Smith, I., Kornhuber, H. H., & Mountcastle, V. B. (1968). The sense of flutter-vibration: Comparison of the human capacity with response patterns of mechanoreceptive afferents from the monkey hand. *Journal of Neurophysiology, 31,* 301–334.

Tapper, D. N. (1965). Stimulus–response relationships in the cutaneous slowly-adapting mechanoreceptor in hairy skin of the cat. *Experimental Neurology, 13,* 364–385.

Taylor, M. M., & Lederman, S. J. (1975). Tactile roughness of grooved surfaces: A model and the effect of friction. *Perception and Psychophysics, 17,* 23–36.

Titchener, E. B. (1916). On ethnological tests of sensation and perception with special reference to the tests of color vision and tactile discrimination described in reports of the Cambridge Anthropological Expedition to Torres Straits. *Proceedings of the American Philosophical Society, 55,* 204–236.

Torebjörk, H. E., & Ochoa, J. L. (1980). Specific-sensations evoked by activity in single identified sensory units in man. *Acta Physiologica Scandinavica, 110,* 445–447.

Torebjörk, H. E., Vallbo, A. B., & Ochoa, J. L. (1987). Intraneural microstimulation in man. Its relation to specificity of tactile sensations. *Brain, 110,* 1509–1530.

Tuckett, R. P., Horch, K. W., & Burgess, P. R. (1978). Response of cutaneous hair and field mechanoreceptors in cat to threshold stimuli. *Journal of Neurophysiology, 41,* 138–149.

Vallbo, Å. B. (1981). Sensations evoked from the glabrous skin of the human hand by electrical stimulation of unitary mechanosensitive afferents. *Brain Research, 215,* 359–363.

Vallbo, Å. B., Hagbarth, K.-E., Torebjörk, H. E., & Wallin, B. G. (1979). Somatosensory, proprioceptive, and sympathetic activity in human peripheral nerves. *Physiological Reviews, 59,* 919–957.

Vallbo, Å. B., & Johansson, R. S. (1978). The tactile sensory innervation of the glabrous skin of the human hand. In G. Gordon (Ed.), *Active touch: The mechanism of recognition of objects by manipulation. A multi-disciplinary approach* (pp. 29–54). Oxford: Pergamon Press.

Vallbo, Å. B., & Johansson, R. S. (1984). Properties of cutaneous mechanoreceptors in the human hand related to touch sensation. *Human Neurobiology, 3,* 3–14.

Vallbo, Å. B., Olausson, H., Wessberg, J., & Kakuda, N. (1995). Receptive field characteristics of tactile units with myelinated afferents in hairy skin of human subjects. *Journal of Physiology, 483,* 783–795.

Vallbo, Å. B., Olausson, H., Wessberg, J., & Norrsell, U. (1993). A system of unmyelinated afferents for innocuous mechanoreception in the human skin. *Brain Research, 628,* 301–304.

Vallbo, Å. B., Olsson, K. Å., Westberg, K.-G., & Clark, F. J. (1984). Microstimulation of single tactile afferents from the human hand. Sensory attributes related to unit type and properties of receptive fields. *Brain, 107,* 727–749.

Van Boven, R. W., & Johnson, K. O. (1994a). A psychophysical study of the mechanisms of sensory recovery following nerve injury in humans. *Brain, 117,* 149–167.

Van Boven, R. W., & Johnson, K. O. (1994b). The limits of tactile spatial resolution in humans: Grating orentation discrimination at the lip, tongue, and finger. *Neurology, 44,* 2361–2366.

Van Doren, C. L. (1989). A model of spatiotemporal tactile sensitivity linking psychophysics to tissue mechanics. *Journal of the Acoustical Society of America, 85,* 2065–2080.

Vega-Bermudez, F., Johnson, K. O., & Hsiao, S. S. (1991). Human tactile pattern recognition: Active versus passive touch, velocity effects, and patterns of confusion. *Journal of Neurophysiology, 65,* 531–546.

Verrillo, R. T. (1962). Investigation of some parameters of the cutaneous threshold for vibration. *Journal of the Acoustical Society of America, 34,* 1768–1773.

Verrillo, R. T. (1963). Effect of contactor area on the vibrotactile threshold. *Journal of the Acoustical Society of America, 35,* 1962–1966.

Verrillo, R. T. (1965). Temporal summation on vibrotactile sensitivity. *Journal of the Acoustical Society of America, 37,* 843–846.

Verrillo, R. T. (1966a). Effects of spatial parameters on the vibrotactile threshold. *Journal of Experimental Psychology, 71,* 570–574.

Verrillo, R. T. (1966b). Specificity of a cutaneous receptor. *Perception and Psychophysics, 1,* 149–153.

Verrillo, R. T. (1966c). Vibrotactile sensitivity and the frequency of response of the Pacinian corpuscle. *Psychonomic Science, 4,* 135–136.

Verrillo, R. T. (1968). A duplex mechanism of mechanoreception. In D. R. Kenshalo (Ed.), *The skin senses* (pp. 139–156), Springfield, IL: Charles C. Thomas.

Verrillo, R. T. (1974). Vibrotactile intensity scaling at several body sites. In F. A. Geldard (Ed.), *Cutaneous communication systems and devices.* Austin, TX: Psychonomic Society.

Verrillo, R. T. (1979). Comparison of vibrotactile threshold and suprathreshold responses in men and women. *Perception and Psychophysics, 26,* 20–24.

Verrillo, R. T. (1982). Effects of aging on the suprathreshold responses to vibration. *Perception and Psychophysics, 32,* 61–68.

Verrillo, R. T. (1991). Measurement of vibrotactile sensation magnitude. In S. J. Bolanowski,

Jr., & G. A. Gescheider (Eds.), *Ratio scaling of psychological magnitude* (pp. 260–275). Hillsdale, NJ: Lawrence Erlbaum.

Verrillo, R. T., & Bolanowski, S. J., Jr. (1986). The effects of skin temperature on the psychophysical responses to vibration on glabrous and hairy skin. *Journal of the Acoustical Society of America, 80,* 528–532.

Verrillo, R. T., & Bolanowski, S. J., Jr. (1993). The perception of mechanical stimuli through the skin of the hand and its physiological bases. In: P. Dario, G. Sandini and P. Aebischer (Eds.), *Robots and Biological Systems: Towards a New Bionics?* (pp. 123–138), Berlin: Springer-Verlag.

Verrillo, R. T., & Capraro, A. J. (1975). Effect of stimulus frequency on subjective vibrotactile magnitude functions. *Perception and Psychophysics, 17,* 91–96.

Verrillo, R. T., & Chamberlain, S. C. (1972). The effect of neural density and contractor surround on vibrotactile sensation magnitude. *Perception and Psychophysics, 11,* 117–120.

Verrillo, R. T., Fraioli, A. J., & Smith, R. L. (1969). Sensation magnitude of vibrotactile stimuli. *Perception and Psychophysics, 6,* 366–372.

Verrillo, R. T., & Gescheider, G. A. (1977). Effect of prior stimulation on vibrotactile thresholds. *Sensory Processes, 1,* 292–300.

Verrillo, R. T., & Smith, R. L. (1976). Effect of stimulus duration on vibrotactile sensation magnitude. *Bulletin of the Psychonomic Society, 8,* 112–114.

Vierck, C. J., Jr., & Cooper, B. Y. (1990). Epicritic sensations of primates. In M. A. Berkley & W. C. Stebbins (Eds.), *Comparative perception, Vol. I: Basic mechanisms* (pp. 29–66). New York: John Wiley.

Vierck, C. J., Jr., & Jones, M. B. (1969). Size discrimination on the skin. *Science, 163,* 488–489.

Vierck, C. J., Jr., & Jones, M. B. (1970). Influences of low and high frequency oscillation upon spatio-tactile resolution. *Physiology and Behaviour, 5,* 1431–1435.

Vierordt, K. H. (1870). Abhängigkeit der ausbildung des raumsinnes der haut von der beweglichkeit der körpertheile. [Dependence of the development of the skin's spatial sense on the flexibility of parts of the body]. *Zeitschrift für Biologie, 6,* 53–72.

Volkmann, A. W. (1858). Über den einfluss der übung auf das erkennen räumlicher distanzen. [On the influence of practice on the recognition of spatial distances]. *Berichte der königlich-sächsischen Gesellschaft der Wissenschaften zu Leipzig, mathematisch-physische Classe, Reports of the Royal Saxon Academy of Sciences at Leipzig, mathematical-physical class, 10,* 38–69.

von Békésy, G. (1939). Über die Vibrationsempfindung. [On the vibration sense]. *Akustische Zeitschrift, 4,* 315–334.

von Békésy, G. (1955). Human skin perception of traveling waves similar to those on the cochlea. *Journal of the Acoustical Society of America, 27,* 830–841.

von Békésy, G. (1957). Neural volleys and the similarity between some sensations produced by tones and by skin vibrations. *Journal of the Acoustical Society of America, 29,* 1059–1069.

von Békésy, G. (1958). Funneling in the nervous system and its role in loudness and sensation intensity on the skin. *Journal of the Acoustical Society of America, 30,* 399–412.

von Békésy, G. (1959). Neural funneling along the skin and between the inner and outer hair cells of the cochlea. *Journal of the Acoustical Society of America, 31,* 1236–1249.

von Békésy, G. (1960). *Experiments in hearing.* New York: McGraw-Hill.

von Békésy, G. (1962). Can we feel the nervous discharges of the end organs during vibratory stimulation of the skin? *Journal of the Acoustical Society of America, 34,* 850–856.

von Békésy, G. (1967). *Sensory inhibition,* New York: McGraw-Hill.

von Euler, C., Franzén, O., Lindblom, U., & Ottoson, D. (1984). *Somatosensory mechanisms.* New York: Plenum.

von Frey, M. (1895). Beiträge zur sinnesphysiologie der haut. *Berichte der königlich-sächsischen Gesellschaft der Wissenschaften zu Leipzig, mathematisch-physische Classe, 47,* 166–184.

von Frey, M., & Kiesow, F. (1899). Über die function der tastkörperchen. *Zeitschift für Psychologie und Physiologie der Sinnesorgane, 20,* 126–163.

von Gierke, H. E., ,Oestreicher, H. L., Franke, E. K., Parrack, H. O., & von Wittern, W. W. (1952). Physics of vibrations in living tissues. *Journal of Applied Physiology, 4,* 886–900.

von Gilmer, B. (1935). The measurement of the sensitivity of the skin to mechanical vibration. *Journal of General Psychology, 13,* 42–61.

Warren, S., Hämäläinen, H. A., Palmer, C. I., & Gardner, E. P. (1991). Transformation of spatio-temporal information by somatosensory neural networks. In O. Franzén & J. Westman (Eds.), *Information processing in the somatosensory system* (pp. 319–328). New York: Stockton.

Weber, E. H. (1834). *De pulsu, resorptione, auditu et tactu.* Leipzig: Koehler.

Weber, E. H. (1846). Der Tastsinn und das Gemeingefül. In R. Wagner (Ed.), *Handwörterbuch der Physiologie, Vol III* (pp. 481–588). Brunswick: Vieweg.

Weber, E. H. (1852). Über den raumsinn und die empfindungskreise in der haut und im auge. *Berichte der königlich-sächsischen Gesellschaft der Wissenschaften zu Leipzig, mathematisch-physiche Classe, 4,* 87–105.

Weddell, G. (1955). Somesthesis and the chemical senses. *Annual Review of Psychology, 6,* 19–136.

Weinstein, S. (1968). Intensive and extensive aspects of tactile sensitivity as a function of body part, sex, and laterality. In D. R. Kenshalo (Ed.), *The skin senses* (pp. 195–218). Springfield, IL: Charles C. Thomas.

Werner, G., & Mountcastle, V. B. (1965). Neural activity in mechanoreceptive cutaneous afferents: Stimulus–response relations, Weber functions, and information transmission. *Journal of Neurophysiology, 28,* 359–397.

Westling, G. K. (1986). Sensori-motor mechanisms during precision grip in man. *Umea University Medical Dissertations,* New Series No. 171, Umea, Sweden.

Wheat, H. E., Goodwin, A. W., & Browning, A. S. (1995). Tactile resolution: Peripheral neural mechanisms underlying the human capacity to determine positions of objects contacting the fingerpad. *Journal of Neuroscience, 15,* 5582–5595.

Whitsel, B. L., Dreyer, D. A., Hollins, M., & Young, M. G. (1979). The coding of direction of tactile stimulus movement: Correlative psychophysical and electrophysiological data. In D. R. Kenshalo (Ed.), *Sensory functions of the skin in humans* (pp. 79–107). New York: Plenum.

Whitsel, B. L., Franzén, O., Dreyer, D. A., Hollins, M., Young, M., Essick, G. K., & Wong, C. (1986). Dependence of subjective traverse length on velocity of moving tactile stimuli. *Somatosensory Research 3,* 185–196.

Willis, W. D. Jr., & Coggeshall, R E. (1991). *Sensory mechanisms of the spinal cord* (2nd ed.). New York: Plenum Press.

Woodward, K. L. (1993). The relationship between skin compliance, age, gender, and tactile discriminative thresholds in humans. *Somatosensory Motor Research, 10,* 63–67.

Zigler, M. J. (1932). Pressure adaptation-time: A function of intensity and extensity. *American Journal of Psychology, 44,* 709–720.

Zigler, M. J. (1935). The experimental relation of the two-point limen to the error of localization. *Journal of General Psychology, 13,* 316–331.

Zotterman, Y. (1939). Touch, pain and tickling: An electro-physiological investigation on cutaneous sensory nerves. *Journal of Physiology, 95,* 1–28.

Zwislocki, J. J. (1960). Theory of temporal summation. *Journal of the Acoustical Society of America, 32,* 1046–1060.

Zwislocki, J. J. (1965). Analysis of some auditory characteristics. In R. D. Luce, R. R. Bush, & E. Galanter (Eds.), *Handbook of mathematical psychology* (Vol. 3, pp. 1–97), New York: Wiley.

Zwislocki, J. J., Adams, W. B., & Kletsky, E. J. (1970). Intensity characteristics of mechanoreceptors. *Journal of the Acoustical Society of America, 47,* 96.

CHAPTER 3

Somatosensory Cortex and Tactile Perceptions

Harold Burton
Robert Sinclair

I. INTRODUCTION

This chapter considers the cerebral cortical areas responsible for low-threshold touch perception in primates. We emphasize cognitive factors in tactile perception, discuss receptive field organization in the cortex, and highlight issues that may benefit from future study. The physiological perspective primarily focuses on the isolated elements responsible for perceptions of well-controlled tactile stimuli.

II. SOMATOSENSORY CORTICAL AREAS

On the basis of anatomical and physiological criteria, approximately 10 parietal cortical areas involve somatosensory processing. These areas display separable cytoarchitecture and form a connected somatosensory network because they share thalamic and cortical connections. All 10 areas respond in some way to cutaneous or deep receptor stimulation and most contain somatotopically organized maps.

A. Anatomical Basis for Inclusion

Based on cytoarchitecture and connections, the parietal cortex is divisible into four anterior areas (e.g., 3a, 3b, 1, and 2), two posterior areas (e.g., 5

Pain and Touch

and 7b), and four lateral regions (e.g., SIIrostral, SIIposterior, retroinsular, granular insula). Collectively these occupy ~11% of a macaque's cortex (Felleman & Van Essen, 1991). An extensive literature describes the criteria for these identifications (Figure 1) (Brodmann, 1994; Jones, 1985; Jones &

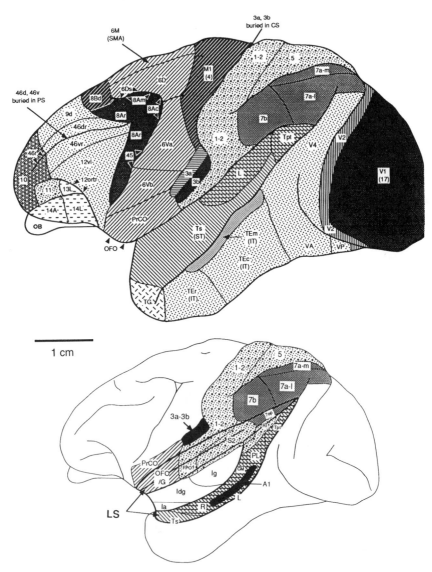

FIGURE 1   Composite presentation of the distribution of lateral cortical areas in the macaque monkey. The lower panel shows cortical areas identified within the depths of the lateral sulcus (LS). (Modified from Preuss & Goldman-Rakic, 1991. Copyright © 1991. Reprinted by permission of Wiley-Liss, Inc., a subsidiary of John Wiley & Sons, Inc.)

Burton, 1976; Jones, Coulter, & Hendry, 1978; Pandya & Seltzer, 1982; Preuss & Goldman-Rakic, 1991; Roberts & Akert, 1963; Sanides & Krishnamurti, 1967).

1. Cytoarchitectonic Boundaries: Anterior, Posterior, and Lateral Areas

Nearly coincident with recognition of the postcentral gyrus as the somatosensory cortex, Brodmann identified three parallel, medial to lateral strips of anatomically distinguishable anterior parietal cortex that he numbered areas 1, 2, and 3. Area 3 contains a high density of small stellate-type cells in an expanded layer 4. Area 1 retains a thick layer 4 but has more large pyramidal cells in supra- and infragranular layers. Area 2 is distinctly homotypical with six clearly defined layers; its cell groups also align radially into palisades across layers. Many of Brodmann's original descriptions were later confirmed (Jones et al., 1978; Powell & Mountcastle, 1959). Area 3 was further subdivided into areas 3a and 3b based on an attenuated layer 4 and increased number of larger pyramidal cells in supra- and infragranular layers in area 3a when compared to adjoining areas 4 and 3b (Jones, 1986; Jones et al., 1978; Jones & Friedman, 1982; Jones & Porter, 1980). The original anterior boundary of area 1 was confirmed; its posterior border and, therefore, the identification of the boundaries of area 2 is debated (Burton, Fabri, & Alloway, 1995; Iwamura, Tanaka, Sakamoto, & Hikosaka, 1993; Jones et al., 1978; Pons, Garraghty, Cusick, & Kaas, 1985b).

Brodmann placed area 5 in posterior parietal cortex (Figure 1). Later clinical and physiological studies show that it is part of the somatosensory cortical regions (Critchley, 1949, 1953; Denny-Brown & Chambers, 1958; Duffy & Burchfiel, 1971; Hyvärinen, 1982; Iwamura, Iriki, & Tanaka, 1994; Iwamura et al., 1993; LaMotte & Acuña, 1978; Mountcastle, Lynch, Georgopoulos, Sakata, & Acuña, 1975; Pons, Garraghty, Cusick, & Kaas, 1985a). Although areas 2 and 5 are difficult to separate, parts of area 5 contain many large, rounded pyramidal cells in layer 5; cells in area 5 often also are arranged radially into palisades throughout the supragranular layers (Jones & Burton, 1976; Jones et al., 1978).

Based on connections with various visual areas, recent studies divide Brodmann's area 7 into several subdivisions (Figure 1). Despite this fractionation, a lateral subportion named area 7b persists in part of Brodmann's original area 40 (Andersen, Asanuma, Essick, & Siegel, 1990; Brodmann, 1994; Felleman & Van Essen, 1991; Preuss & Goldman-Rakic, 1991; Vogt & Vogt, 1919). We discuss below evidence that this region is somatosensory.

Several somatosensory cortical areas occupy portions of the parietal operculum and neighboring inferior parietal lobule (Burton, 1986; Johnson, 1990; Jones & Burton, 1976; Roberts & Akert, 1963). The nomenclature for these areas varies (Figure 1). Brodmann originally identified two areas that bordered on the lateral sulcus. He considered area 43 (called subcentral area) a part of the parietal cortex that extends around the inferior end of the central sulcus and the adjoining parietal–frontal operculum. His area 43 reached to the insula. He noted that the architecture of area 43 was like that in the postcentral cortex (i.e., had a prominent layer 4). However, today the region included within area 43 encompasses (a) the surface cortex for the face regions of areas 3 and 1 that course around the central sulcus and (b) separate regions along the parietal operculum (Jones & Burton, 1976; Roberts & Akert, 1963). Posterior to area 43 Brodmann identified an area 40 that extends over much of the inferior parietal lobule in monkeys. The most anterior and inferior portion of Brodmann's area 40 should be included in current interpretations of lateral parietal somatosensory cortex (Burton et al., 1995). This would include portions of area 7b on the surface and the SII region along the parietal operculum. The layering pattern in the opercular parts of this region is atypical for parietal cortex because the usual six layers are obscured by fusion of layers 5 and 6 (Burton, 1986; Jones & Burton, 1976; Roberts & Akert, 1963).

Posterior insula was originally recognized as displaying a distinct granular layer 4 (Brodmann, 1994). This region was later named the granular insula (Ig). Several findings suggest Ig probably is part of the somatosensory cortical areas. Cutaneous stimuli elicit responses and 2-DG labeling in this region (Juliano, Hand, & Whitsel, 1983; Robinson & Burton, 1980b; Schneider, Friedman, & Mishkin, 1993). Ig connects with thalamic nuclei and other areas of the cortex with somatosensory roles (Friedman & Murray, 1986; Jones, 1985).

2. Thalamic Connections

The connections between each somatosensory cortical area and different thalamic nuclei form another defining basis for separate areal designations (Jones, 1984; Jones, 1985; Lin, Merzenich, Sur, & Kaas, 1979). Briefly, portions of the ventroposterior nuclei that receive terminal projections from the medial lemniscus connect with all of the anterior parietal regions. The central core of the ventral posterior lateral nucleus (VPL), which receives the bulk of medial lemniscal projections, sends dense projections to area 3b. Area 1 receives connections from a broader expanse of VPL, including the central core. Area 2 principally receives projections from the dorsal and rostral shell of VPL. Lateral parietal regions receive some projections from these same nuclei. A greater density of connections to this cortex comes from thalamic nuclei located ventral and posterior to VPL (ventroposterior inferior; posterior nucleus; suprageniculate nucleus). These thalamic nuclei receive somatosensory information principally (but not exclusively) through spinothalamic pathways (Apkarian & Shi, 1994; Burton, 1984; Burton & Carlson, 1986; Burton & Craig, 1983; Burton & Jones, 1976; Craig, Bushnell, Zhang, & Blomqvist, 1994; Friedman & Murray, 1986; Krubitzer & Kaas, 1992; Stevens, London, & Apkarian, 1993). The anterior pulvinar nucleus and neighboring portions of the lateral posterior nucleus project to posterior parietal areas. These thalamic nuclei receive few direct ascending spinal-trigeminal projections (Jones, 1985).

3. Ipsilateral Intracortical Connections

Each of the somatosensory cortical areas forms dense reciprocal connections with its nearest and next nearest neighbor (Burton & Fabri, 1995; Felleman & Van Essen, 1991; Jones et al., 1978; Young, 1992). The connection patterns for many of these follow hierarchical rules similar to those proposed for the visual cortex (Figure 2). Connections from more anterior areas in the

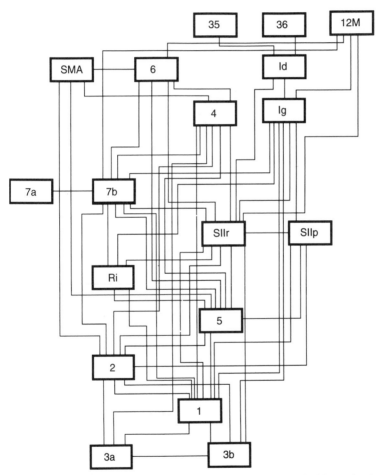

FIGURE 2   Intracortical connections according to previously described hierarchical rules (Felleman & Van Essen, 1991). Areas 3b and 3a form the lowest rung of the hierarchy. This figure emphasizes connections of parietal and somatosensory areas. Supplementary motor area, SMA; dysgranular insula, Id; granular insula, Ig. Assigned interconnections modified from previous reports according to recently summarized findings (Burton & Fabri, 1995; Burton et al., 1995).

parietal cortex mainly end with greater density in middle cortical layers in more posterior and lateral parietal areas (i.e., display a "feed-forward" pattern) (Felleman & Van Essen, 1991; Friedman, Murray, O'Neill, & Mishkin, 1986); lateral parietal areas project with a similar pattern to posterior parietal and insular regions (Burton et al., 1995; Friedman et al., 1986). Most return connections are to upper and lower cortical layers (i.e., "feedback" pattern). However, examples of reciprocal patterns of connections that originate from both supra- and infragranular layers show a far more complicated arrangement of feed-forward, feedback, and lateral projections between two areas (Burton & Fabri, 1995; Lund, Yoshioka, & Levitt, 1993). In addition, distant connections beyond the next nearest neighbor pass to frontal and temporal cortex from areas 2, 5, 7b, SII, and insula (Burton et al., 1995; Carmichael & Price, 1994; Cavada & Goldman-Rakic, 1989; Jones et al., 1978; Jones & Powell, 1970; Matelli, Camarda, Glickstein, & Rizzolatti, 1986; Mesulam & Mufson, 1982; Mufson & Mesulam, 1982; Preuss & Goldman-Rakic, 1989; Stepniewska, Preuss, & Kaas, 1993). These more distant connections suggest a likely contribution of somatosensory information to higher cognitive and motor control centers.

B. Physiological Basis for Inclusion

Early clinical and experimental findings identified the role of the postcentral gyrus in cutaneous perceptions (Barenne, 1916; Campbell, 1905; Head, 1920; Penfield & Boldrey, 1937). Pioneering studies of Woolsey and colleagues revealed details about the extent and topographical organization of body representation in this cortex (Woolsey, Marshall, & Bard, 1942) and confirmed the extension of somatosensory cortical areas outside the postcentral region (Adrian, 1941; Woolsey, 1943; Woolsey, 1958; Woolsey & Fairman, 1946; Woolsey & Walzl, 1981; Woolsey & Wang, 1945). The interpretation, salience, and role of proposed somatosensory cortical areas in the modern literature relate to receptive field structure and conditions needed to establish responsiveness to cutaneous stimulation.

Preservation of peripheral-like specificity across well-defined receptive fields within postcentral cortical regions supports analogous classifications of these central neurons with peripheral counterparts (Mountcastle, 1984; Mountcastle, Talbot, Sakata, & Hyvärinen, 1969; Paul, Merzenich, & Goodman, 1972; Powell & Mountcastle, 1959; Sur, Wall, & Kaas, 1984). Such tight physiological links arise from equally precise subcortical connections. Cortical areas containing readily activated cell populations with peripheral-like mimicry are by definition somatosensory cortical regions (Kaas, 1983). Exceptions to this notion are the cutaneous, low-threshold-driven responses in motor cortex.

Each of the somatosensory parietal areas responds in varying degrees to stimulation of cutaneous or proprioceptive-kinesthetic receptors. Responses in anterior parietal areas 3b and 1 are obvious, reproducible, and readily delimited to well-defined receptive fields under a variety of physiological conditions. Because of these characteristics, it is frequently possible to identify the particular classes of peripheral mechanoreceptors that send information to these cortical areas (Friedman & Jones, 1981; Iwamura et al., 1993; Mountcastle, 1984; Paul et al., 1972; Powell & Mountcastle, 1959). There is evidence that some peripheral submodalities may dominate activity throughout a vertical column of cells in area 3b and possibly area 1 (Paul et al., 1972; Powell & Mountcastle, 1959; Sur et al., 1984). However, the changing character of responses across the cortical layers makes it difficult to prove this hypothesis conclusively despite the evident selectivity of cells isolated within the middle cortical layers (Favorov, Diamond, & Whitsel, 1987; Favorov & Whitsel, 1988a,b; Iwamura, Tanaka, Sakamoto, & Hikosaka, 1983a,b, 1985b). Many studies provided stimulus–response characterizations from neurons in these areas (see below).

Inclusion in the family of somatosensory cortical areas becomes more difficult where the predominant cell populations lack peripheral-like capacities. In these regions physiological identifications often include responsiveness when a body part is touched, manipulated, involved, engaged, and so on. (Burton, 1986; Iwamura & Tanaka, 1978; Iwamura, Tanaka, & Hikosaka, 1980; Iwamura et al., 1983b; Iwamura et al., 1993; Sinclair & Burton, 1993a). Receptive field delimitations vary from equivalent to those observed in the postcentral gyrus to those without clear borders.

Some recordings from SII showed nearly comparable consistent responsiveness to cutaneous stimulation (Burton & Sinclair, 1990; Hsiao, O'Shaughnessy, & Johnson, 1993; Robinson & Burton, 1980a,c; Sinclair & Burton, 1993a), but many investigations reported greater difficulty in evoking responses in these cortical areas, especially in anesthetized animals (Burton, 1986).

More than 70% of isolated neurons in the granular insula responded to innocuous somatosensory stimuli; these mostly responded to stimulation anywhere on the body and often included bilateral receptive fields (Robinson & Burton, 1980b,c; Schneider et al., 1993). Many Ig neurons might respond better to noxious stimuli because this region receives connections from the thalamic projection target for specific nociceptor neurons in the marginal layer of the spinal cord (Craig et al., 1994).

Studies of somatosensory activation in posterior parietal cortex reported even fewer correlations between responses and classification schemes useful in anterior parietal areas. Cells in area 5, for example, responded to different spatial positions or rotations of the limb (Duffy & Burchfiel, 1971; Ferraina & Bianchi, 1994; Mountcastle et al., 1975; Sakata, Takaoka, Kawarasaki, & Shibutani, 1973). Descriptions of these responses as more complex often means less identifiable mimicry of single classes of peripheral receptors and large, poorly defined receptive fields. For example, activity might depend on placing the arm in a particular spatial orientation to move it through some specified part of visual space to reach a target (Ferraina & Bianchi, 1994, p. 178). Similarly, neurons in portions of area 7 (ventral intraparietal area VIP and 7b) combine visual and somatosensory responsiveness that may create a body-centered coordinate system in visual space (Colby & Duhamel, 1991; Colby, Duhamel, & Goldberg, 1993; Duhamel, Colby, & Goldberg, 1991; Hyvärinen, 1981, 1982; Hyvärinen & Shelepin, 1979).

1. Maps in Areas 3b and 1

Early recordings from the postcentral gyrus to cutaneous stimulation of the contralateral body surface showed a medial to lateral map of the body that is nondermatomal but largely coherent (Figure 3A). The map appears somatotopic (Kaas, 1983; Woolsey, 1958) despite distortions introduced by enlarged representations for body parts with high peripheral innervation densities (i.e., fingers, peri-oral regions, and toes). This view supports notions that perception of body space reflects activity across one homogeneous and continuous two-dimensional representation (Werner & Whitsel, 1968). However, several early findings foreshadowed later reports of separate and complete maps of the body in areas 3b and 1 (Figure 3B) (Kaas, Nelson, Sur, Lin, & Merzenich, 1979; Merzenich, Kaas, Sur, & Lin, 1978; Nelson, Sur, Felleman, & Kaas, 1980; Pons, Wall, Garraghty, Cusick, & Kaas, 1987). These included reports of differing percentages of cells responding to cutaneous and deep receptors in these cortical areas (Dreyer, Loe, Metz, & Whitsel, 1975; Powell & Mountcastle, 1959). Later studies expanded to the whole body the initial observation of replicate receptive field representations for the digits in areas 3b and 1 in different species of primates (Paul et al., 1972). The highly detailed maps provide mirror reversed representations mostly of the cutaneous surface of the body (Figure 3B). In the digit zones of the primary somatosensory cortex (SI) of macaques one finds segregation of receptive field representations in two axes. Medial to lateral regions of cortex represent the ulnar to radial axis of the hand and digits, and the rostro-caudal dimension of the map includes the proximal distal axis of the digits. In area 3b fingertips are anterior; in area 1 they are posterior (Figure 4). These findings suggest that perception of a single body space may arise from a distributed network involving several maps. Important unresolved issues are understanding the separate contributions of each area's map and the cohesion of these distinct responses into unified perceptions of touched objects or body parts.

The selective contribution of each map to a body image remains unknown. Data from comparative studies showed that most nonprimate spe-

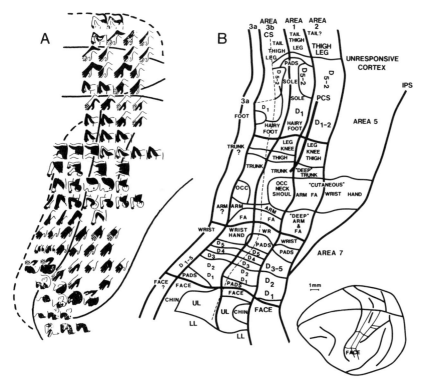

FIGURE 3   Distribution of cutaneous responsive loci across the medial to lateral extent (top to bottom of drawings) of the postcentral gyrus. (A) Drawings of selected body regions placed at sites of recorded maximum surface evoked potentials to tapping the receptive field drawn in black. The dashed, medial-lateral line shows the fundus of the central sulcus. Parallel solid line shows the gyral crown of the same sulcus (modified from Woolsey, 1958). (B) Summary drawing of the somatotopic organization across anterior parietal cortical areas 3a, 3b, 1, and 2 (collectively named SI) in macaque monkeys. Drawing on lower right shows the location of the SI region on a dorsolateral view of the hemisphere with some included somatotopic boundaries for reference to the enlarged image to the left. CS, central sulcus; D, digit; FA, forearm; IPS, intraparietal sulcus; LL, lower lip; OCC, occiput; PCS, postcentral sulcus; UL, upper lip; WR, wrist. (Modified from Pons et al., 1985b. Copyright © 1985. Reprinted by permission of Wiley-Liss, Inc., a subsidiary of John Wiley & Sons, Inc.)

cies have only a single, somatotopical representation that may be analogous to the predominant cutaneous sensitivity of area 3b in higher primates (Kaas, 1983). Therefore, cutaneous tactile discriminations in these animals, like spatial localization, may simply involve a single neural space. In rats, however, cortical processing engages several regions with less responsiveness to tactile stimuli and with different connections. This matrix of cortex surrounds the main cutaneous representation (Chapin & Lin, 1984; Fabri &

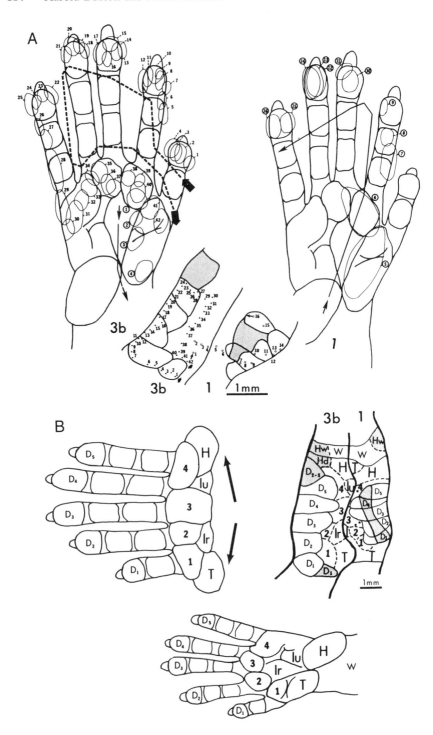

Burton, 1991; Killackey, Koralek, Chiaia, & Rhoades, 1989). Several older primate species also have pockets of duplicate representations devoted largely to cutaneous receptors from glabrous and hairy skin parts of the fingers and hand (Carlson, 1985, 1990). Carlson and others hypothesized that these examples of partial multiple areas emerge from evolutionary pressures "associated with advances in the power or complexity of cortical processing" (Carlson, 1990, p. 460). Thus each area may relate to some developmental specialization associated with improved sensory capacities of somatosensory organs like the hands (Carlson, 1985, 1990; Carlson & Nystrom, 1994; Kaas, 1983). However applicable these ideas are to the hand, they do not explain multiple representations in higher primates for parts of the body that show no unique structural and functional demands. For example, proximal limb representations are multiple only in primates.

Perceptions of body space are hypothesized to reflect the structure of cortical maps. Evidence for this idea comes from several observations in humans. First, electrical stimulation of selected parts of the postcentral gyrus evoked sensations attributed to a body site whose map is at the stimulated focus (Penfield & Boldrey, 1937). Second, modern imaging techniques revealed topographically distributed maps across the postcentral gyrus that follow stimulated sites on the body where subjects report localized sensations (Burton, Videen, & Raichle, 1993; Fox, Burton, & Raichle, 1987). Third, patients with limb amputations can localize sensations to a phantom and to stimulation of neighboring intact body sites. Neuroimaging in some of these cases shows conjoint activity in the cortical region that would have served the amputated limb and the intact site (Ramachandran, 1993). Together these data show that spatial localization on the body correlates with site activation in the SI map.

2. Map in Area 3a

Recordings from several species discovered responses evoked from muscle spindle receptors at the anterior border of a cutaneous representation in

FIGURE 4 Organization of areas 3b and 1 activated by cutaneous stimulation of the hand and finger. Data obtained from owl monkeys. (A) Receptive fields for single or small clusters of neurons from recordings in area 3b (left) or area 1 (right). A reconstruction of the SI region for the hand and fingers appears between drawings of the receptive fields. Numbered penetrations match those for corresponding receptive fields in area 3b (dots) and in area 1 (open circles). (B) Summary of the glabrous hand representation in areas 3b and 1. The map (upper left) distorts the hand (below). For example, adjacent skin regions for the hypothenar (H) and thenar (T) pads lie separated by enlarged representations for the pads at the base of each digit. The representation of the dorsal surfaces (shaded areas in drawing on upper right) of the digits, hand (Hd) and wrist (Hw) also extend into segregated, discontinuous medial and lateral segments of the map (upper right). The area 1 representation is a mirror reversal of the map in area 3b. Distal tips of the digits lie anterior in area 3b and posterior in area 1. (Modified from Merzenich et al., 1978. Copyright © 1978. Reprinted by permission of Wiley-Liss, Inc., a subsidiary of John Wiley & Sons, Inc.)

anterior parietal cortex (Jones & Porter, 1980; Lucier, Rüegg, & Wiesendanger, 1975; C. B. Phillips, Powell, & Wiesendanger, 1971; Wiesendanger, 1973). Brodmann had not distinguished a separate cortical area between sensory and motor strips along the central sulcus of primates. Later debates on the definition of the boundaries of a distinct somatosensory cortical area in this location sometimes exist even in the views of the same authors (Friedman & Jones, 1981; Jones, 1985; Jones et al., 1978; Jones & Porter, 1980; Mountcastle, 1984; Phillips et al., 1971; Powell & Mountcastle, 1959; Vogt & Vogt, 1919). Two observations eliminated much of the earlier ambiguity surrounding the identification of area 3a. First, it was shown that a separate rostral part of VPL projects to a narrow stripe of cortex near the fundus of the central sulcus (Friedman & Jones, 1981). The particular thalamic region receives projections from the medial lemniscus and responds to stimulation of muscle spindle receptors. Furthermore, recognition that the cytoarchitecture along area 3a changes from medial to lateral in the respective lower limb to face representations dispels much earlier confusion regarding the appropriate anatomical criteria for defining this cortex (Friedman & Jones, 1981; Jones et al., 1978; Jones & Porter, 1980). No mapping studies have described details about the topography of muscle spindle representation in area 3a of primates except to confirm a parallel organization with area 3b (Figure 3B). Evidence from tracing intracortical connections with more reliably defined cutaneous maps supports a likely medial to lateral topography that mirrors the better known cutaneous maps (Burton & Fabri, 1995). As muscle spindle activity contributes significantly to proprioceptive sense of limb position (Goodwin, McCloskey, & Matthews, 1972), it is likely area 3a also has a role in these perceptions. Direct demonstrations of this involvement do not exist.

3. Maps in Areas 2 and 5

Area 2 contains another somatotopic map of the contralateral body that is also parallel to those in areas 3b and 1 (Figure 3B). Receptive fields in area 2 are larger, responses to cutaneous stimulation are less frequent, and proprioceptive-kinesthetic information (e.g., deep inputs) predominates (Iwamura & Tanaka, 1978; Iwamura et al., 1980; Iwamura, Tanaka, Sakamoto, & Hikosaka, 1985a; Iwamura et al., 1993; Pons et al., 1985b; Powell & Mountcastle, 1959). In the distal digit representations, especially, the map in area 2 appears mirror reversed from the sequence in area 1. Several studies describe additional responsiveness to somatosensory stimulation further posterior in the depths of the intraparietal sulcus (Duffy & Burchfiel, 1971; Iwamura et al., 1994; Iwamura et al., 1993; Mountcastle et al., 1975; Pons et al., 1985a). However, the boundaries and organization for responses in areas 2 and 5 differ between studies. This possibly reflects interpretive difficulties in assigning receptive fields where most cells respond to stimulation of deep receptors or integrate across submodalities and where anesthetic status dramatically modifies responsiveness. In more posterior parietal locations responses to cutaneous and proprioceptive-kinesthetic stimuli become more intertwined, sometimes even in the same cell. Many cells have bilateral receptive fields and often responded only to changes in multijoint limb projections (Duffy & Burchfiel, 1971; Ferraina & Bianchi, 1994; Iwamura et al., 1994; Iwamura & Tanaka, 1978; Iwamura et al., 1980; Iwamura et al., 1985a; Iwamura et al., 1993; Mountcastle et al., 1975; Sakata et al., 1973). The representation in area 5 still aligns with the medial to lateral receptive field sequences found in anterior parietal cortex, but there is much greater integration across body regions (i.e., there are no individual digit regions).

4. Maps in Lateral Parietal Areas

Contemporaneous with descriptions of somatosensory activation in what became known as the primary somatosensory cortex (Woolsey et al., 1942), Adrian found another parietal region in cats that responded to stimulating the forelimb claws (Adrian, 1941). Woolsey later reported responses in a second representation (SII) from a greater variety of cutaneous receptors located in the face and other parts of the body in several mammalian species, including primates (Burton, 1986; Woolsey, 1943; Woolsey, 1958; Woolsey & Fairman, 1946; Woolsey & Walzl, 1981; Woolsey & Wang, 1945). The SII region is considerably smaller than SI. Activation from cutaneous receptors in the SII region varies across species, being easier to elicit in cats and less in primates in whom anesthesia depresses responsiveness. SII responses to low- or high-intensity tactile stimuli occur in humans (Burton et al., 1993; Coghill et al., 1994; Hari et al., 1993). Stimulation of deep or high threshold receptors evoked responses in some cells in monkeys but these were not studied systematically (Robinson & Burton, 1980c).

Most studies reported one map of the body surface in SII, although with lower resolution, greater receptive field overlaps, and reduced distortions (Burton, 1986). The presence of bilateral receptive fields in SII prompted the suggestion that this region serves bimanual transfers of information and creation of a fused body image across the midline (Burton, 1986; Manzoni, Conti, & Fabri, 1986; Whitsel, Petrucelli, & Werner, 1969). The presence of bilateral receptive fields in posterior parietal cortex (Iwamura et al., 1994) suggests that this hypothesis about the role of SII may be more appropriately shared with area 5. In addition, recent investigations revealed two mirror-reversed representations in the originally described SII region of several different mammals (Burton et al., 1995; Fabri & Burton, 1991; Krubitzer & Calford, 1992; Krubitzer, Calford, & Schmid, 1993; Krubitzer et al., 1995). Some bilateral representation exists in both parts, so it is difficult to ascribe bimanual transfers to just one SII region.

The map in the more anterior–lateral part of the SII region represents the

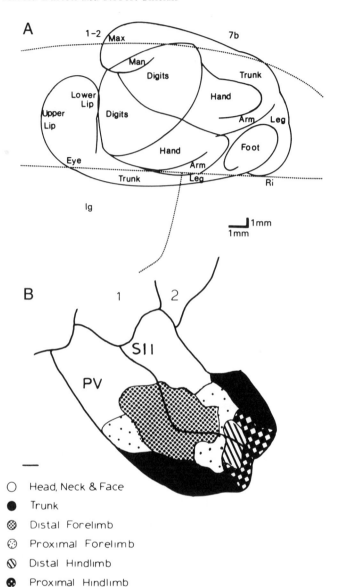

○ Head, Neck & Face
● Trunk
⊗ Distal Forelimb
⊙ Proximal Forelimb
◐ Distal Hindlimb
◑ Proximal Hindlimb

FIGURE 5   Organization of two separate body maps in the SII region of macaques. (A) This drawing summarizes a map based on the distribution of cortical connections with selected body representations in areas 3b and 1 of SI. (Modified from Burton et al., 1995.) (B) This map compiles the distribution of multineuron responses to cutaneous stimulation of indicated body regions across parallel penetrations through the parietal operculum. (Modified from Krubitzer, Clarey, Tweedale, Elston, & Calford, 1995.)

body in a supine position. It lies upright in the posterior region, and the two join along the distal extremities (Figure 5). Compared to most parts of SI, there is less detailed representation of individual body parts throughout the SII region. Nevertheless small pockets of cells within the more anterior map have cutaneous receptive fields confined to individual digits (Robinson & Burton, 1980a,c); most cells even in the distal digit zone have large receptive fields compared to those characterizing cells in the middle layers of areas 3b and 1.

The evidence is scant that somatosensory maps found beyond the postcentral gyrus region also contribute to a body image. Electrical stimulation near the lateral sulcus in humans elicits sensations from different body regions (Penfield & Rasmussen, 1950). However, more recent neuroimaging data in humans do not yet reveal somatotopic maps along the parietal operculum (Burton et al., 1993). Patients or monkeys with lesions involving this cortex appear to retain a sense of touch to different body regions despite compromised tactile object recognition (Carlson & Burton, 1988; Caselli, 1991, 1993).

Cortex lying lateral to SII also responds to somatosensory stimulation. This includes the posterior third of the insula known as the Ig (Jones & Burton, 1976) and an adjoining posterior segment of cortex named the retroinsular area, Ri (Jones & Burton, 1976; Robinson & Burton, 1980b; Sanides & Krishnamurti, 1967). Cells in both regions responded to innocuous cutaneous stimuli (Robinson & Burton, 1980c). However, the source of thalamocortical projections, especially to the insula, suggests a possible role in pain perceptions (Craig et al., 1994; Stevens et al., 1993). Ri has some topographic organization, but Ig has none (Burton et al., 1995; Robinson & Burton, 1980b).

III. SOME CONSIDERATIONS OF RECEPTIVE FIELDS

A. Changing Dimensions across Cortical Areas

1. Receptive Field Size

Several studies surveyed more than one anterior parietal area using the same recording, stimulation and anesthesia methods in a single recording session (Favorov & Whitsel, 1988a,b; Favorov & Diamond, 1990; Iwamura et al., 1993; Merzenich et al., 1978, 1984; Pons et al., 1987). Most findings agree despite procedural differences between studies. These comparisons consider responses to manually applied cutaneous stimulation over well-defined receptive fields. Stimulation consisted of light stroking or pressure with small probes like artist brushes, thin nylon filaments, dowel sticks, or wisps of cotton. Recordings from the middle cortical layers found responses evoked from the smallest receptive fields. In these layers more recording sites from cortex devoted to overrepresented structures (e.g., distal digits and face) displayed smaller receptive fields than in most other representations. Sites in area 3b had the smallest receptive fields; remarkably, receptive fields for these were often only single-integer multiples of those observed subcortically. Thus, some cells in area 3b responded to stimulation of ellipses covering 1–2 mm^2 of skin on a finger pad (Figure 4A). Recordings from area 1 generally showed activation from larger receptive fields (Figure 4A). The smallest of these still covered a fraction of a single finger pad, others extended over adjoining skin, and some included disjoint receptive fields from adjacent digit tips. Sites in area 2 responding to skin stimulation have the largest receptive fields of the anterior parietal areas. Receptive fields in area 2 still include body segments (i.e., neighboring pair of fingers, hand, and fingers, lower arm, leg, trunk, or trigeminal divisions of the head) (Figure 6) (Iwamura et al., 1993; Pons et al., 1985b). Some cells in area 1–2 responded best when tactile stimuli moved across their receptive fields (see below).

Few recording sites in posterior parietal cortex revealed responses from similarly contained receptive fields (Duffy & Burchfiel, 1971; Hyvärinen, 1981, 1982; Hyvärinen & Shelepin, 1979; Leinonen, Hyvärinen, Nyman, & Linnankoski, 1979; Leinonen & Nyman, 1979; Robinson & Burton, 1980c; Sakata et al., 1973), although there were exceptions to this observation (Iwamura et al., 1994). Despite the absence of distinct boundaries, most receptive fields in areas 5 and 7b are larger. Some in area 5 are bilateral, and most cover a major body region (i.e., arm, including the hand and digits, all fingers on one hand, much of the head, all of the upper trunk, etc.). Effective stimulation of the receptive fields for neurons in these posterior areas often involved more integrated stimulation procedures than simply touching the skin. Responses in area 5 occurred only to particular shapes in association with particular limb positions. They also related to a spatial coordinate that placed the touched region within a visual domain (Ferraina & Bianchi, 1994) (e.g., upper visual quadrant for receptive fields on the face). Some neurons in area 7 also showed convergence of visual and tactile sensitivity where the receptive fields in both modalities tend to include congruent space (Duhamel et al., 1991; Robinson & Burton, 1980b). Responses appeared when there was some projected visual alignment to targeted skin locations. However, there are difficulties in interpreting these results because there were no eye-movement records. The results are ambiguous as the responses could signal body, visual, or eye spatial coordinates; they could also relate to visual or tactile modalities. Additional studies need to establish the modal qualities of area 7b responses.

Recordings from SII, especially without anesthesia and in the hand and digit representations, found neurons with cutaneous-activated receptive fields whose size matched that in areas 1 and 2 (Burton & Sinclair, 1990;

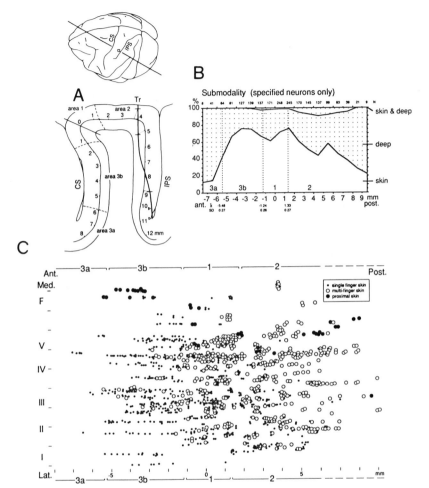

FIGURE 6 Summary of a survey showing receptive field and submodality distribution for single neurons recorded across areas 3b, 1, and 2 in awake macaques. (A) Lines through a dorsolateral view of the left hemisphere show the sagittal orientation and location of penetrations that entered the SI region. The sagittal section displays anterior (negative) to posterior (positive) distance from the center of area 1 (0-mm position) as measured along layer 4. Tr (penetration track) shows the position of a typical recording site through the anterior bank of the intraparietal sulcus (IPS) and in area 2. (B) Proportional submodality distribution across 16 mm of the anterior parietal cortex. Percentages based on cells responding to cutaneous (skin) or proprioceptive (deep) type stimulation. (C) Medial (Med.) to lateral (Lat.) and anterior (Ant.) to posterior (Post.) distribution of cells with different receptive field sizes that covered some part of one finger (small open circles), multiple fingers (large open circles), or some part of the hand and palm (filled circles). I. II, III, IV, V, fingers 1–5, respectively; F, forearm. (Modified from Iwamura et al., 1993. Reprinted with permission of Springer-Verlag.)

Robinson & Burton, 1980c; Sinclair & Burton, 1993a). Some SII neurons only responded while the monkey actively engaged in a tactile discrimination task but otherwise had no demonstrable receptive fields to passive stimuli. During a haptic task, however, responses can be evoked from restricted skin areas. In contrast, cells recorded from the insula often have receptive fields that include the whole body (Robinson & Burton, 1980b,c; Schneider et al., 1993).

2. Factors Effecting Receptive Field Size

Receptive fields in the cortex derive from anatomical factors such as proximity, density, and type of thalamic and cortical projections and physiological factors dictating the spatial and temporal balance between various types of synaptic activity. For example, in recordings from the most anterior parts of SI in monkeys, receptive field sizes increase following pharmacological disruption of local inhibitory circuits (Alloway & Burton, 1991). These circuits vary between and within cortical areas and, therefore, affect the definition of receptive fields in different places.

Recordings in different layers within one cortical area revealed varying receptive field sizes that can overlap a single point on the body (Favorov et al., 1987; Favorov & Whitsel, 1988a,b). Consequently, contrasting small receptive fields across areas is meaningless without matching recording depths. For example, multidigit receptive fields common to recordings from the middle layers of area 1 have been discovered within the infragranular layers of area 3b (Iwamura et al., 1985b; Iwamura, Tanaka, Sakamoto, & Hikosaka, 1985c).

Anesthetic conditions influence responses of cortical cells and by that modify possible delineations of receptive fields in different studies (G. Duncan, Dreyer, McKenna, & Whitsel, 1982). Small receptive fields found in anesthetized animals may therefore underestimate spatial convergence in a region of cortex. Anesthesia affects also vary considerably by cortical area. Most recordings from areas 3b and 1 found responses to cutaneous stimulation in anesthetized animals, but such consistent activation required awake animals in most other somatosensory cortical areas.

Overrepresentation of skin regions with high peripheral innervation densities greatly distorts cortical maps and distributions of receptive field sizes (i.e., comparatively smaller receptive fields for overrepresented structures). The disproportionate representation of highly innervated structures provides a neural magnification factor for the hands, feet, and mouth. Two-point limens that depend on preservation of minimal spatial resolution on the skin (Phillips & Johnson, 1981a,b) are also lowest for these same structures.

Cortical areas display varying sensitivity and modality specificity to passively applied stimuli; areas 3b and 1 are most responsive to innocuous cutaneous stimuli, areas 3a and 2 respond best to stimulation of deep receptors, and most other somatosensory areas display a range of sensitivity (Figure 6). Therefore, comparisons of receptive field dimensions between areas may confound responses arising from noncomparable stimulations.

3. Sources of Receptive Field Size Differences between Cortical Areas

Receptive fields change when considering responses just to cutaneous stimuli within different cortical regions representing the same body part. In anterior parietal regions receptive fields increase in size in recordings from anterior to posterior locations. The source of these changes in receptive field size is unknown. One possibility is differences in divergence of thalamocortical projections. Recent findings of greater spread of projections to area 3b suggest that these connections may contribute to some receptive field expansions to adjacent skin regions (e.g., adjoining digits) (Rausell & Jones, 1995). Similar evidence of even greater divergence has not been reported for the thalamocortical projections to other somatosensory areas.

Another way to increase receptive field size is through interareal projections; multiple regions from within an area with smaller receptive fields converge in the next area (DeFelipe, Conley, & Jones, 1986; Jones & Powell, 1970). Undoubtedly this happens, but most studies of interareal connections show the densest connections between similar somatotopic representations (i.e., a digit region in area 3b projects mostly to the same digit region in area 1) (Burton & Fabri, 1995). Similar homotopical connections occur between portions of SI and SII (Burton et al., 1995). There also are no reports of double-labeled cells in areas 3b, 2, or SII that show possible divergence of projections to two adjacent digit representations in area 1.

Intracortical connections may be yet another means to enlarge receptive fields. These connections extend across components of the somatotopic map within each of the anterior parietal areas. For example, an area 1 single-digit representation connects with much of the neighboring digit representation in area 1; the range of these nonsomatotopical projections increases in successive steps away from area 3b (Lund et al., 1993). Connections within an area are thus less homotopical than those between cortical areas (Alloway & Burton, 1985; Burton & Fabri, 1995; Fabri & Burton, 1991; Lund et al., 1993). Similar extensions in posterior and lateral parietal areas in combination with some convergence between areas could be responsible for the observed differences in receptive field size across somatosensory cortex.

4. Receptive Field Size and Tactile Discriminations

There is little data on how areas with different average receptive field sizes contribute to tactile perceptions. As in the peripheral nervous system, one might suggest that psychophysical thresholds for spatial resolution of gap sizes, letter heights, or two-point separations depend on the spatial resolution provided by the density and distribution of peripheral mechanoreceptors in the skin (K. Johnson & Hsiao, 1992; Phillips, Johnson, & Hsiao, 1988). Generalizing, one might therefore suggest that cortical regions with the smallest receptive fields are necessary and sufficient for such spatially restrictive perceptions. However, the high spatial resolution obtained from area 3b cells with the smallest receptive fields is less than that of most peripheral receptors. These cells are close to the input layers of this cortex, whereas the larger receptive fields of cortical output layers provide even worse spatial resolution. Furthermore, some detection thresholds are even below the spatial resolution of peripheral fiber innervation patterns (e.g., distinguishing texture differences or detecting small asperities (Lamb, 1983a,b; LaMotte & Whitehouse, 1986)). Therefore, preservation of psychophysical limens attributable to the spatial array of peripheral receptors where cortical receptive fields fail to mimic this resolution requires central enhancement mechanisms.

There is evidence that excitatory and inhibitory connections lead to enhanced signal-to-noise ratios at several levels in the somatosensory system (Mountcastle, 1984). Several aspects of the response evoked from the central point in a cell's receptive field aid this process: greater firing rates from the center, a response pattern with greater temporal uniformity, and shorter latency responses from the center compared to less intensely stimulated regions from the surround (Gardner & Spencer, 1972; K. Johnson & Hsiao, 1992; Mountcastle, 1984). It may be inappropriate to conclude that a cortical area with only small receptive fields, like 3b, uniquely serves perceptions requiring high spatial resolution because excitatory–inhibitory mechanisms can enhance the central zone of stimulation in cortical regions with normally large receptive fields. Lesion studies showed that area 3b must be present for tactile discriminations of texture, size, and shape. However, detection of vibrotactile stimuli occurred following destruction of most of the postcentral gyrus (Carlson, 1990; LaMotte & Mountcastle, 1979; Randolph & Semmes, 1974). In addition, damage only to lateral parietal areas with only large receptive fields still interfered with tactile object recognition and isolated texture and size discriminations despite preservation of the small receptive fields in area 3b (Carlson & Burton, 1988; Murray & Mishkin, 1984a).

Differences in receptive field size across the cortical areas may suggest different qualitative processes. Areas with smaller receptive fields are specific for object features. Regions with larger receptive fields may provide for object invariance despite haptic explorations that bring the same object into contact with different parts of the skin. Areas with larger receptive fields provide for predictive perceptual completion of object features when the whole cannot be processed at once. This process is then analogous to visual identification of partially obscured objects. Thus, loss of stereognosis following area 3b lesions may arise from loss of the ability to process the detailed features of touched objects. Lesions of SII might lead to problems with object recognition because object constancy cannot be maintained while the hand explores the object using a variety of manipulative procedures.

IV. ATTRIBUTES REPRESENTED IN CORTICAL ACTIVITY PATTERNS

A. Neural Metrics for Temporal Patterns of Skin Stimulation

1. Flutter-Vibration Sensations

Vibratory stimuli evoke two distinct perceptions in humans. Frequencies below 40 Hz feel like a flutter. This changes to a vibratory hum for frequencies above this range. The greatest sensitivity to vibration is for frequencies of 250 Hz (Mountcastle, Steinmetz, & Romo, 1990b). The amplitude threshold for detecting vibration is severalfold lower than for flutter. Several sources of evidence suggest that sensations of flutter and vibration arise from the activity of two populations of peripheral mechanoreceptors and their associated projection targets in the cortex. Rapidly adapting (RA) afferent fibers, thought to innervate Meissner (MEI) corpuscles in the skin, respond best below 40 Hz in the range of the sense of flutter. Pacinian corpuscles (PCs) respond best at 250 Hz. These two receptor classes respond maximally to different ranges of frequency of mechanical stimulation that in concert cover human and monkey vibrotactile sensitivity. Slowly adapting afferents (SAI—Merkel) are also sensitive to low frequency vibration, but probably contribute to a sense of pressure rather than pitch (Torebjörk, Schady, & Ochoa, 1984; Valbo & Johansson, 1984).

Additional psychophysical results support the notion of segregated channels for flutter and vibration sensations. Local skin anesthesia eliminates the sense of flutter and dissociates it from retained vibratory sensitivity (Gescheider, 1976; Gescheider, O'Malley, & Verrillo, 1983; Gescheider & Verrillo, 1979). Masking stimuli in the frequency range of flutter raise detection thresholds for flutter but not for higher frequency vibration stimuli. Higher frequency masking stimuli affect only vibratory detection. In addition, masking occurs when separate, or even contralateral, skin sites receive mask and test stimuli (Gescheider, Verrillo, & Van Doren, 1982). This suggests the phenomenon of masking depends on central mechanisms.

2. Neural Metrics in SI for Flutter Vibration

Knowledge of the central neural substrates of vibrotactile perception mostly comes from studies of response patterns recorded in SI (LaMotte &

Mountcastle, 1975; LaMotte & Mountcastle, 1979; Mountcastle, Steinmetz, & Romo, 1990a,b; Mountcastle et al., 1969). Different neuron populations within this cortex have response properties that suggest they receive information from MEI or PC receptors. One class of neurons, particularly in area 3b, entrains to the temporal period of low-frequency (<100 Hz) vibratory stimuli in the same range that maximally excites the afferent fibers associated with MEI corpuscles (LaMotte & Mountcastle, 1975; Mountcastle et al., 1990a,b). Most reports of responses in SI to higher frequency stimuli, like those best processed by PCs, show shifts in the average rates of activity with different frequencies. There have been reports of temporally encoded response patterns to higher frequencies in recordings from SII (Burton & Sinclair, 1991). However, activity in SI appears necessary for vibrotactile sensations because lesions of areas 3b and 1 led to elevated frequency-discrimination thresholds in monkeys (LaMotte & Mountcastle, 1979). Some recovery of flutter-discrimination performance occurred with retraining, but the ability to distinguish vibration frequencies was permanently lost following SI lesions. LaMotte and Mountcastle concluded that activity of flutter-vibration-sensitive neurons in areas 3b and 1 contributes to frequency discriminations. They further argued that frequency discrimination requires the cortical processing of vibrotactile stimuli to represent the stimulus period. Detection of the presence of vibration, they suggested, requires no processing beyond the subcortical level.

Frequency (pitch) and intensity (amplitude or loudness) are the primary dimensions of vibrotactile perception in humans (Melara & Day, 1992). Mountcastle and colleagues (LaMotte & Mountcastle, 1975, 1979; Mountcastle, LaMotte, & Carli, 1972; Mountcastle et al., 1969, 1990a,b; Talbot, Darian-Smith, Kornhuber, & Mountcastle, 1968) studied psychophysical and neural response functions associated with the detection and discrimination of amplitudes and frequencies of vibratory stimuli in humans and monkeys. They assessed detection and discrimination thresholds in both species and matched psychophysical with neural functions to determine possible neural metrics for each parameter responsible for the perception of vibrotactile stimuli.

Psychometric functions for amplitude discriminations for vibrotactile stimuli resemble cumulative normal curves. The amplitude difference limen (DL) for monkeys and humans was about 10% of the standard over a 17–30 dB range above detection threshold (LaMotte & Mountcastle, 1975). K. Johnson found that the number of activated fibers increased with the amplitude of a vibrotactile stimulus (K. Johnson, 1974). According to his data, the neural code for vibrotactile intensity is spatial and reflects the size of the population of active peripheral neurons. The relationship between number of cortical cells and a cumulative normal curve for stimulus amplitude remains unknown. Similarly, there is little information about the smallest cell population needed for detection thresholds.

Perceived intensity also varies with vibration frequency (Goff, 1967; LaMotte & Mountcastle, 1975; Verrillo, 1962, 1970, 1971). Below 250 Hz, a lower frequency must be presented with greater amplitude to equal the perceived intensity of a higher frequency stimulus. These differences possibly reflect the sensitivities of the peripheral receptor channels for low- and high-frequency vibrotaction. For example, the amplitude judged equal to 30 Hz at 48 μm or 129 μm varied about 50 μm from 20–40 Hz (LaMotte & Mountcastle, 1975). There have been no studies of responses from cortical neurons to find the characteristics associated with equal subjective intensity for different stimulus frequencies.

Frequency DLs are 1.8 Hz for humans and 2.7 Hz for monkeys against a standard of 30 Hz. In humans, the DLs for higher frequencies are 9.75 Hz for 100 Hz and 12.4 Hz for 200 Hz (calculated from published Weber fractions) (Mountcastle et al., 1990a). Vibration amplitudes must be a minimum of 8 dB above detection threshold for frequency discriminations. Similar to results with auditory stimuli, an "atonal" interval separates detection of the onset of vibration from the amplitudes needed to identify frequencies (LaMotte & Mountcastle, 1975; Mountcastle, 1984). The ratio of the discriminable frequency difference to a standard frequency (Weber fractions) is nearly constant (~20%) over the range of 5–200 Hz in monkeys and humans (Mountcastle et al., 1990a). Therefore, the just noticeable difference (JND) in pitch increases with frequency. The similarities between results from humans and monkeys suggest analogous processes for perceiving flutter vibration. This provides for confident generalizability from neurophysiological studies of monkey somatosensory cortex using vibrotactile stimuli to likely neural processing in humans.

Early comparisons of psychophysical and neurophysiological data suggested that the tuning points of both classes of afferent fibers correspond to human and monkey frequency-discrimination thresholds (Mountcastle et al., 1972). However, at these same threshold amplitudes, there was no corresponding sharp change in the tuning of cortical cells (Figure 7). Instead, a minimum discharge appeared in SI neurons at amplitudes matched to detection thresholds (LaMotte & Mountcastle, 1975; Mountcastle, 1984). Increasing amplitudes above detection thresholds for flutter-type stimuli resulted in a decrease in the dispersion of discharges within a stimulus cycle (increased phase locking) in cortical neurons and a graded decrease in the JND in frequency to minimum levels (LaMotte & Mountcastle, 1975; Mountcastle et al., 1990a). As amplitude increased into the atonal interval, cortical cells, especially sensitive to lower frequencies, discharged more frequently and with greater synchrony to the stimulus period (Figure 7). The degree of phase locking in the neuronal firing pattern progressively increased with stimulus amplitude. The variation in mean firing rate in entrained neurons also decreased with these stimulus amplitude shifts. As amplitude rises, Fourier transformations of the discharge train show greater

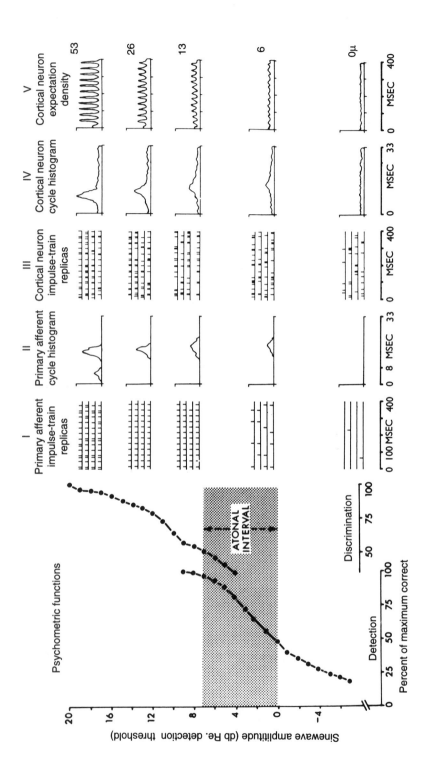

I. Primary afferent impulse-train replicas

II. Primary afferent cycle histogram

III. Cortical neuron impulse-train replicas

IV. Cortical neuron cycle histogram

V. Cortical neuron expectation density

Psychometric functions

Sinewave amplitude (db Re. detection threshold)

ATONAL INTERVAL

Detection

Discrimination

Percent of maximum correct

power for the stimulus frequency (Mountcastle et al., 1990a,b). These changes in neural activity with amplitude parallel the effects comparable changes in stimulus amplitude have on increased frequency-discrimination performance toward an arbitrary threshold level (e.g., 75%). These results clearly show that sequences of an entrained neural discharge, with intervals equal to the cycle lengths of the stimuli over a wide range of stimulus amplitudes, signal the frequency of vibrotactile stimuli. Thus the difference in interspike intervals serves as the neural code for frequency (Mountcastle, 1984; Mountcastle et al., 1990a,b).

This conclusion, however, applies only to frequencies sensed as flutter. For higher vibration frequencies, most reported SI cells only show changes in the average discharge rates for stimulus amplitudes associated with behavioral threshold shifts (Mountcastle et al., 1969). In contrast, peripheral receptors show comparable degrees of entrained discharge patterns at amplitudes that match psychophysical performance (Mountcastle et al., 1972). This entrainment of peripheral fiber responses does not reappear in areas 3b and 1 of SI. As lesions of SI disrupt frequency discriminations for flutter and vibration, the processes responsible for both probably reside within this cortex. One hypothesized mechanism is that discriminating higher frequencies relies on changes in the average level of activity in selected responsive neurons in SI (Mountcastle, 1984; Mountcastle et al., 1990b; Mountcastle et al., 1969). The next section discusses other possibilities.

3. Flutter Vibration beyond SI

The discharge intervals of many SI cells reflect the stimulus frequencies for perceived flutter but not for vibration. However, as with other features of tactile stimuli (see following sections), the link with peripheral response patterns weakens at higher levels of the somatosensory system. Only some

←——————————————————————————————

FIGURE 7 Comparisons between psychophysical measurements and neural responses from mechanoreceptor afferent fibers and SI cortical neurons to different amplitudes of 30-Hz vibrotactile stimuli. The psychometric functions on the left display the stimulus amplitudes needed by monkeys to detect the presence of a vibration (curve on the left) or discriminate between two closely spaced frequencies (curve on the right) when the base stimulus vibrated at 30 Hz. Shaded rectangle (atonal interval) shows distribution of stimulus amplitudes where performance was above chance for both tasks but not >75% correct for discrimination performance. Columns I through V show representative analyses of mechanoreceptor and cortical neuron responses to 30-Hz sinusoidal vibrations at amplitudes scaled in microns (listed on the far right) to match the sine wave amplitudes used for the psychophysical tests. Columns I and III show single trial examples of the firing patterns for different stimulus amplitudes. Columns II and IV contain examples of summated histograms folded over the period interval for the 30-Hz stimulus. The expectation density plots (auto-correlation functions) in column V display the probability that the cortical neuron firing pattern sequentially replicates the temporal consistency of the stimulus. (Modified from LaMotte & Mountcastle, 1975.)

cells in SII respond with high temporal fidelity (i.e., frequency phase locking) to vibrotactile stimuli. Most SII cells failed to follow vibrotactile frequencies above 10 Hz; a few followed frequencies up to 75 Hz. In contrast to previous recordings from SI, a small sample of SII cells displayed PC-like response patterns in showing sustained activation to higher frequency stimuli. These cells discharged at a maximum of 100 Hz. Some cells in this latter group continued to respond with temporally entrained discharge patterns even for low amplitude ($<$ 30 μm) stimuli with frequencies above 100 Hz. They displayed no further increases in discharge rate for stimulus frequencies up to 300 Hz, although the average interspike intervals follow stimulus periods (Burton & Sinclair, 1991). These cells could signal the presence of high-frequency vibration without distinguishing pitch. Thus SII provides a poor literal representation of vibrotactile frequency and only a modest rate code transformation of frequency. Some less obvious feature extraction of pitch remains a possibility.

Finding some SII cells with discharge patterns that entrain to higher frequencies introduces several speculations. First, all frequency discriminations may depend on some form of temporal encoding. Sensations of flutter and vibration might then be processed in different cortical areas. However, as responses in SII depend on intact connections with SI (Burton, Sathian, & Dian-Hua, 1990; Pons, Garraghty, Friedman, & Mishkin, 1987; Pons, Garraghty, & Mishkin, 1992) and as all areas of SI connect with SII (Burton et al., 1995; Friedman et al., 1986; Pons & Kaas, 1985), reports of no entrained responses to high-frequency stimuli in SI may be incomplete. Most previous studies concentrated on recording from areas 3b and 1. It is still possible that some area 2 cells follow high-frequency stimuli with PC-like fidelity. This may be likely, as PC receptors are prevalent in subcutaneous locations, and area 2 predominantly represents innervation of noncutaneous structures.

If the temporal information in vibrotactile stimuli penetrates from the periphery to the cortex, and if perception of these stimuli arises from direct access to a neural metric of stimulus period, then how does the brain dissociate entrained activity from different cell populations? SI neurons with SA-type response patterns entrain nearly like RAs to vibratory stimuli (e.g., (Freeman & Johnson, 1982). Yet, evidence suggests SAI afferents signal changes in pressure rather than frequency (Torebjork et al., 1984; Valbo & Johansson, 1984). If discharge period is an epiphenomena in one set of neurons (i.e., SAs), one might worry about its perceptual importance in another neural sample (i.e., RAs). Alternately, higher somatosensory areas may extract information about frequency as a feature represented in nonentrained activity patterns in a neural network. Thus response patterns of nonentrained neurons are transforms of stimulus temporal information. Alternatively, perception of the temporal characteristics resides within the entrained activity of SI neurons and does not arise from later stage process-ing.

No other parietal cortical areas have been studied using parametrically manipulated vibrotactile stimuli. Neurons in area 7b, Ri area, and insula display heightened discharges to vibrotactile stimuli (Robinson & Burton, 1980c). Positron emission tomography (PET) studies in humans (Burton et al., 1993; Coghill et al., 1994) and 2-DG studies in monkeys (Juliano et al., 1983) confirm that vibrotactile stimuli influence multiple cortical areas be-yond SI and SII. Most cells in these regions respond phasically, and there-fore the discharge pattern cannot entrain stimulus frequencies. The source of activation is unknown. If it were through cortical connections from SI or SII, the evident absence of sustained responses to persistent vibration sug-gests stimulus-frequency representation is either absent or encoded in a format that does not replicate stimulus temporal order.

4. Perceiving Tactile Motion

A sensation of motion arises when an object moves across a region of skin, when a part of skin is drawn across a stationary object, or when a stimulus moves between points on the skin. Thus, the sensory experience arises from receptors in one patch of skin (and associated central representation with this receptive field) or from a distribution of elements active in temporal and spatial sequence. Proposed neural mechanisms responsible for motion sen-sations from the skin are varied and, in most instances, inconclusive expla-nations. One possible and intuitively appealing notion considers perceived motion is the greatly enhanced responses of low-threshold cutaneous me-chanoreceptors and their associated central neurons to moving a stimulus across receptive fields (Costanzo & Gardner, 1980; Essick & Whitsel, 1985a,b; Gardner & Costanzo, 1980; Hyvärinen & Poranen, 1978a; Whitsel, Roppolo, & Werner, 1972). The simple perception of motion then results because the peaks of higher average activity shift across neural space, be-tween neurons with overlapping receptive fields, and within the so-matotopic maps of SI cortex (Gardner & Costanzo, 1980). At some unde-fined higher level, changes in the parameters of movement may be detected by processes that interpret the sequential changes in peak activity in the somatotopic maps.

This hypothesis is insufficient because a perception of movement occurs when the same area of skin passes over or is passed over by an object. In this circumstance of limited shifts in activated skin and, therefore, regions across the somatotopic maps, the question then becomes one of defining the least detectable shifts across a somatotopic map. Alternatively, perceived move-ments lacking spatial shifts across the skin could involve processes like skin stretch, which activate a sense of movement with less than 1-mm stretches

(Gould, Vierck, & Luck, 1979; Johansson, 1978; Johansson & Vallbo, 1983). Furthermore, the hypothesis lacks an *a priori* reference for activity levels; it also provides no mechanism for distinguishing different sources of enhanced responses. Many neurons show high-frequency, phasic bursts of discharges to skin indentations. Punctate vibrotactile stimuli similarly evoke high firing rates, but these do not create illusions of movement, thus, a hypothesis of motion sense from spatial changes across a neural map fails to consider the metric(s) that associate the altered mean firing rates to a moving stimulus as opposed to other sources of increased activity.

Another potential mechanism able to signal movement is activation of selectively tuned neurons that only respond to moving tactile stimuli (Cos-

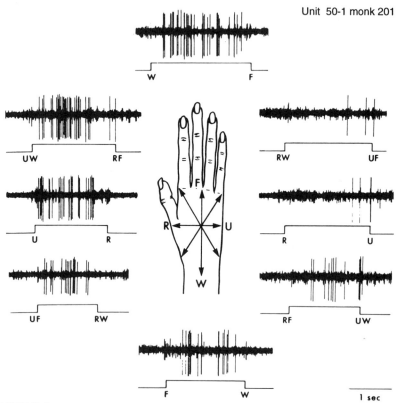

FIGURE 8   Example tuning properties of different direction-sensitive neurons recorded from SI. Panel on the left shows typical responses from a cell with unidirectional tuning. Here brushing a 10-mm edges across the hand dorsum, from the ulnar side of the wrist (UW) to the radial side of the fingers (RF), evoked the largest response. Moving the stimulus in the mirror opposite direction (RW to UF) caused the smallest responses. Polar plots summarize this cell (E in panel on the right) and examples from 8 other cells. The plots in A–D are from broadly tuned cells; E–F show moderate tuning; and G–I display cells with narrow tuning. (Modified from Costanzo & Gardner, 1980.)

tanzo & Gardner, 1980; Essick & Whitsel, 1985a,b; Hyvärinen & Poranen, 1978a,b; Whitsel et al., 1972). Polar plots of the average firing rates for some of these cells show preferential higher average firing rates for selected directions of movement across their receptive fields (Figure 8). These studies described only small percentages of SI neurons with such invariant direction-specific features (range of 3–27%). Another group, called variant direction cells, responded preferentially to opposed directions of movement where the polarization pointed toward or away from a limb joint (Essick & Whitsel, 1985a,b, 1993). Both classes of cells might provide for the simple perception of moving stimuli exclusively or in addition to "simple" cells that respond to any movement of suprathreshold mechanical stimuli in their receptive fields. Movement sensation is then a consequence of activity in these feature-specific populations or some higher stage of processing able to interpret the responses of these cells. With the activation of tuned cells,

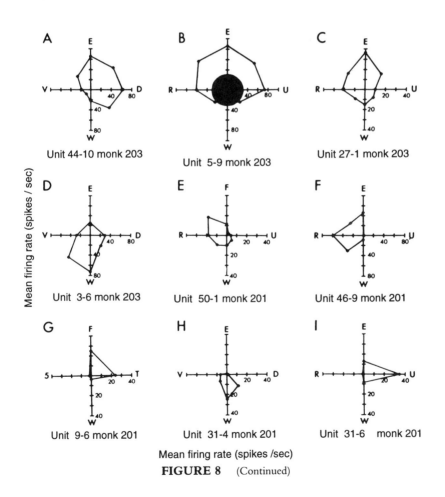

Mean firing rate (spikes /sec)

FIGURE 8 (Continued)

perceptions of movement might be expected to intertwine with knowledge of the direction of movements (see section IV.A.7, Perceiving Stimulus Direction).

Several studies found that only posterior parts of the somatosensory cortex (i.e., area 2 or posterior area 1) contained examples of movement-sensitive neurons; these regions receive cortical projections from more anterior areas. In addition, many directionally selective cells were recorded in supra- and infragranular layers of the cortex. Implied is that cells in these layers are distant from the main thalamocortical projection zones and thus respond after intracortical processing (Whitsel et al., 1972). Cutting the dorsal columns eliminates a monkey's ability to discriminate movement directions on the limbs below the lesion (Vierck, 1974) and removes movement-selective responses in SI cortex (Dreyer, Schneider, Metz, & Whitsel, 1975).

Some findings from the peripheral nervous system are at variance with the idea that motion sense comes from activation of feature-selective neurons. First, the tenet that feature selection for attributes of movement only appears after central processing is unrepresentative. Most classes of low-threshold mechanoreceptors already display highly reproducible response differences to distinct directions and velocities of motion. This selection arises from exquisite peripheral receptor sensitivity to variations in the waves of compression, force, and stretch produced by moving a tactile stimulus across the skin (Edin, Essick, Trulsson, & Olsson, 1995; Essick & Edin, 1995; Ray, Mallach, & Kruger, 1985; Srinivasan & LaMotte, 1987). It is still conjecture, however, that the precise patterns seen in peripheral responses persist across the spatial and temporal asynchrony of multiple synapses to converge on the larger receptive fields of cortical neurons (Edin et al., 1995). Furthermore, the isomorphic preservation of the mechanical stresses and strains across the path of a moving stimulus does not appear necessary to explain psychophysical performance with moving stimuli. Careful matches between psychophysics of motion detection and average firing rates across a population of peripheral receptors suggest that this broadly filtered average response provide a sufficient measure of the attributes of movements (Edin et al., 1995). Important, however, is that these average rates, even from separate receptors on the skin, converge centrally where the particular peripheral receptor directional responses from local skin mechanics are unlikely to reappear (Essick, 1991).

Additional recordings from SI cortex suggest that perceiving motion may emerge from more general mechanisms and attributes of responses to moving stimuli because feature selectivity for stimulus motion is not restricted to a small population of cells. Percentages of over 40% of recorded cells in anterior areas (e.g., areas 3b and 1) responded to real or apparent movements (Gardner, Palmer, Hamalainen, & Warren, 1992; Ruiz, Crespo, & Romo, 1995; Warren, Hamalainen, & Gardner, 1986a). Thus, stimulus motion engaged many rather than selected populations of cells.

One possible mechanism for distinguishing the source of raised firing rates in a more general population of cortical cells is differences in the temporal pattern of discharges. Gardner and colleagues found that enhanced responses to simulated moving stimuli were more continuous for the combinations of temporal and spatial stimuli that produced a sense of apparent motion. Virtual motion sensations occur when the interval between stimulating rows of an OPTACON (Optical-to-Tactile Conversion) device shortens to less than 20 ms (Gardner et al., 1992). Low-frequency (25 Hz) shifts of stimuli across the rows produced phase-locked response patterns in nearly 40% of the sample of SI neurons. Higher rates of interrow sequential stimulation (50 and 100 Hz) evoked progressively less oscillatory firing patterns. Humans exposed to the same stimulus patterns reported feeling a more pronounced sense of apparent motion with the higher frequencies and pulsatile, punctate stimuli at lower frequencies of stimulation. Thus, changes in temporal characteristics of cortical neurons to the same stimulus patterns show a central mechanism of low-pass filtering of incoming information. The decay time of in-field inhibition produced by the initial excitatory responses may set the wavelength of this filter (Gardner et al., 1992). A sensation of movement then corresponds to the onset of continuous activity in a population of SI cells. Local inhibitory circuits mediate the transition into this pattern of responding.

A temporal hypothesis that proposes a shift from oscillatory to continuous firing patterns may be limited to perception of apparent motion produced from displaced stimuli. With continuous stimulation the attributes of velocity, stimulus direction, and traverse length have complex influences on perceived movements (see following sections). For example, slow and fast movements across a single patch of skin evoke motion sensations and, although the traverse is within the cell's receptive field, different rates of continuous neural activity rather than a low-pass filtered transition of firing patterns. In this circumstance measures based on mean firing rates (or some derivative statistic) provide a closer fit to the sensations. A sense of movement across continuous patches of skin similarly evokes a summed, continuous discharge in many cortical neurons. There is no evidence showing how "central neural responses elicited by continuously moving stimuli can be approximated by the presentation of *two* spatially and temporally separated punctate stimuli" (Essick, McGuire, Joseph, & Franzen, 1992, p. 183). The particular temporal pattern discovered with simulated motion is also misleading for circumstances where stationary stimuli, such as vibrations or airpuffs directed at one spot, produce continuous firing rates.

5. Perceiving the Length of a Stimulus Path

Studies of perceived stimulation distance across the skin (e.g., traverse length) relate to the duration of stimulation. Thus, estimates of traverse length are inverse functions of stimulus velocity for brushing stimuli (Essick, Franzén, McMillian, & Whitsel, 1991; Whitsel et al., 1986). Temporal factors also account for the mechanisms responsible for apparent motion sensations from successive, pulsatile vibrotactile taps and electrocutaneous shocks or cutaneous stimulation through separated apertures (Essick et al., 1991; Essick et al., 1992). For example, studies of the phenomena known as the Tau effect or cutaneous saltation "have shown that the temporal patterning of stationary tactile pulses greatly influences subjects' perception of the spatial locus of those pulses, and of the distance between them" (Essick et al., 1991, p. 345). As the interval between successive pulses decreases, the sensed distance separating these shrinks and migrates toward the location of succeeding pulses in the series. At an interval of 20 ms the two stimuli feel coincident. In another study, subjects perceived a smooth, continuous movement when a brushing stimulus spanned two sequential, slit-like openings over the skin. This occurred only when the stroke velocity increased appropriately for different-spaced openings so that the delay between apertures was ~30 ms (Essick et al., 1992). In these same studies, "the perceived length of skin traversed . . . decreased with the time interval [sic] between initial and final skin contact, following a time course grossly comparable to that characteristic of the cutaneous saltation phenomenon" (Essick et al., 1991, p. 348).

These critical interstimulus intervals suggest a likely common mechanism for producing spatial compressions and apparent motion. Changes in the receptive fields of cortical neurons to differences in the timed sequence of stimulating the receptive field may be responsible for perceived spatial compression. Gardner and colleagues showed that the number of OPTACON rows over which a cell responds decreased when they shortened the interval between successive rows of stimulation (Gardner, Hamalainen, Palmer, & Warren, 1989; Gardner, Palmer, Hamalainen, & Warren, 1992). They suggested that the cell's receptive field might shrink because it does not recover in time from in-field inhibition produced by the first stimulus. Using a similar argument, higher velocity stimuli yield estimates of shorter traverse lengths because the leading edge of stimulation within the receptive field arrives before that portion of the map recovers from in-field inhibition produced from previously stimulated skin.

6. Perceiving Stimulus Velocity

Perceptions of stimulus velocity, like those for traverse lengths, are related to stimulus durations. Estimates of stimulus velocity, however, compress the physical dimension and relate to it with a power function whose mean exponent is 0.33 for cutaneous stimulation using continuous or discontinuous brushing stimuli across constant traverse lengths (Essick et al., 1988; Essick et al., 1991; Essick, Franzen, & Whitsel, 1988; Essick & Whitsel, 1988). Based on finding a direct proportionality between estimates of stimulus duration and velocity, these investigations suggested that "subjects' estimates of stimulus velocity are based on temporal information, i.e., on the time elapsed between stimulus onset and offset" (Essick et al., 1991, p. 349). Additional factors must contribute to estimates of velocity because stimulation across different traverse lengths does not affect perceived velocities. Essick and colleagues suggested two mechanisms to explain the relative lack of influence of traverse lengths (Essick et al., 1991; Essick & Whitsel, 1988). First, different stimulus durations from varying path lengths could be nullified by a within-fiber code based on the known changes in average firing rates of individual mechanoreceptors by different stimulus velocities (e.g., higher rates with faster stimuli). A second alternative for which there are no data is that velocity perceptions come from neural mechanisms that process the time for raised activity to move across neural space. Another alternative is that the necessary information for estimating stimulus velocity comes from a range of stimulus durations that have a fixed maximum; the system ignores information from long traverse lengths (e.g., greater durations).

7. Perceiving Stimulus Direction

Perceiving the direction of tactile motion involves interactions among parameters of stimulus velocity, body region stimulated, and length of skin traversed (Essick, 1991). The relationship between directional sensitivity and stimulus velocity is nonlinear. For a given traverse length and stimulus location, the function is a distorted inverted U. "The effects of velocity on directional sensitivity . . . can be described by curves belonging to a subset of the gamma function family which pass through the origin, increase with increasing velocity to a maximum . . . and then decrease asymptotically to zero" (Essick, 1991, p. 336). Increased traverse lengths compensate for declines in sensitivity with higher velocities.

The distribution of peripheral innervation density (i.e., higher on the hand and face than on the trunk and arm) correlates with several factors that effect directional sensitivity. For example, traverse lengths are shorter and optimal stimulus velocities are slower for maximal direction sensitivities at distal skin regions like the digits compared to the arm and shoulder. Traverse length and optimal velocity increase linearly for stimulus locations that progress up the arm from the digits (e.g., respectively, traverse length and optimal velocity are 0.33 cm and 5.4 cm/sec on the digits and 1.75 cm and 18.6 cm/sec on the arm). However, maximum directional sensitivity is similar for all body regions studied with appropriately selected traverse length and stimulus velocity (Essick, 1991). The greatest resolution of direction (i.e., shortest traverse length) is in highly innervated structures. Sensitivity also increases with longer traverses. These results suggest perception of stimulus direction requires recruitment of some minimum sample of low-threshold mechanoreceptors and associated population of cortical neurons. This explains why shorter traverse lengths reach this level where innervation densities are greater. Differences in optimal velocities for different body regions suggest that directional sensitivity also might involve timing of successive activation within a minimum population of activated structures. The critical factor for best perception of stimulus direction is an optimum temporal frequency of stimulation (Essick, 1991). Thus, faster optimal velocities are necessary to reach the requisite numbers with the right timing where innervation densities are lower.

Collectively, these psychophysical observations suggest that neural models based on the timing of responses may underlie direction sensitivity. However, there are only a few examples of how the cortex processes this (or any other) metric for stimulus direction. Most published evidence describes feature-selective cells whose firing rates vary with the coordinate direction of movement across the receptive field (Figure 8) (Costanzo & Gardner, 1980; Essick & Whitsel, 1985a,b; Essick & Whitsel, 1993; Hyvärinen & Poranen, 1978a; Whitsel et al., 1972). As in the visual system, mean firing rates presumably carry information about stimulus direction (Costanzo & Gardner, 1980; Gardner & Costanzo, 1980; Hyvärinen & Poranen, 1978a; Whitsel et al., 1972). Accordingly, perceiving different directions involves heightened activity among such feature-specific cells. Responses from these selected cells are the percept or the information for other brain regions that decipher the different populations. A problem with these ideas is lack of evidence of how some optimal temporal frequency of stimulation translates to mean firing rates. A possibility is that stimulation of successive receptive fields leads to undefined and undetected temporal summation at central synapses. Another problem with mean firing rate codes is that variability of the discharge train rises with firing rates. Comparisons based on maximal firing rates are, therefore, not always associated with optimal distinctions for opposed directions of movement.

Essick and colleagues proposed a signal-detection measure (i.e., d') as an alternative index for assessing directional information in the discharges of selective feature cells. The estimated neural d' (named delta'e) is calculated from the difference in mean firing rates obtained from opposed directions of motion divided by the average of the standard deviations for these two directions of movement (Essick & Whitsel, 1985a). Thus, changes in variance with different average firing rates nullified with a d' measure.

Several observations support the hypothesis that populations of tuned cells have a role in the perception of movement direction. First, changes in velocity and traverse length that influence direction sensitivities (i.e., d' from signal-detection statistics) in humans cause comparable changes in direction detectability estimated from neural activity using delta'e (Essick & Whitsel, 1985a,b; Essick & Whitsel, 1993). For example, psychophysical studies (see above) showed that direction sensitivity from a region of skin was a gamma-shaped function of velocity and that traverse length was indirectly proportional to direction sensitivities. Several neurons with a single-direction polarization and with responses to different velocity stimuli showed gamma functions like those observed in the psychophysics studies. The cell's maximum delta'e was for velocities of 5 cm/sec, close to the maximum sensitivity of human observers. Lower sensitivities occurred for higher and lower velocity stimuli in the cell discharge rates and in human observers. Longer traverse lengths made these distinctions more evident (e.g., improved direction sensitivities). Mean firing-rate differences, however, did not fit the psychophysical data (Essick & Whitsel, 1985b).

Despite the comparable sensitivities to direction found with the neural delta'e and human d' values, this neural measure has problems that limit its generalization. First, the delta'e index is calculated from responses to opposed directions of movement. During a task, discrimination follows presentation of a stroke in one direction. Therefore, a predictive difference index must derive from single-direction trials. Reference to different mean firing rates for an opposed direction of movement within a single discrimination trial would have to incorporate memory of the firing rates for opposed directions. Second, if delta'e is only a measurement and mean firing rates of a population of directionally selective neurons is the basis of discriminations, then the problems with mean firing rates discussed above still apply. Third, there are very few examples of direction invariant neurons in the literature. Although technical problems explain some of this (Essick & Whitsel, 1985a), the results are scarce for what is a very general and easily learned tactile discrimination. Fourth, an ability to distinguish any direction of movement requires populations of directionally invariant neurons that encompass different vector directions. Such populations remain hypothetical in the literature, especially for receptive fields involving the digit tips. Finally, there is no evidence regarding the interpretation of a delta'e index by cortical regions that decide the direction of movement. Are the distinctions in adjusted mean firing rates the percept? If not, then what kind of information does a delta'e index convey to later stages of processing?

Another approach to the issue of a metric for direction considers changes in mean firing rates over time in a population of SI cells. Ruiz and colleagues calculated the coefficients of the Karhunen–Loeve transform from sequential small time segments of smoothed average firing rates for different directions of movement (Ruiz et al., 1995). They showed that the value of the coefficient for the first principal component systematically varies with the direction of movements across a cell's receptive field. More cells in area 3b than area 1 showed this result (62 vs. 38%). The responses from cells with slowly adapting responses were more likely to display such orderly variations in activity over time (i.e., in area 3b, 55.5%, and in area 1, 63.6%). This measure derives from the matrix of correlations between the firing rates (adjusted for mean activity) determined in each of a series of selected intervals across the duration of a moving stimulus. Its size, therefore, reflected the extent of temporal covariance of neuronal activity during strokes across the receptive field. The interesting aspect of this measure is that its magnitude may relate potential sequential relatedness of neural activity as a stimulus successively moves across receptor space. Thus, the amount of temporal covariance measures how much of a cell's discharge pattern models optimum temporal frequency of stimulation for a particular direction. Ruiz and colleagues found that the first coefficient varies for different directions, and most cells displayed maximum positive and negative values for different directions. Unlike differences in mean firing rates for opposed directions but collapsed across time, this quotient considers activity changes with time.

By summing the magnitudes of the first coefficients for a population of cells for different movement directions, they created direction vectors (Ruiz et al., 1995). The angle of each vector closely matched the direction of movement across the fingertip. This result implies that direction perceptions may arise from such sums across a population of SI cells. Additional modeling showed that a population of only ~100 cells would have to retain close temporal linkages of their respective discharges to yield discriminable vectors.

There are several attractive features in this model: (a) It considers all potential directions of movement without requiring unique feature specific neurons. (b) It incorporates information about changes in neural activity over the time for movements. (c) It suggests a specific summed neural signal that could be the percept for selected directions of movement.

The model also raises some concerns. Most of the cortical cells (e.g., SA, area 3b cells recorded near layer IV) whose response patterns satisfied the constraints of the measurements also display the highest fidelity reproduction of peripheral mechanoreceptor transduction events for spatial forms (K. Johnson & Hsiao, 1992). The responses of such cells probably follow changes in the distribution of stress and strains for different movement directions. This may explain remarkable similarities in the average discharge patterns for recordings from the peripheral mechanoreceptors and SI neurons in studies that used very different stimulators (Edin et al., 1995; Ruiz et al., 1995). Thus, the model may apply only for cells with such isomorphic transformations. The next stage of processing in cells that lack such peripheral-like responses is still a puzzle.

Another issue is modeling a realistic synaptic circuit that produces different amounts of temporal covariance across a discharge train. The calculated quotient reflects levels of correlation between different parts of the matrix created from the discharge rates per trial for multiple movement directions. The preceding discussion suggested that greater temporal covariance arose from sequential relationships in firing rates because these might represent successive stimulation of the receptive field. However, the specific correlations have not been specified. For example, high levels of temporal covariance could arise from moving through excitatory–inhibitory sequences at the beginning and end of a stroke rather than sequentially through the entire movement.

Most neural models for representing attributes of motion remain inconclusive. This is partly due to difficulties in parsing the contribution of spatial, temporal, or feature selectivity aspects of central neuron responses to the attributes of motion (i.e., movement velocity, direction, and traverse length). Models based on simulated motion from pulsatile stimuli may not account for the mechanical and temporal complexities produced by continuous stimulation of the skin. None of the developed models adequately considers motion sensations from back and forth stimulation over the same patch of skin.

B. Evidence for Material (Micromorphic) Components of Objects

1. Texture

One specialization of the somatosensory system of primates may be for the discrimination of surface texture with the hand (Taylor, Lederman, & Gibson, 1973). For instance, touch may be superior to vision for discriminations of very smooth textures (Heller, 1989; Lederman, Thorne, & Jones, 1986). Furthermore, sighted and congenitally blind observers equivalently discriminate textures examined only by touch. This suggests that, unlike three-dimensional form, texture may not be automatically "recoded" into a visual format (Heller, 1989).

The sense of roughness is the intensive aspect of texture studied most often. Suggested neural codes for roughness emphasize temporal, spatial, intensive, or modal properties of neural activation.

a. Temporal Codes for Roughness

An early theory held that roughness depends on the sense of vibration (Katz, 1989). With the hand stationary, perception of surface roughness quickly fades. Katz suggested that moving the skin over a textured surface creates vibrations, as might result when micromorphic irregularities in surface height strike the papillary ridges of the fingertips. One possibility is the timing of activation of tactile receptors replicates the temporal information in such sequential skin compressions.

Several psychophysical experiments weaken the credibility of a temporal code for roughness. Taylor and Lederman (1975) found that wide variations in scanning velocity (and, to a lessor degree changing force) had virtually no influence on the perceived roughness of a tactual grating. These experiments verified the common perceptual experience that the texture of a homogeneous surface feels constant although the hand moves about in a continually varying pattern producing shifting receptor activation. Velocity insensitivity in magnitude estimates of roughness is particularly embarrassing for a temporal code for roughness, which might predictably change with the rate at which surface microstructures strike cutaneous receptors. When Taylor and Lederman independently varied the groove and ridge widths of tactual gratings, they found that human magnitude judgment of roughness increased as a function of larger groove widths, but remained unaffected or even decreased with larger ridge widths (Taylor & Lederman, 1975). The temporal period of the ridges striking the skin varies with either of these parameters. Rather than a temporal code, Taylor and Lederman proposed a spatial-intensity mechanism in which perceived roughness is proportional to the volume of skin falling down into gaps between high points on a surface.

Temporal coding of surface features such as roughness requires that stimulus-related periodicities in peripheral afferent fiber discharges persist in the activity patterns of central neurons. Examples of periodic changes were observed in the discharge bursts of many cells in areas 3b and 1 of SI when the receptive fields of these cells touched grating surfaces with differing spatial-temporal period of ridges and grooves (Figure 9) (Sinclair & Burton, 1991a, 1993b). However, the duration of the bursts to each ridge increased with groove width. The pattern of discharges within each burst showed no obvious temporal structure related to the stimulus. SI neurons discharged to rising and falling force levels as surface structures such as the edges of ridges compressed the skin during stroking (Sinclair & Burton, 1993b). These results suggest cortical neurons receive information from peripheral mechanoreceptors sensitive to skin curvature changes at the edges of surface structures such as grating ridges. Sensitivity to edges in cortical cells is consistent with descriptions of peripheral afferent fiber activation to step indentations (Srinivasan & LaMotte, 1987). For gratings, larger groove dimensions result in greater skin distortion as the skin falls deeper into the gap. The increases in discharge burst duration and average firing rates in many cortical cells possibly reflect these skin distortions. Timing information related to spatial-temporal period, though present in the discharge, has

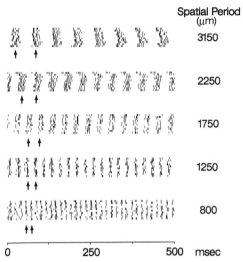

Spatial Period
(μm)

3150

2250

1750

1250

800

0   250   500 msec

FIGURE 9   An SI neuron's response patterns to rubbing different embossed gratings across its receptive field. Surfaces with constant ridge widths (Rw) of 250 μm but various groove widths (Gw) stroked at constant velocity of 50 mm/sec using mechanically controlled horizontal stroking. Spatial period equals Rw + Gw. Two arrows at the bottom of each group of ten raster lines mark the interval of the spatial-temporal period for the stimulating surface. The burst discharge duration increased with spatial period while the interval between bursts accurately replicated the spatial-temporal period of stimulation.

dubious value in predicting surface roughness. The frequency of these discharge bursts displayed the temporal period of ridges striking a point on the skin, but bore no relation to the groove width (Taylor & Lederman, 1975; Taylor et al., 1973). In addition, temporal fidelity in neural discharges to the stimulus period of tactual gratings decreased in recordings from different levels: it was most evident in VPL, seen in some SI cells and was essentially absent in SII (Sinclair & Burton, 1991a, 1993a; Sinclair, Sathian, & Burton, 1991). This suggests a neural transformation from thalamus through cortex that accentuates nontemporal aspects of texture.

A critical experiment by Lamb (1983b) proved that discriminations of at least some textured surfaces are not based on differences in temporal patterns. Lamb had human subjects discriminate raised dot surfaces that varied in interdot spacing in the direction perpendicular or parallel to finger motion. Stimulus temporal period varied only for the latter surfaces. Lamb found that subjects discriminated both types of surfaces equivalently—both those varying in temporal period and those with constant temporal period. In a parallel neurophysiological study in monkeys, Lamb (1983a) found that increments in average firing rates in peripheral afferent fibers correlated with increments in dot spacing in either direction relative to hand motion.

These results show that an intensity code based on mean firing rate could account for human ability to discriminate the roughness of dot patterns.

b. Intensity Code

The average firing rates of cells at many levels of the somatosensory system predict the roughness of dot patterns and tactual gratings with surface elements spaced between .5 and 3 mm (Darian-Smith, 1982, 1984; Darian-Smith, Goodwin, Sugitani, & Heywood, 1985; Darian-Smith, Sugitani, & Heywood, 1982; Sinclair & Burton, 1991a, 1993a; Sinclair et al., 1991). An example of the SI average firing rates in monkeys trained to rub distal fingertips over gratings of varying groove width is shown in Figure 10 (Sinclair & Burton, 1991a). The graph shows the relation between average firing rates (AFR) and groove width by the strength of correlation (*z*-axis) between these factors for a sample of >80 cells. The result simulates average firing rate for a SI neuronal population to tactual gratings. AFR functions were standardized for each cell to equalize basal firing rate and to aid generalizations concerning the shape of these response functions. This strategy produces a continuum of response functions from positive to negative that

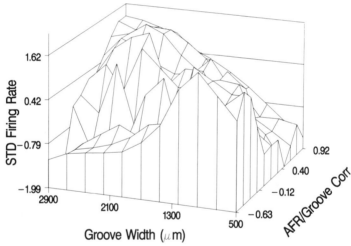

FIGURE 10   Synthesized SI neuron population response for different gratings rubbed across the receptive fields. This graph plots standardized values of signal-detection theory (SDT) firing rate (AFR) on the ordinate against groove width (abscissa). AFR was standardized by converting each cell's firing rates to Z-scores (based on mean and standard deviation of firing rates per cell). The *z*-axis shows sorting by the correlation between AFR and groove width. The functions for cells with maximum positive correlation values appear toward the rear and those with maximum negative correlations are toward the front. The response range forms a continuum from high-positive to high-negative correlations to roughness. No discrete subgroups are visible. (Data analyzed from previously published results, Sinclair & Burton, 1991a.)

is "saddle" shaped. Cells with SA and RA response patterns responded similarly in this task and data from both are in this graph. As shown toward the rear of the graph, the AFRs of many cells increased as a linear function of grating groove width. Functions toward the center show that responses of many cells saturated at larger spacings (>~2200 μm). Shown in the forefront, the responses of a few neurons decreased as a function of groove width. A slice through this graph from front to back at any groove width could resemble the brain's view of the entire population of SI cells discharging during one stroke. Cells with steep positive or negative response functions to groove width most reliably represent the continuum of roughness touched by the monkeys. These cells could provide information sufficient to account for discrimination of surface roughness. The variety of response functions shown in Figure 10 may illustrate stages of neural processing in different cortical layers.

c. Spatial Code for Surface Texture and Spatial Form

Taylor and Lederman suggested a sense of roughness might relate to the volume of skin falling into the grooves of a tactual grating (Taylor & Lederman, 1975). Later descriptions of afferent fiber responses (LaMotte & Srinivasan, 1987a,b,c; LaMotte & Whitehouse, 1986; Srinivasan & LaMotte, 1987) and studies of cortical responses to edges (Sinclair & Burton, 1993a) support the idea that skin distortions provide a critical component to texture information. Johnson and colleagues extended this idea by contending roughness depends on distributed activation of receptors in the area of skin contacting a surface. This latter suggestion is appealing because exploring the spatially distributed surfaces of natural objects engages multiple receptors.

In two closely related papers, Johnson and colleagues examined several candidate neural metrics in peripheral receptors for representing roughness (Connor, Hsiao, Phillips, & Johnson, 1990; Connor & Johnson, 1992). They concluded that neural codes based on average firing rate, firing rate variance, and temporal variations in firing rate failed to account for human magnitude estimation functions of the same surfaces (see below). Instead, they argued that roughness depends on spatial form represented in the discharge patterns of a population of cutaneous mechanoreceptors activated by surface contact at distributed points over an area of skin. Johnson and colleagues reasoned that the topological organization of the tactile receptor sheet produces an initial cortical representation of a spatial stimulus as an isomorphic, spatial pattern of neural activity.

Spatial event plots (SEPs) from single-neuron responses elegantly showed this isomorphism for embossed patterns such as letters of the alphabet or raised dots scanned across the skin (K. Johnson, Phillips, Hsiao, & Bankman, 1991; Phillips et al., 1988). The spatial information apparent in the SEPs constructed from some area 3b neurons with SA responses closely reproduced the stimulus patterns (Figure 11). The SEPs from area 1 cells were less structured. Many neurons in both areas responded best to vertical leading and trailing edges. This spatial filtering may arise partly from skin and receptor mechanics or from spatially structured excitatory and inhibitory networks in the cortex. Gaps in SEPs corresponding to components of some letters correlated well with confusion matrices generated by human subjects identifying the same letter patterns (K. Johnson & Phillips, 1981; Loomis, 1982; Phillips, Johnson, & Browne, 1983). For instance, many neurons clearly represented the crescent but failed to respond to the hori-

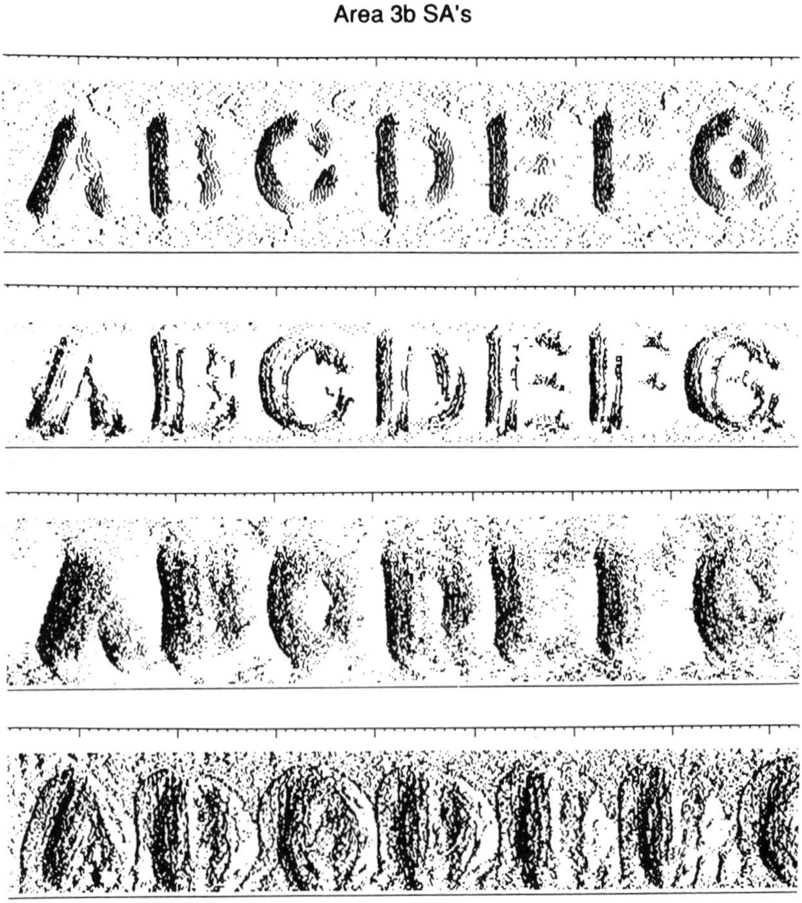

Area 3b SA's

FIGURE 11   Spatial event plots (SEPs) created from the discharge patterns of five slowly adapting neurons recorded in area 3b of awake macaque monkeys. "Letter height: 8.0 mm., Scanning velocity: 50 mm/sec with the exception of panel 4, for which the letter height and scanning velocity were 6.0 mm and 20 mm/sec respectively. Contact force, 60 g" (Johnson et al., 1991, p. 311)

zontal bar in the letter *G*, which human observers often confused with the letter *C* (Figure 11). In contrast, the letter *C* was never confused with *G*. Isomorphic neural representations were most common in layer 4 and upper layer 3 in area 3b (DiCarlo, Hsaio, & Johnson, 1994). Degraded or distinctly nonisomorphic neural patterns appeared in recordings from cells in supra- and infragranular cortical layers. These results suggested processing of spatial form may go from layer 4 to supra- and infragranular layers and that this leads to transformation of spatial information (K. Johnson & Hsiao, 1992; Johnson et al., 1991; Phillips et al., 1983). They speculated that the final nonisomorphic representation may be a map of stimulus features independent of size and locus of stimulation (Johnson et al., 1991).

The application of studies of spatial form to roughness perception stemmed from the finding that human estimates of the roughness of raised dot surfaces, an idealized spatial pattern, yield an inverted U-shaped relationship as a function of dot-to-dot spacing (Connor et al., 1990; Connor & Johnson, 1992). This function has a maximum at 3–3.5 mm for dot diameters of .5 to 1.2 mm. An influential finding was that mean response rates of SA and RA afferent fibers in monkeys similarly followed inverted-U functions, but the peak for these was at dot spacings about 1 mm less than the human psychometric functions. Connor and colleagues concluded that this disparity in the peak of the psychometric and neurometric functions discounted mean rate as the neural code for roughness. Instead, they proposed a spatial variation metric, computed on the SEPs of SAI afferents, provided the best fit to the human psychometric data. Spatial variation reflects the degree of coincident activation of neighboring receptors in the skin area contacting a surface.

The spatial variation metric was computed from peripheral afferent fiber response patterns. However, the receptive fields of these receptors converge on central neurons where isomorphic representations are soon lost. They suggested that the distributed array of peripheral activation collectively engages antagonistic excitatory and inhibitory projections on a central neuron from neighboring cutaneous receptors (Connor & Johnson, 1992). In the simplest case excitatory and inhibitory inputs from neighboring receptors converge on a central neuron. With no contact, all receptors are inactive and cortical cells are silent. When all receptors are active due to uniform compression by a dot on a surface, excitation and inhibition from each receptor cancels out and the central cell remains quiet. However, the central neuron responds maximally when only the excitatory receptor strikes a raised dot and neighboring receptors are over the surrounding gap. This model of spatial convergence would enhance responses to edges or contours, points of greatest skin distortions. The consequence of this model is to translate spatial variation into a rate code or intensity function in cortical neurons across the whole range of human roughness estimates.

A spatial variation metric predicts that central neurons receiving convergent information from spatially distributed receptors would change their average firing rate correspondingly with the shape of magnitude estimation functions for dot spacings larger than 3 mm, (i.e., they would appear as gamma-type functions) (Connor et al., 1990; Connor & Johnson, 1992). Johnson and colleagues suggest integration over a few millimeters, and implicitly spatial convergence need only occur over the area of skin being stimulated during texture exploration (e.g., the fingertips). Unfortunately, data from central neurons currently is available only for surface spacings up to 3–3.5 mm, the peak in the human psychometric function. In agreement with the model, many cortical cells show progressively higher firing rates over this range of spacings (Sinclair & Burton, 1991a, 1993a). Response functions of these cells might peak at the 3 to 3.5 mm spacing predicted by the human psychophysics. Saturated response functions seen in SI suggest the possibility of a response inversion at wider spacings (Figure 10; Sinclair & Burton, 1990). However, response functions of many cells plateau at 2 mm or above (Figure 10). The latter resemble response functions of many peripheral afferent fibers (Goodwin, John, Sathian, & Darian-Smith, 1989) perhaps due to close ties to the cortical input layers. Studies using wider spacings may clarify this issue.

An attractive feature of the spatial variation explanation of roughness perception is its power to reconcile perceptual constancy. Humans demonstrate constant magnitude estimates despite large variations in velocity and force of strokes across touched surfaces (Lederman, 1974; Lederman & Taylor, 1972; Loomis & Lederman, 1986). These variations produce confounding information from the receptor sheet as stimulation parameters follow the ever-changing motion of the hand. Individual peripheral receptors render ambiguous information about surface dimensions from which the proposed spatial variation mechanism extracts invariant spatial information. Temporal coincidence of activation would affect the exact discharge pattern of individual receptors, but motion over a uniform surface would produce an average activation across a population of receptors converging on central neurons. Consistent with this prediction many SI and SII cells respond only to spacings and not the applied force or velocity of strokes across grating surfaces (Sinclair & Burton, 1991a, 1993a). This type of independent response to roughness is not found in recordings from peripheral afferent fibers (Goodwin et al., 1989; Goodwin & Morley, 1987a,b; Sathian, Goodwin, John, & Darian-Smith, 1989) or thalamus (Sinclair & Burton, 1991b), suggesting a convergence mechanism selectively sensitive to spatially distributed variations in surface dimensions.

Some phenomena remain unexplained by a spatial variation mechanism of roughness perception. The idea of spatial variation, and the simpler notion of volume of skin distortion, as the effective stimulus for roughness fails to explain enhanced roughness discrimination through intermediate surfaces. Interposing a piece of paper between finger and surface improves discrimination of surface roughness (Lederman, 1978). However, the raised ridges of a grating would still transmit a temporally modulated (and spatially blurred) signal through the paper and activate receptors crudely as a function of spacing. It remains unclear why an intermediate surface enhances roughness perceptions unless this method activates more receptors or brings the stimulus into the response range of other receptor classes (e.g., PCs).

The spatial variation model emphasizes the role of SA afferents in representing roughness. However, roughness discrimination extends to surface spacings below the spatial resolution of SAs. Possibly the spatial variation mechanism might lose resolution but still respond differentially to these very smooth surfaces. Alternately, discriminations may be based on the RA receptor population, which responds differentially in this very smooth texture range. Enterprising primates may use a variety of strategies to identify surface roughness depending on the contrived laboratory task or natural environmental requirements. SAs, if the basis of a spatial variation code, may provide an estimate of surface roughness unbiased by movement patterns. Finer discriminations, such as between grades of silk, may depend on RA activation. Stereotyped movements may be essential for such fine discriminations because they fall below the resolution of the movement-independent spatial variation mechanism.

V. COGNITIVE ASPECTS OF TOUCH

A. Tactile Memory

1. Evidence for a Working Tactile Memory

Short-term memory (e.g., working, operant, sensory, primary, or provisional memory) refers to the retention of some sensory information for a brief time. Long-term tactile memory might be thought of as the stored "templates" of familiar objects. There is no physiological data on long-term tactile memory. However, building on a large body of psychophysical evidence, some initial studies have examined neural correlates of short-term memory for information obtained by touch.

Studies using Braille letters, three-dimensional shapes, spatial position of punctate stimulation, and vibrotactile stimuli provide data on tactile short-term memory. Results of these studies support the proposition that tactile short-term memory partially depends on a continuing "trace" of the original sensation that gradually decays over a period of about 10–15 sec after stimulation ends (Millar, 1975). Performance on tactile-discrimination tasks, in which a delay separates two stimuli, decrements over this period

(Gilson & Baddeley, 1969; Sinclair & Burton, 1995; Sullivan & Turvey, 1974). However, subjects remember a series of stimuli presented to the same skin site though each subsequent item replaces the previous lingering sensory memory (Millar, 1978). The detrimental effects of interfering with mnemonic rehearsal in delayed discrimination tasks, some psychophysical data suggest, is that tactile sensory information temporarily persists in condensed form beyond the initial sensory "image" (Gilson & Baddeley, 1969; Millar, 1978; Sinclair & Burton, 1996; Sullivan & Turvey, 1974). For instance, Sinclair and Burton (Sinclair & Burton, 1996) had human subjects identify the higher of two vibrotactile frequencies separated by delays of 0.5–30 sec. Percent correct and d' decreased and reaction time increased as a function of length of delay. After delays of 5 sec, performance decreased more when subjects counted backwards during the delay interval. This change in decay rate beyond 5-sec delays with versus without interference suggests the involvement of two mnemonic mechanisms in which retained information decays at different rates. In this interpretation, information retained by one mechanism, commonly called sensory memory, decays rapidly in the first 5 sec following stimulation. In this preattentive form of short-term memory the observer can afterward attend to the sensory trace even if distracted during stimulation. A second, more durable mechanism is responsible for higher performance levels out to 30-sec delays without distraction. This latter mechanism may rely on voluntary condensation and rehearsal of sensory information, an attentive process susceptible to interference by distracting tasks like backwards counting.

The embryonic level of current investigations of the neural substrates of tactile memory provides little empirical data to evaluate mnemonic mechanisms suggested by psychophysical evidence. However, several potential somatosensory areas likely to exhibit tactile memory effects are suggested by physiological studies and anatomical connection patterns.

Mishkin proposed that SII might serve a role in somesthesis for the tactile recognition of objects (Mishkin, 1979) based on its connections with parietal cortex and through the insula to frontal cortex (Felleman & Van Essen, 1991; Friedman et al., 1986; Mesulam & Mufson, 1982; Preuss & Goldman-Rakic, 1989). He noted the similarity of this connection sequence to that of inferior temporal (IT) cortex, which connects to parahippocampal and frontal areas and is implicated in the visual discrimination of objects (Friedman et al., 1986; Mishkin, 1979). The reciprocal connections of SII to all parts of SI, motor areas 4 and 6, proprioceptive-kinesthetic area 5, visual-tactual area 7b, orbital frontal areas, and, via insular projections, to frontal and parahippocampal areas (Burton et al., 1995; Carmichael & Price, 1994; Felleman & Van Essen, 1991; Friedman et al., 1986; Jones et al., 1978; Mesulam & Mufson, 1982; Mufson & Mesulam, 1982; Pons & Kaas, 1986; Preuss & Goldman-Rakic, 1989; Stepniewska et al., 1993), make SII optimal to pass integrated sensory information obtained during object exploration on to higher cognitive areas. Additionally, through its motor connections (Matelli et al., 1986; Rizzolatti et al., 1987; Rizzolatti, Gentilucci, & Matelli, 1985; Stepniewska et al., 1993), SII may reconcile various manipulatory and sensory information to produce a constant, unitary object percept (Sinclair & Burton, 1993a).

Support for a role of SII in memory and tactile object recognition comes from case studies of human parieto-temporal lesions involving the parietal opercular region. Patients show impaired recognition of tactual objects (tactile agnosia) without more rudimentary perceptual dysfunction (Caselli, 1991, 1993; Greenspan & Winfield, 1992). Lesions of SII in visual-somatosensory anatomical similarities make SII and the insula likely sites for memory processes related to tactile object recognition. In turn, insular and SII projection patterns to frontal cortex areas 12 and 46 suggest that frontal cortex plays a more direct role in tactile object recognition (Carmichael & Price, 1994; Felleman & Van Essen, 1991; Friedman et al., 1986; Mesulam & Mufson, 1982; Mufson & Mesulam, 1982; Preuss & Goldman-Rakic, 1989).

Fuster and colleagues described sustained poststimulus activity related to a tactile stimulus in a delayed discrimination task (Koch & Fuster, 1989; Zhou & Fuster, 1992). Cells in SI, area 5, and lateral area 7 show transient and sustained responses during the delay before presentation of a comparison object. These responses correlate with the identity of three-dimensional sample objects (spheres vs. cubes, or vertically vs. horizontally ridged cylinders). These valuable studies show selective changes related to delayed discriminations of tactile objects. However, they suffer methodological shortcomings. First, due to uncontrolled manual object palpation, and the complex, multifeatured nature of the stimuli, specific attributes of the test objects are not identifiable in the pattern of activity during stimulus presentation. Second, nothing clearly separates stimulus from response features. Manipulatory movements needed for palpating the objects rather than specific sensory events related to the sample object may selectively engage the proprioceptive–kinesthetic cortical areas studied.

B. Attention

1. Theories

Attention involves allocating scarce neural resources for selective processing of sensory, motor, or goal-directed behavior (Allport, 1989; Desimone & Duncan, 1995; Kahneman, 1973; Posner, 1978; Posner & Dehaene, 1994; Posner & Petersen, 1990). Directing attention to some stimulus location, feature, or modality generally leads to improved accuracy and speed of detection or discrimination of that stimulus. Attention differs from vigilance and arousal, two other states that also lead to changes in performance. Being aroused or engaged in a task to achieve optimal performance uses resources, but these processes lack the selectivity associated with attention mechanisms.

Two theories encompass current ideas about attention. These postulates derive mainly from studies of attention in the visual system. We consider below whether analogous notions apply to tactile attention. One theory attributes attention to an active, high-speed, serial process of "searchlight-like" selectivity that uses a top-down mechanism to scan and amplify attended events (Crick, 1984; Olshausen, Anderson, & Van Essen, 1993; Posner & Dehaene, 1994; Posner & Petersen, 1990). According to Posner and colleagues the selection process for attention is multifaceted. A mechanism for shifting attention involves disengagement from an attended location or object, movement to a new focus and engagement (Corbetta, Miezin, Dobmeyer, Shulman, & Petersen, 1991; Corbetta, Miezin, Shulman, & Petersen, 1993; Posner & Presti, 1987; Posner, Walker, Friedrech, & Rafal, 1984). Symptoms of tactile neglect following strokes in the right parietal cortex may involve disruption of the normal disengagement mechanisms. Posterior parietal cortex may shift or allow shifts in the center of attention within specific spatial coordinates. The circuits responsible for shifts possibly have control neurons that serve to direct the spatial focus of higher stage neurons (Olshausen et al., 1993). Sustained attention to some object may involve portions of frontal cortex (Posner & Petersen, 1990). These notions of attention are primarily concerned with the spatial location of visual attention.

A second theory considers attention a more passive mechanism, for example, "an emergent property of slow, competitive interactions that work in parallel across the visual field" (Desimone & Duncan, 1995, p. 217; Niebur & Koch, 1994). Attention may influence any stage of processing where neurons with sensitivity to isolated stimulus features have receptive fields that include multiple, competing stimuli (e.g., targets and distractors). For stimuli lacking automatic pop-out characteristics (Desimone & Duncan, 1995), competitive advantage, however, still arises from top-down influences. A source may be working memory that amplifies by matches between stimulus-evoked activity in a target cortex and attention templates projected into that region (Desimone & Duncan, 1995). This theory of attention mostly emphasizes selective processes in object identification.

Both theories invoke ideas of amplification of pooled activity levels for selected features or locations within a receptive field. As shown in PET studies in humans or single-cell studies in monkeys, the cortical loci showing gains in activity frequently are selectively responsive to the attended stimulus features (Corbetta et al., 1991; Corbetta et al., 1993; Desimone & Duncan, 1995; Motter, 1993, 1994a,b; Posner & Dehaene, 1994). The mechanism of gain control may be through controlling synaptic connections on a cell or modifying the total activity of a group of cells. The net effect is biasing the activity of a subset of cells (Anderson & Van Essen, 1987; Niebur & Koch, 1994; Olshausen et al., 1993). Improved signals for targets over the noise of distractors can arise through several mechanisms: suppressing responses to unattended stimuli within the receptive field, enhancing activity to attended stimuli, combining enhancement and suppression, general background suppression, shifting receptive field location for maximal responses or, theoretically, improved temporal correlation between activated detectors (Desimone & Duncan, 1995; Motter, 1993, 1994a; Posner & Dehaene, 1994; D. Robinson, McClurkin, Kertzman, & Petersen, 1991; Whang, Burton, & Shulman, 1991). Suppression and enhancement effects may be serially linked. Thus, suppression early in processing attenuates unattended stimuli, whereas later stage enhancement of task-related activity amplifies attended information (Posner & Dehaene, 1994).

Although few studies have investigated tactile attention, several observations show evidence of shifting attentive focus, limited-capacity processing, and mechanisms of selectivity that resemble findings in visual attention. Introspectively, stimulus selectivity is a common experience, as humans ignore the plethora of tactile stimuli normally impinging on the body. Selectivity is also evident when switching between subjective and objective awareness while rubbing anything. Patients with parietal lobe lesions also neglect somatosensory stimuli located contralateral to the infarcts (Critchley, 1949, 1953; Denny-Brown & Chambers, 1958; Moscovitch & Behrmann, 1994). As in studies of visual neglect (Posner et al., 1984), the deficit in these patients appears as an inability to shift attention into the space that is on the contralateral side. Even in body space that is ipsilateral to the lesion, there is neglect in the contralateral half of that space (Moscovitch & Behrmann, 1994). Thus, a right-sided lesion leads to detection failures for any left-sided locations when two simultaneous stimuli occur. If both stimuli are on an ipsilateral body site (e.g., right arm or hand), patients have greater difficulty with the contralateral (left) side of that structure. Neglect of the perceived contralateral space changes with 180-degree rotation of the limb. For example, patients with a right posterior parietal stroke neglect the radial side of the palm with the hand pronated and the ulnar side of the back of the hand with the hand supinated (Moscovitch & Behrmann, 1994).

Most tactile pattern recognition tasks also provide evidence of limited-capacity resources in the somatosensory system and therefore suggest a need for tactile attention processes. For example, in Braille reading optimal processing is serial and selectively confined to the receptors from a single fingerpad. Occasionally two hands are better than one for pattern recognition or for divided attention tasks (Evans & Craig, 1991, 1992; Sathian & Burton, 1991). Usually, tactile information from multireceptive fields only establishes the crude outlines of three-dimensional, novel objects. Explorations focused on stimulation of the receptors on a single finger normally reveal the more detailed material properties of objects like surface texture, contours, and compliance (Klatzky, Lederman, & Reed, 1989).

2. Psychophysical Studies

Attention demands of a task can be proved by cuing selections to improve detection, discrimination, or capture time of a stimulus in a distracting background (Posner, 1978). In these experiments stimulus parameters are identical across tasks, and only a cue informs the subject where to shift attention. Many studies of visual attention assess reaction times to cued (selected) locations. Levels of correct performance with valid and invalid or neutral cues provide similar evidence of attention (Posner, 1978; Sathian & Burton, 1991; Whang et al., 1991). Only a few studies of tactile attention have used similar cuing paradigms.

Several experiments showed that performance in tasks with multiple tactile stimuli is minimally or not affected by probability cuing when subjects detect the presence of an increase in vibratory amplitude (Butter, Buchtel, & Santucci, 1989; Shiffrin, Craig, & Cohen, 1973; Whang et al., 1991), changes in roughness (Sathian & Burton, 1991), or the occurrence of light taps (Posner, 1978). Even tactile distractor stimuli presented simultaneously or sequentially at two or more other skin locations do not disrupt performance at the site stimulated with a change in a tactile feature. One interpretation offered for these findings is that touch is an alerting sense that requires little or no attention (Posner, 1978).

There is, however, some evidence from cross-modal tasks that touch also engages attention processes. For example, studies with visual and tactile stimuli found that valid cuing to the location of a tactile vibration produces the shortest reaction times for detecting this stimulus. Reaction times are longer when a neutral cue divides attention between touch and vision. An invalid cue that shifts attention to vision prolongs reaction times to the tactile stimulus even more. The authors suggested "that the ability to detect the stimulus had been enhanced . . . when the vibrotactile stimulus was expected and attention was directed toward that modality" (Post & Chapman, 1991, p. 154).

This demonstration of possible "top-down" influences on reaction times to the appearance of vibrotactile stimuli is consistent with similar enhancements using only visual stimuli (Posner, 1978; Posner & Petersen, 1990). By argument of similarity, these results suggest that a process of tactile attention might provide for the enhanced performance. However, positive results with cross-modal stimuli might also suggest that selection of vision inhibits touch. Thus, slowed reaction with neutral or invalid cuing do not show selective enhancement through tactile attention but predominance of one modality over the other. This form of modulation may still selectively allocate resources for optimal processing but it is not of the same form as has previously been discussed for within-modality attention. Slowed reaction times also may reflect the different levels of difficulty in detecting visual as opposed to tactile stimuli.

Evidence of tactile attention was found in another study that used multiple tactile stimuli. Subjects detected a cued location where there was no change in stimulus amplitude (e.g., target stimulus), while several other sites received tactile distractor stimuli with amplitude changes. Thus, cuing provides advantage in detecting targets of no vibratory amplitude shifts or no changes in surface roughness at one finger; three other finger pads received brief vibratory amplitude increments or roughness changes (Sathian & Burton, 1991; Whang et al., 1991). The different effects of cuing, when target stimuli have or do not have a particular tactile feature, suggest analogies with descriptions of selective, serial search strategies in the visual system (Treisman & Gormican, 1988; Treisman & Souther, 1985). Thus, there is automatic or preattentive processing of transient tactile changes that are alerting (i.e., they are presence features that "pop-out") (Sathian & Burton, 1991; Whang et al., 1991). The absence of a feature (e.g., constant vibratory amplitude or uniform roughness) requires serial searches that place demands on scarce resources.

These asymmetrical findings with present and absent tactile stimulus features are interpretable in terms of pooling information from an array of feature detectors as previously proposed for visual stimulus arrays (Treisman & Gormican, 1988; Treisman & Souther, 1985; Whang et al., 1991). The relative weighing of inputs from different feature detectors (i.e., spatial location) represents the allocation of attention resources. Simulating these results with a signal-detection model more successfully duplicated the observed differences than a previously proposed Weber fraction model. The signal-detection model provides a better fit for two key observations: (a) no effect of cuing on performance with different intensities of transient amplitude changes at all locations, and (b) the detectability of stimuli lacking a feature change (Whang et al., 1991). The model emphasizes both the mean and variances of the responses of feature detectors in the pooled distribution of activity and, from this, the probability of detecting the presence or absence of a feature. Where pooled activity converges onto one neuron with a large receptive field, the model results are consistent with observed suppression of responses when attending away from the optimal stimulus in the array (Moran & Desimone, 1985; Whang et al., 1991).

Cuing also aids performance that involves discrimination of particular tactile features (i.e., vibratory frequencies or patterns of roughness (Posner, 1978; Sathian & Burton, 1991). Selective detection of such features possibly improves object recognition. For example, valid cuing improves detecting the location exposed to a selected direction of texture gradients (increasing or decreasing roughness) in the presence of opposite gradients as distractors at other locations. Cuing also helps identify the location of an absolute texture (smooth or rough) in the presence of the opposite uniform texture (Sathian & Burton, 1991). Attention demands in these experiments suggest a competitive model based on spatial selection in the presence of multiple target stimuli. As previously hypothesized for visual stimuli, competition between targets causes interference and requires selective allocation of resources (Desimone & Duncan, 1995; Duncan, 1980). This allocation may lead to suppression or enhancement of neural responses to target features in cued locations (Motter, 1993; Motter, 1994a).

3. Neurophysiological Studies of Response Modulation

Response enhancements for cued events correspond with the neural expression of attention in several studies using visual stimuli (Posner & Dehaene, 1994). Findings of task-related neural activity changes do not necessarily show modulation based on attention. One example is lability of responses that provide hints of "cognitive" influences. As in the visual cortex, response consistency decreases in "higher" somatosensory cortical areas. In SII we observed cells with no responses to passive stimulation but vigorous activity while a monkey actively rubbed the same region of skin to distinguish the roughness of touched surfaces (Sinclair & Burton, 1993a). Some cells in SII also responded when monkeys touched particular features of discriminated objects even when sensitivity to these features could not be attributed to differences in physical parameters of the effective surfaces. No comparable selectivity between passive and active modes of stimulation appeared in a sample of SI cells studied while the same monkeys performed in the identical paradigm (Sinclair & Burton, 1991a). Other examples of selectively enhanced responses to somatosensory sensory stimuli have been reported from recordings in posterior parietal cortex, especially area 7b, areas 1 and 2, and granular insula (Dong, Chudler, Sugiyama, Roberts, & Hayashi, 1994; Hyvärinen, 1982; Iwamura et al., 1983b, 1985b; Robinson & Burton, 1980b,c; Schneider et al., 1993). Response increments frequently correlated with when the monkey appeared to engage some object or stimulus delivered by the experimenters. As attention was not controlled, these studies only revealed a potential substrate for dynamic changes in neural activity during a tactile task.

Hyvärinen and colleagues described differences between task and no-task conditions (Hyvärinen, Poranen, & Jokinen, 1980; Poranen & Hyvärinen, 1982). Neurons in SI and SII had enhanced responses during blocks of behaviorally relevant trials when monkeys received a reward for detecting the cessation of a constant amplitude vibrotactile stimulus. Less activity occurred to the same stimulus on blocks of irrelevant trials with no rewards and no required behavior. Average firing rates were greater in approximately 16% of recorded SI cells during task relevant trials. Multicell recordings in SII showed more widespread enhancement of responses when studied with the same task–no-task paradigm (Poranen & Hyvärinen, 1982). In these studies task performance led to immediate reward only for correct performance during behaviorally relevant blocks. Successful detection required vigilance during the vibration interval, but it also engaged the expectancy of the animal in a different reward condition. Response gains during task trials might reflect sustained attention to enhance processing of the vibrotactile stimulus. As task trials also were rewarded, altered firing rates might also correlate with heightened arousal as the animal expected an immediate reward after the vibration interval only on these trials. This study therefore failed to separate the effects of arousal and vigilance from selective attention. However, consistent with recent findings, a greater incidence of response enhancements appeared in SII, a "higher" somatosensory cortical area.

Evidence for response suppression during some aspect of a behavior has also been described at cortical and subcortical levels of the somatosensory system. Response suppression of unattended stimuli has been described as a component expression of selective attention in the visual system (e.g., Moran & Desimone, 1985). Greater effects occurred at progressively higher levels (Sinclair & Burton, 1993a). Most examples were response reductions to cutaneous stimuli before or during a movement (e.g., "sensory gating"). Chapin and colleagues found low-threshold, cutaneous evoked responses only during one phase of the walking cycle (Chapin & Woodward, 1981, 1982a,b, 1986). Several groups showed suppression of responses to cutaneous stimuli from noninvolved skin regions during stimulus-cued arm or finger movements (Chapman, Jiang, & Lamarre, 1988; Jiang, Chapman, & Lamarre, 1991; Nelson, Smith, & Douglas, 1991; Schmidt, Schady, & Torebjörk, 1990). Thresholds for detecting stimulation at effected skin locations rose during intervals of suppression or "gating" (Coquery, 1978; Dyhre-Poulsen, 1978). In contrast, no suppression of responses occurred from the skin regions moved over surfaces in active tactile discrimination tasks (Post, Zompa, & Chapman, 1994; Sinclair & Burton, 1991a, 1993a; Sinclair et al., 1991). Collectively, these examples of "gating" suggest processes can improve signal-to-noise ratios by suppressing information from selected body locations. Without cuing to particular target sites, however, there is no way to judge the appropriateness of these movement-associated gain controls for attention.

Currently, only one published study presents evidence of "top-down" influences in somatosensory areas SI and SII during sustained vigilance in two different tasks (Hsiao et al., 1993). In trials of a cross-modal task a monkey matched tactile and visual stimuli. It did a visual-detection task in alternating blocks of trials. The visual stimuli differed between tasks. In the cross-modal task, the monkey obtained rewards for correctly matching a visually displayed letter and a felt embossed letter, passively rubbed over a finger pad. Different raised letters were continuously scanned across the skin while the visual display was constant. For the visual task the monkey attended to three simultaneous lighted target squares to detect dimming in one. The same embossed letters rubbed the finger pad, but the tactile stimuli were irrelevant during the visual task. The analysis compared responses to the same tactile stimulus letters when these were relevant or irrelevant to obtaining rewards. Neurons in SI (20 of 41) and SII (55 of 69) responded with significantly enhanced firing rates whenever the tactile stimuli matched the visually presented letter. Responses were lower during presentation of the same letters during the visual-detection task (Figure 12). Performance accuracy did not correlate with any neural response differences, suggesting sustained effects that related to the different demands in each task. All effected neurons in SI and most in SII (40 of 69) showed significantly higher firing rates to the tactile stimuli during the relevant cross-modal task. Fifteen SII cells displayed lower activity during the tactile task.

These results concur with studies of visual attention in showing a type of signal amplification (e.g., enhancement or suppression of firing rates) while processing tactile stimuli. Several differences in experimental design make it difficult to compare these observations to findings on visual attention. First, cuing in this study did not shift attention between matched paradigms. The monkey obviously could not match felt letters to viewed ones without dividing attention between modalities. In this study there is only one target or distractor tactile stimulus at a time inside a cell's receptive field. There is no competition for scarce resources between simultaneously presented tactile stimuli. Second, in this study the monkey learned and discriminated a paired association in the cross-modal task but simply detected in the visual task. Discrimination and detection tasks made different demands on memory and, therefore, were likely to engage different processes that cannot be dissociated in the results. Third, selective shifts of attention in this study differed from processes examined in visual attention (Desimone & Duncan, 1995; Posner & Dehaene, 1994; Posner & Petersen, 1990). The monkey "selects vision" and not some target stimulus when performing the behaviorally irrelevant tactile task. Do the lower activity levels in most neurons suggest that performing visual tasks engaged inhibitory mechanisms equivalent to movement-triggered sensory gating (i.e., vision inhibits somatosensory activity)? Whatever form of modulation represented by these results, they do suggest "cognitive" processes associated with vigilance (e.g., sustained attention) can modulate somatosensory cortical responses. As in

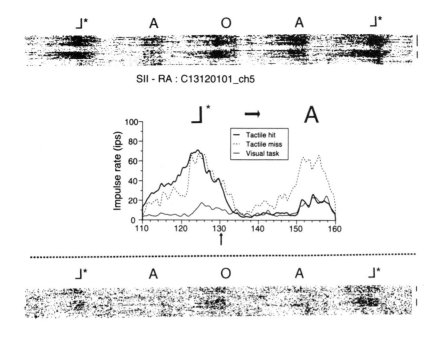

⌐*   A   O   A   ⌐*

SII - RA : C13120101_ch5

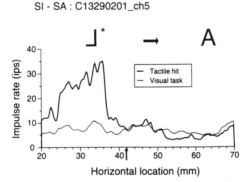

SI - SA : C13290201_ch5

FIGURE 12   Single neuron responses from SII (upper) and SI (lower) during multiple alternations between two tasks. The tactual task had the monkey indicate a match between the visual display of a letter, the cued stimulus, and the touched embossed equivalent letter, the "hit." The visual task had the monkey detect target light dimming while the now irrelevant, but same tactile letters still rubbed across the receptive field of the cell. Short vertical lines to the right of the rasters mark trials of tactual tasks; the visual task occurred during remaining trials. Raster dots show the times of action potentials with respect to the spatial pattern (letters) illustrated above the raster lines. The sequence of stimulation progresses from right to left. For the illustrated examples during the tactual task, the cued letter was *L*. Lower half of each panel shows stimulus histograms. "The abscissa in each panel represents circumferential distance from the join in the pattern (15 mm = 1 s). The ordinate represents mean impulse rate (histogram bins 0.2-mm wide, triangular smoothing over ± 3 bins). The thick solid, dash, and thin solid lines represent impulse rates during and following hits and misses and during the visual task, respectively. The arrow on each abscissa represents the mean location of the monkey's response (twist) to the target letters. Top arrows represent scanning direction" (Hsiao et al., 1993, p. 446). (Modified from Hsiao et al., 1993.)

other examples cited above, this has a greater effect on hierarchically later stages like SII.

4. Neuroimaging Studies of Attention in Humans

Corbetta and colleagues used PET to find cortical foci of increased cerebral blood flow, especially in posterior parietal cortex, when subjects attend to one of the two simultaneously presented features of different visual stimuli (i.e., color vs. shape, color vs. motion, motion vs. shape). They also found feature-related shifts in the location of significant regions of interest in extrastriate cortex. In addition, foci near the anterior cingulate correlated with sustaining attention, whereas those in posterior parietal cortex corresponded to shifting attention from one stimulus to another (Posner & Petersen, 1990; Posner et al., 1984). These findings suggest that visual attention might be a distributed process that engages different cortical areas (Posner & Dehaene, 1994).

No PET studies of tactile attention used multifeature somatosensory stimuli. However, a distributed array of foci with increased blood flows can be identified when subjects sustain attention in doing somatosensory detections of painful stimuli. The foci include anterior cingulate, postcentral gyrus, and parietal operculum (Coghill et al., 1994). Like the studies with visual stimuli, the regions include early-stage sensory cortical areas that may determine the features of an attended stimulus and higher, cognitive areas whose role is unknown for somatic stimuli.

The results from another PET study involving somatosensory stimuli suggest that vigilance may affect global background levels of activity. In this study subjects attend to an anticipated somatosensory stimulus throughout a scan session (Drevets et al., 1995). Although no stimulus occurred during the scan, blood flow measurements showed significant decreases in blood flow over portions of somatosensory cortex representing parts of the body not attended to during the scan. Thus, blood flow decreased bilaterally over the hand and face regions of SI in subjects expecting tactile stimulation of the toes, over the foot and face regions of SI when subjects expected a painful shock to the finger, and over the hand region when subjects anticipated stimulation of the face. Blood flow did not increase in the region of expected stimulation. These topographically focused blood flow changes suggest that tactile spatial selection may involve control of background activity within specific topographic representations of the body. The decreases in blood flow for portions of sensory cortex representing unattended locations may enhance signals in cortical sites for attended receptive fields (Drevets et al., 1995).

VI. CONCLUDING COMMENTS

Touch is unique among sensory systems because through it we experience varied stimulation of body regions and, with the same peripheral receptors and associated central neural structures, we identify external objects. The sense of touch is both objective and subjective. Katz recognized this dichotomy in *The World of Touch* more than 75 years ago when he summarized a rich literature on the phenomenology of touch (Katz, 1989). The perceptions of the material and geometric attributes of objects form objective reality. *(What did I touch?)* Objects have perceived material features of texture, compliance and elasticity, weight, or temperature. Objects also have geometric properties of volume, shape or size (Klatzky, Lederman, & Reed, 1987; Klatzky et al., 1989; Lederman, Browse, & Klatzky, 1988; Lederman & Klatzky, 1987, 1990; Lederman, Klatzky, & Reed, 1993; Loomis, Lederman, Wake, & Fujita, 1993; Reed, Lederman, & Klatzky, 1990). These features characterize the reality of object recognition. Thus, a real-world, physical thing exists under outstretched fingertips. The percept is a table not fingers.

The sensations evoked from touching also form the subjective or internal aspects of touch. *(Where and how was I touched?)* A touch experience includes perceived spatial dimensions of localization, pressure, indentation, and stimulus direction across the skin and temporal dimensions of motion or vibrations. Some phenomena of touch like oiliness, wetness, or stickiness combine subjective and objective features, thus highlighting the complexity in trying to parse the duality of touch perceptions. In contrast, vision, audition, and chemical senses normally serve perceptions of stimuli that originate within objects. Hue is inherently an object's property; touch is both something at a location on the skin or is an object with particular features. This duality is apparent from the simple procedure of running fingers from one hand across any other body part or over an object and alternatively attending to the touched or touching (Katz, 1989).

Much of the research summarized in this and chapter 2 by Greenspan and Bolanowski show detailed knowledge about many of these phenomena. We know much about the neural basis for tactile thresholds, discrimination sensitivities, magnitude estimates of touch sensations from vibrations, textures, moving stimuli, simple forms, and so on. Many fundamental issues for these simple percepts remain unresolved. For example, we still know little about the specific roles played by each cortical cytoarchitectonic subdivision along the postcentral gyrus. We know nothing about the functioning of the network of cortical areas tied to the initial processing in SI. What information passes between cortical layers or between cortical areas? What do higher cortical areas add to low-level threshold detections or magnitude estimates? What is the role of SII?

To "know" one's car key in the dark involves cognitive processes of attention, memory, discrimination, and categorization. This recognition task also means earlier processes of learning with possible visual-to-tactile paired associations. There is scant knowledge on these neural mechanisms in the sense of touch. We even do not know whether early-stage tactile cognitive processing occurs within the somatosensory cortical areas that readily respond to mechanical stresses and strains transduced by peripheral receptors. Is it a key because an isomorphic representation appears in area 3b? The variety of phenomena inherent in touch sensations is matched by the diversity of exploratory procedures used by humans to obtain information about novel objects (Klatzky et al., 1987; Klatzky et al., 1989; Lederman et al., 1988; Lederman & Klatzky, 1987, 1990; Lederman et al., 1993; Reed et al., 1990). These behaviors isolate specified aspects of the material and geometric properties of objects. A striking gap exists in our understanding of these actions in reference to the detailed, single-neuron analyses described above. A major goal for future studies will be integrating haptic exploration and models of the neural networks associated with gathering information about touching and being touched. Crucial to these studies must be attempts to dissociate and then reintegrate proprioceptive and kinesthetic with tactile information.

Acknowledgments

Preparation of this chapter supported by funds from National Institutes of Health grant NS 31005.

References

Adrian, E. D. (1941). Afferent discharges to the cerebral cortex from peripheral sense organs. *Journal of Physiology (London)*, *100*, 159–191.

Alloway, K. D., & Burton, H. (1985). Homotypical ipsilateral cortical projections between somatosensory areas I and II in the cat. *Neuroscience*, *14*, 15–35.

Alloway, K. D., & Burton, H. (1991). Differential effects of GABA and bicuculline on rapidly- and slowly-adapting neurons in primary somatosensory cortex of primates. *Experimental Brain Research*, *85*, 598–610.

Allport, A. (1989). Visual attention. In M. I. Posner (Ed.), *Foundations of cognitive science* (pp. 631–682). Cambridge, MA: The MIT Press.

Andersen, R. A., Asanuma, C., Essick, G., & Siegel, R. M. (1990). Corticocortical connections of anatomically and physiologically defined subdivisions within the inferior parietal lobule. *Journal of Comparative Neurology*, *296*, 65–113.

Anderson, C. H., & Van Essen, D. C. (1987). Shifter circuits: A computational strategy for dynamic aspects of visual processing. *Proceedings of the National Academy of Sciences (USA)*, *84*, 6927–6301.

Apkarian, A. V., & Shi, T. (1994). Squirrel monkey lateral thalamus. I. Somatic nociresponsive neurons and their relation to spinothalamic terminals. *Journal of Neuroscience*, *14*, 6779–6795.

Barenne, J. G. D. d. (1916). Experimental researches on sensory localization in the cerebral cortex. *Quarterly Journal of Experimental Physiology*, *9*, 355–390.

Brodmann, K. (1994). *Brodmann's "Localisation in the cerebral cortex."* (L. J. Garey, Trans.). London: Smith-Gordon.

Burton, H. (1984). Corticothalamic connections from the second somatosensory area and neighboring regions in the lateral sulcus of macaque monkeys. *Brain Research*, *309*, 367–372.

Burton, H. (1986). Second somatosensory cortex and related areas. In E. G. Jones & A. Peters (Ed.), *Cerebral cortex, sensory-motor areas and aspects of cortical connectivity* (Vol. 5, pp. 31–98). New York: Plenum.

Burton, H., & Carlson, M. (1986). Second somatic sensory cortical area (SII) in a prosimian primate, *Galago crassicaudatus*. *Journal of Comparative Neurology, 168,* 200–220.

Burton, H., & Craig, A. D. (1983). Spinothalamic projections in cat, raccoon and monkey: A study based on anterograde transport of horseradish peroxidase. In G. Macchi, A. Rustioni, & R. Spreafico (Eds.), *Somatosensory integration in the thalamus: A re-evaluation based on the new methodological approaches* (pp. 17–41). Amsterdam: Elsevier Science Publishers.

Burton, H., & Fabri, M. (1995). Ipsilateral intracortical connections of physiologically defined cutaneous representations in areas 3b and 1 of macaque monkeys: Projections in the vicinity of the central sulcus. *Journal of Comparative Neurology, 355,* 508–538.

Burton, H., Fabri, M., & Alloway, K. (1995). Cortical areas within the lateral sulcus connected to cutaneous representations in areas 3b and 1: A revised interpretation of the second somatosensory area in macaque monkeys. *Journal of Comparative Neurology, 355,* 539–562.

Burton, H., & Jones, E. G. (1976). The posterior thalamic region and its cortical projection in new and old world monkeys. *Journal of Comparative Neurology, 168,* 249–299.

Burton, H., & Robinson, C. J. (1981). Organization of the SII parietal cortex: Multiple somatic sensory representations within and near the second somatic sensory area of cynomolgus monkeys. In C. N. Woolsey (Ed.), *Cortical Sensory Organization* (pp. 67–119). Clifton, NJ: Humana Press.

Burton, H., Sathian, K., & Dian-Hua, S. (1990). Altered responses to cutaneous stimuli in the second somatosensory cortex following lesions of the postcentral gyrus in infant and juvenile macaques. *Journal of Comparative Neurology, 291,* 395–414.

Burton, H., & Sinclair, R. J. (1990). Second somatosensory area in macaque monkeys. I. Neuronal responses to controlled, punctate indentations of glabrous skin on the hand. *Brain Research, 520,* 262–271.

Burton, H., & Sinclair, R. J. (1991). Second somatosensory cortical area in macaque monkeys: 2. Neuronal responses to punctate vibrotactile stimulation of glabrous skin on the hand. *Brain Research, 538,* 127–135.

Burton, H., Videen, T. O., & Raichle, M. E. (1993). Tactile vibration activated foci in insular and parietal opercular cortex studied with positron emission tomography: Mapping the second somatosensory area in humans. *Somatosensory and Motor Research, 10,* 297–308.

Butter, C. M., Buchtel, H. A., & Santucci, R. (1989). Spatial attentional shifts: Further evidence for the role of polysensory mechanisms using visual and tactile stimuli. *Neuropsychologia, 27,* 1231–1240.

Campbell, A. W. (1905). *Histological studies on the localization of cerebral function.* London: Cambridge University Press.

Carlson, M. (1985). The significance of single or multiple cortical areas for tactile discrimination in primates. In A. W. Goodwin & I. Darian-Smith (Eds.), *Hand function and the neocortex* (pp. 1–16). Berlin: Springer-Verlag.

Carlson, M. (1990). The role of somatic sensory cortex in tactile discrimination in primates. In E. G. Jones & A. Peters (Eds.), *Cerebral cortex* (Vol. 8B, pp. 451–486). New York: Plenum Publishing.

Carlson, M., & Burton, H. (1988). Recovery of tactile function after damage to primary or secondary somatic sensory cortex in infant macaca mulatta. *Journal of Neuroscience, 8,* 833–859.

Carlson, M., & Nystrom, P. (1994). Tactile discrimination capacity in relation to size and organization of somatic sensory cortex in primates: I. old-world prosimian, *galago;* II. new-world anthropoids, *saimiri* and *cebus. Journal of Neuroscience, 14,* 1516–1541.

Carmichael, S. T., & Price, J. L. (1994). Architectonic subdivision of the orbital and medial prefrontal cortex in the macaque monkey. *Journal of Comparative Neurology, 346,* 366–402.

Caselli, R. J. (1991). Rediscovering tactile agnosia. *Mayo Clinic Proceedings, 66,* 129–142.

Caselli, R. J. (1993). Ventrolateral and dorsomedial somatosensory association cortex damage produces distinct somesthetic syndromes in humans. *Neurology, 43,* 762–771.

Cavada, C., & Goldman-Rakic, P. S. (1989). Posterior parietal cortex in rhesus monkey: I. Parcellation of areas based on distinctive limbic and sensory corticocortical connections. *Journal of Comparative Neurology, 287,* 393–421.

Chapin, J. K., & Lin, C.-S. (1984). Mapping the body representation in the SI cortex of anesthetized and awake rats. *Journal of Comparative Neurology, 229,* 199–213.

Chapin, J. K., & Woodward, D. J. (1981). Modulation of sensory responsiveness of single somatosensory cortical cells during movement and arousal behaviors. *Experimental Neurology, 72,* 164–178.

Chapin, J. K., & Woodward, D. J. (1982a). Somatic sensory transmission to the cortex during movement: Gating of single cell responses to touch. *Experimental Neurology, 78,* 654–669.

Chapin, J. K., & Woodward, D. J. (1982b). Somatic sensory transmission to the cortex during movement: Phasic modulation over the locomotor step cycle. *Experimental Neurology, 78,* 670–684.

Chapin, J. K., & Woodward, D. J. (1986). Distribution of somatic sensory and active-movement neuronal discharge properties in the MI-SI cortical border area in the rat. *Experimental Neurology, 91,* 502–523.

Chapman, C. E., Jiang, W., & Lamarre, Y. (1988). Modulation of lemniscal input during conditioned arm movements in the monkey. *Experimental Brain Research, 72,* 316–334.

Coghill, R. C., Talbot, J. D., Evan, A. C., Meyer, E., Gjedde, A., Bushnell, M. C., & Duncan, G. H. (1994). Distributed processing of pain and vibration by the human brain. *Journal of Neuroscience, 14,* 4095–4108.

Colby, C. L., & Duhamel, J. R. (1991). Heterogeneity of extrastriate visual areas and multiple parietal areas in the macaque monkey. [Review]. *Neuropsychologia, 29,* 517–37.

Colby, C. L., Duhamel, J. R., & Goldberg, M. E. (1993). The analysis of visual space by the lateral intraparietal area of the monkey: The role of extraretinal signals. [Review]. *Progress in Brain Research, 95,* 307–16.

Connor, C. E., Hsiao, S. S., Phillips, J. R., & Johnson, K. O. (1990). Tactile roughness: Neural codes that account for psychophysical magnitude estimates. *Journal of Neuroscience, 10,* 3823–3836.

Connor, C. E., & Johnson, K. O. (1992). Neural coding of tactile texture: Comparison of spatial and temporal mechanisms for roughness perception. *Journal of Neuroscience, 12,* 3414–3426.

Coquery, J.-M. (1978). Role of active movement in control of afferent input from skin in cat and man. In G. Gordon (Ed.), *Active touch* (pp. 161–169). New York: Pergamon.

Corbetta, M., Miezin, F. M., Dobmeyer, S., Shulman, G. L., & Petersen, S. E. (1991). Selective and divided attention during visual discriminations of shape, color, and speed: Functional anatomy by positron emission tomography. *Journal of Neuroscience, 11,* 2383–2402.

Corbetta, M., Miezin, F. M., Shulman, G. L., & Petersen, S. E. (1993). A PET study of visuospatial attention. *Journal of Neuroscience, 13,* 1202–1226.

Costanzo, R. M., & Gardner, E. P. (1980). A quantitative analysis of responses of direction-sensitive neurons in somatosensory cortex of awake monkeys. *Journal of Neurophysiology, 43,* 1319–1341.

Craig, A. D., Bushnell, M. C., Zhang, E.-T., & Blomqvist, A. (1994). A thalamic nucleus specific for pain and temperature sensation. *Nature, 372,* 770–773.

Crick, F. (1984). Function of the thalamic reticular complex: The searchlight hypothesis. *Proceedings of the National Academy of Sciences (USA), 81,* 4586–4590.

Critchley, M. (1949). The phenomenon of tactile inattention with special reference to parietal lesions. *Brain, 72,* 538–561.

Critchley, M. (1953). *The parietal lobe.* London: Edward Arnold.

Darian-Smith, I. (1982). Touch in primates. *Annual Review of Psychology, 33,* 155–194.

Darian-Smith, I. (1984). The sense of touch: Performance and peripheral neural processes. In I. Darian-Smith (Ed.), *Handbook of physiology, Section 1: The nervous system, Vol III. Sensory processes, Part 1* (pp. 739–788). Bethesda: American Physiological Society.

Darian-Smith, I., Goodwin, A., Sugitani, M., & Heywood, J. (1985). Scanning a textured surface with the fingers: Events in sensorimotor cortex. In A. W. Goodwin & I. Darian-Smith (Eds.), *Hand function and the neocortex experimental brain research supplement* (Vol. 10, pp. 17–43). New York: Springer-Verlag.

Darian-Smith, I., Sugitani, M., & Heywood, J. (1982). Touching textured surfaces: Cells in somatosensory cortex respond both to finger movement and to surface features. *Science, 218,* 906–909.

DeFelipe, J., Conley, M., & Jones, E. G. (1986). Long-range focal collateralization of axons arising from corticocortical cells in monkey sensory-motor cortex. *Journal of Neuroscience, 6,* 3749–3766.

Denny-Brown, D., & Chambers, R. A. (1958). The parietal lobe and behavior. *Research Publication Association of Nervous and Mental Diseases, 36,* 35–117.

Desimone, R. (1992). The physiology of memory: Recordings of things past. *Perspectives, 258,* 245–246.

Desimone, R., & Duncan, J. (1995). Neural mechanisms of selective visual attention. *Annual Review of Neuroscience, 18,* 193–222.

Desimone, R., & Ungerleider, L. G. (1989). Neural mechanisms of visual processing in monkeys. In I. Voller & J. Gratman (Eds.), *Handbook of neuropsychology* (Vol. 2, pp. 267–299). New York: Elsevier Science Publishers.

DiCarlo, J. J., Hsaio, S. S., & Johnson, K. O. (1994). Transformation of tactile spatial form within a cortical column in area 3b of the macaque. *Society of Neuroscience Abstracts, 20,* 1387.

Dong, W. K., Chudler, E. H., Sugiyama, K., Roberts, V. J., & Hayashi, T. (1994). Somatosensory, multisensory, and task-related neurons in cortical area 7b (PF) of unanesthetized monkeys. *Journal of Neurophysiology, 72,* 542–564.

Drevets, W. C., Burton, H., Videen, T. O., Snyder, A. Z., Simpson, J. R., & Raichle, M. E. (1995). Blood flow changes in human somatosensory cortex during anticipated stimulation. *Nature, 373,* 249–252.

Dreyer, D. A., Loe, P. R., Metz, C. B., & Whitsel, B. L. (1975). Representation of head and face in postcentral gyrus of the macaque. *Journal of Neurophysiology, 38,* 714–33.

Dreyer, D. A., Schneider, R. J., Metz, C. B., & Whitsel, B. L. (1975). Differential contributions of spinal pathways to body representation in postcentral gyrus of *Macaca mulatta. Journal of Neurophysiology, 37,* 119–145.

Duffy, F. H., & Burchfiel, J. L. (1971). Somatosensory system: Organizational hierarchy from single units in monkey area 5. *Science, 172,* 273–275.

Duhamel, J.-R., Colby, C. L., & Goldberg, M. E. (1991). Congruent representations of visual and somatosensory space in single neurons of monkey ventral intraparietal cortex (area VIP). In J. Paillard (Ed.), *Brain and space* (pp. 223–236). Oxford: Oxford University Press.

Duncan, G. H., Dreyer, D. A., McKenna, T. M., & Whitsel, B. L. (1982). Dose- and time-dependent effects of ketamine on SI neurons with cutaneous receptive fields. *Journal of Neurophysiology, 47,* 677–699.

Dyhre-Poulsen, P. (1978). Perception of tactile stimuli before ballistic and during tracking movements. In G. Gordon (Ed.), *Active touch* (pp. 171–176). New York: Pergamon.

Edin, B. B., Essick, G. K., Trulsson, M., & Olsson, K. A. (1995). Receptor encoding of moving tactile stimuli in humans. I. Temporal pattern of discharge of individual low-threshold mechanoreceptors. *Journal of Neuroscience, 15,* 830–847.

Essick, G. K. (1991). Human capacity to process directional information provided by tactile stimuli which move across the skin: Characterization and potential neural mechanisms. In F. Franzén & J. Westman (Eds.), *Information processing in the somatosensory system: Proceedings of an international symposium at the Wenner-Gren Center* (Vol. 57, pp. 329–340). London: The Macmillan Press.

Essick, G. K., Afferica, T., Aldershof, B., Nestor, J., Kelly, D., & Whitsel, B. (1988). Human perioral directional sensitivity. *Experimental Neurology, 100,* 506–23.

Essick, G. K., & Edin, B. B. (1995). Receptor encoding of moving tactile stimuli in humans. II. The mean response of individual low-threshold mechanoreceptors to motion across the receptive field. *Journal of Neuroscience, 15,* 848–864.

Essick, G. K., Franzén, O., McMillian, A., & Whitsel, B. (1991). Utilization of temporal and spatial cues to judge the velocity and traverse length of a moving tactile stimulus. In F. Franzén & J. Westman (Eds.), *Information processing in the somatosensory system: Proceedings of an international symposium at the Wenner-Gren Center* (Vol. 57, pp. 341–352). London: The Macmillan Press.

Essick, G. K., Franzén, O., & Whitsel, B. L. (1988). Discrimination and scaling of velocity of stimulus motion across the skin. *Somatosensory and Motor Research, 6,* 21–40.

Essick, G. K., McGuire, M., Joseph, A., & Franzen, O. (1992). Characterization of the percepts evoked by discontinuous motion over the perioral skin. *Somatosensory and Motor Research, 9,* 175–84.

Essick, G. K., & Whitsel, B. L. (1985a). Assessment of the capacity of human subjects and S-I neurons to distinguish opposing directions of stimulus motion across the skin. *Brain Research Reviews, 10,* 187–212.

Essick, G. K., & Whitsel, B. L. (1985b). Factors influencing cutaneous directional sensitivity: A correlative psychophysical and neurophysiological investigation. *Brain Research Reviews, 10,* 213–230.

Essick, G. K., & Whitsel, B. L. (1988). The capacity of human subjects to process directional information provided at two skin sites. *Somatosensory and Motor Research, 6,* 1–20.

Essick, G. K., & Whitsel, B. L. (1993). The response of SI directionally selective neurons to stimulus motion occurring at two sites within the receptive field. *Somatosensory and Motor Research, 10,* 97–113.

Evans, P. M., & Craig, J. C. (1991). Tactile attention and the perception of moving tactile stimuli. *Perception and Psychophysics, 49,* 355–364.

Evans, P. M., & Craig, J. C. (1992). Response competition: A major source of interference in a tactile identification task. *Perception and Psychophysics, 51,* 199–206.

Fabri, M., & Burton, H. (1991). Ipsilateral cortical connections of primary somatic sensory cortex in rats. *Journal of Comparative Neurology, 311,* 405–424.

Favorov, O., Diamond, M. E., & Whitsel, B. L. (1987). Evidence for a mosaic representation of the body surface in area 3b of the somatic cortex of cat. *Proceedings of the National Academy of Sciences (USA), 84,* 6606–6610.

Favorov, O., & Whitsel, B. L. (1988a). Spatial organization of the peripheral input to area 1 cell columns. I. The detection of "segregates." *Brain Research Reviews, 13,* 25–42.

Favorov, O., & Whitsel, B. L. (1988b). Spatial organization of the peripheral input to area 1 cell columns. II. The forelimb representation achieved by a mosaic of segregates. *Brain Research Reviews, 13,* 43–56.

Favorov, O. V., & Diamond, M. E. (1990). Demonstration of discrete place-defined columns—segregates—in the cat SI. *Journal of Comparative Neurology, 298,* 97–112.

Felleman, D. J., & Van Essen, D. C. (1991). Distributed hierarchical processing in the primate cerebral cortex. *Cerebral Cortex, 1,* 1–47.

Ferraina, S., & Bianchi, L. (1994). Posterior parietal cortex: functional properties of neurons in area 5 during an instructed-delay reaching task within different parts of space. *Experimental Brain Research, 99,* 175–8.

Fox, P. T., Burton, H., & Raichle, M. E. (1987). Mapping human somatosensory cortex with positron emission tomography. *Journal of Neurosurgery, 67*, 34–43.

Freeman, A. W., & Johnson, K. O. (1982). A model accounting for effects of vibratory amplitude on responses of cutaneous mechanoreceptors in macaque monkey. *Journal of Physiology (London), 323*, 43–64.

Friedman, D. P., & Jones, E. G. (1981). Thalamic input to areas 3a and 2 in monkeys. *Journal of Neurophysiology, 45*, 59–85.

Friedman, D. P., & Murray, E. A. (1986). Thalamic connectivity of the second somatosensory area and neighboring somatosensory cortical fields in the lateral sulcus of the monkey. *Journal of Comparative Neurology, 252*, 348–373.

Friedman, D. P., Murray, E. A., O'Neill, J. B., & Mishkin, M. (1986). Cortical connections of the somatosensory fields of the lateral sulcus of macaques: Evidence for a corticolimbic pathway for touch. *Journal of Comparative Neurology, 252*, 323–347.

Gardner, E. P., & Costanzo, R. M. (1980). Neuronal mechanisms underlying direction sensitivity of somatosensory cortical neurons in awake monkeys. *Journal of Neurophysiology, 43*, 1342–1354.

Gardner, E. P., Hamalainen, H. A., Palmer, C. I., & Warren, S. (1989). Touching the outside world: Representation of motion and direction within primary somatosensory cortex. In J. S. Lund (Ed.), *Sensory processing in the mammalian brain neural substrates and experimental strategies* (pp. 49–65). New York: Oxford University Press.

Gardner, E. P., & Palmer, C. I. (1989). Simulation of motion on the skin: I. Receptive fields and temporal frequency coding by cutaneous mechanoreceptors of optacon pulses delivered to the hand. *Journal of Neurophysiology, 62*, 1410–1436.

Gardner, E. P., Palmer, C. I., Hamalainen, H. A., & Warren, S. (1992). Simulation of motion on the skin. V. Effect of stimulus temporal frequency on the representation of moving bar patterns in primary somatosensory cortex of monkeys. *Journal of Neurophysiology, 67*, 37–63.

Gardner, E. P., & Spencer, W. A. (1972). Sensory funneling. II. Cortical neuronal representation of patterned cutaneous stimuli. *Journal of Neurophysiology, 35*, 954–977.

Gescheider, G. A. (1976). Evidence in support of the duplex theory of mechanoreception. *Sensory Processes, 1*, 68.

Gescheider, G. A., O'Malley, M. J., & Verrillo, R. T. (1983). Vibrotactile forward masking: Evidence for channel independence. *Journal of the Acoustical Society of America, 74*, 474.

Gescheider, G. A., & Verrillo, R. T. (1979). Vibrotactile frequency characteristics as determined by adaptation and masking procedures. In D. R. Kenshalo (Ed.), *Sensory functions of the skin* (pp. 183–203). New York: Plenum Press.

Gescheider, G. A., Verrillo, R. T., & VanDoren, C. L. (1982). Prediction of vibrotactile masking functions. *Journal of the Acoustical Society of America, 72*, 1421–1426.

Gilson, E. Q., & Baddeley, A. D. (1969). Tactile short-term memory. *Quarterly Journal of Experimental Psychology, 21*, 180–184.

Goff, G. D. (1967). Differential discrimination of frequency of cutaneous mechanical vibration. *Journal of Experimental Psychology, 74*, 294–299.

Goodwin, A. W., John, K. T., Sathian, K., & Darian-Smith, I. (1989). Spatial and temporal factors determining afferent fiber responses to a grating moving sinusoidally over the monkey's fingerpad. *Journal of Neuroscience, 9*, 1280–1293.

Goodwin, G. M., McCloskey, D. I., & Matthews, P. B. C. (1972). Proprioceptive illusions induced by muscle vibration: Contribution by muscle spindles to perception. *Science, 175*, 1382–1384.

Goodwin, A. W., & Morley, J. W. (1987a). Sinusoidal movement of a grating across the monkey's fingerpad: Representation of grating and movement features in afferent fiber responses. *Journal of Neuroscience, 7*, 2168–2180.

Goodwin, A. W., & Morley, J. W. (1987b). Sinusoidal movement of a grating across the monkey's fingerpad: Effect of contact angle and force of the grating on afferent fiber responses. *Journal of Neuroscience, 7,* 2192–2202.

Gould, W. R., Vierck, J., C.J., & Luck, M. M. (1979). Cues supporting recognition of the orientation or direction of movement of tactile stimuli. In D. R. Kenshalo (Ed.), *Sensory functions of the skin of humans* (pp. 63–78). New York: Plenum Press.

Greenspan, J. D., & Winfield, J. A. (1992). Reversible pain and tactile deficits associated with a cerebral tumor compressing the posterior insula and parietal operculum. *Pain, 50,* 29–39.

Hari, R., Karhu, J., Hamalainen, M., Knuutila, J., Salonen, O., Sams, M., & Vikman, V. (1993). Functional organization of the human first and second somatosensory cortices: a neuromagnetic study. *European Journal of Neuroscience, 5,* 724–734.

Head, H. (1920). *Studies in neurology.* London: Oxford University Press.

Heller, M. A. (1989). Texture perception in sighted and blind observers. *Perception and Psychophysics, 45,* 49–54.

Horster, W., & Ettlinger, G. (1987). Unilateral removal of the posterior insular or of area SII: Inconsistent effects on tactile, visual and auditory performance in the monkey. *Behavioral Brain Research, 26,* 1–17.

Hsiao, S. S., O'Shaughnesy, D. M., & Johnson, K. O. (1993). Effects of selective attention on spatial form processing in monkey primary and secondary somatosensory cortex. *Journal of Neurophysiology, 70,* 444–447.

Hyvärinen, J. (1981). Functional mechanisms of the parietal cortex. *Advances in Physiological Sciences, 16,* 35–49.

Hyvärinen, J. (1982). Posterior parietal lobe of the primate brain. *Physiological Reviews, 62,* 1060–1129.

Hyvärinen, J., & Poranen, A. (1978a). Movement-sensitive and direction and orientation-selective cutaneous receptive fields in the hand area of the post-central gyrus in monkeys. *Journal of Physiology (London), 283,* 523–537.

Hyvärinen, J., & Poranen, A. (1978b). Receptive field integration and submodality convergence in the hand area of the post-central gyrus of the alert monkey. *Journal of Physiology (London), 283,* 539–556.

Hyvärinen, J., Poranen, A., & Jokinen, Y. (1980). Influence of attentive behavior on neuronal responses to vibration in primary somatosensory cortex of the monkey. *Journal of Neurophysiology, 43,* 870–882.

Hyvärinen, J., & Shelepin, Y. (1979). Distribution of visual and somatic functions in the parietal associative area 7 of the monkey. *Brain Research, 169,* 561–564.

Iwamura, Y., Iriki, A., & Tanaka, M. (1994). Bilateral hand representation in the postcentral somatosensory cortex. *Nature, 369,* 554–6.

Iwamura, Y., & Tanaka, M. (1978). Postcentral neurons in hand region of area 2: Their possible role in the form discrimination of objects. *Brain Research, 150,* 662–666.

Iwamura, Y., Tanaka, M., & Hikosaka, O. (1980). Overlapping representation of fingers in the somatosensory cortex (area 2) of the conscious monkey. *Brain Research, 197,* 516–520.

Iwamura, Y., Tanaka, M., Sakamoto, M., & Hikosaka, O. (1983a). Functional subdivisions representing different finger regions in area 3 of the first somatosensory cortex of the conscious monkey. *Experimental Brain Research, 51,* 315–326.

Iwamura, Y., Tanaka, M., Sakamoto, M., & Hikosaka, O. (1983b). Converging patterns of finger representation and complex response properties of neurons in area 1 of the first somatosensory cortex in the conscious monkey. *Experimental Brain Research, 51,* 327–337.

Iwamura, Y., Tanaka, M., Sakamoto, M., & Hikosaka, O. (1985a). Comparison of the hand and finger representation in areas 3, 1, and 2 of the monkey somatosensory cortex. In M. Rowe & W. O. Willis (Eds.), *Development, organization, and processing in somatosensory pathways* (pp. 239–245). New York: Liss.

Iwamura, Y., Tanaka, M., Sakamoto, M., & Hikosaka, O. (1985b). Diversity in receptive field properties of vertical neuronal arrays in the crown of the postcentral gyrus of the conscious monkey. *Experimental Brain Research, 58,* 400–11.

Iwamura, Y., Tanaka, M., Sakamoto, M., & Hikosaka, O. (1985c). Vertical neuronal arrays in the postcentral gyrus signaling active touch: A receptive field study in the conscious monkey. *Experimental Brain Research, 58,* 412–20.

Iwamura, Y., Tanaka, M., Sakamoto, M., & Hikosaka, O. (1993). Rostrocaudal gradients in the neuronal receptive field complexity in the finger region of the alert monkey's postcentral gyrus. *Experimental Brain Research, 92,* 360–368.

Jiang, W., Chapman, C. E., & Lamarre, Y. (1991). Modulation of the cutaneous responsiveness of neurones in the primary somatosensory cortex during conditioned arm movements in the monkey. *Experimental Brain Research, 84,* 342–354.

Johansson, R. S. (1978). Tactile sensibility in the human hand: Receptive field characteristics of mechanoreceptive units in the glabrous skin area. *Journal of Physiology (London), 281,* 101–123.

Johansson, R. S., & Vallbo, Å. B. (1983). Tactile sensory coding in the glabrous skin of the human hand. *Trends in Neuroscience, 6,* 27–32.

Johnson, J. I. (1990). Comparative development of somatic sensory cortex. In E. G. Jones & A. Peters (Eds.), *Cerebral cortex* (Vol. 8B, pp. 335–449). New York: Plenum.

Johnson, K. O. (1974). Reconstruction of population response to a vibratory stimulus in quickly adapting mechanoreceptive afferent fiber population innervating glabrous skin of the monkey. *Journal of Neurophysiology, 37,* 48–72.

Johnson, K. O., & Hsiao, S. S. (1992). Neural mechanisms of tactual form and texture perception. *Annual Review of Neuroscience, 15,* 277–350.

Johnson, K. O., & Phillips, J. R. (1981). Tactile spatial resolution. I. Two-point discrimination, gap detection, grating resolution, and letter recognition. *Journal of Neurophysiology, 46,* 1177–1191.

Johnson, K. O., Phillips, J. R., Hsiao, S. S., & Bankman, I. N. (1991). Tactile pattern recognition. In O. Franzén & J. Westman (Eds.), *Information processing in the somatosensory system* (pp. 305–318). London: The MacMillan Press.

Jones, E. G. (1984). Organization of the thalamocortical complex and its relation to sensory processes. In I. Darian-Smith (Ed.), *Handbook of physiology. Section I. The nervous system, sensory processes. Part 2* (Vol. III, pp. 149–212). Bethesda: American Physiological Society.

Jones, E. G. (1985). *The thalamus.* New York: Plenum.

Jones, E. G. (1986). Connectivity of the primate sensory-motor cortex. In E. G. Jones & A. Peters (Eds.), *Cerebral cortex, sensory-motor areas and aspects of cortical connectivity* (Vol. 5, pp. 113–183). New York: Plenum Press.

Jones, E. G., & Burton, H. (1976). Areal differences in the laminar distribution of thalamic afferents in cortical fields of the insular, parietal and temporal regions of primates. *Journal of Comparative Neurology, 168,* 197–248.

Jones, E. G., Coulter, J. D., & Hendry, S. H. C. (1978). Intracortical connectivity of architectonic fields in the somatic sensory, motor and parietal cortex of monkeys. *Journal of Comparative Neurology, 181,* 291–348.

Jones, E. G., & Friedman, D. P. (1982). Projection patterns of functional components of thalamic ventrobasal complex on monkey somatosensory cortex. *Journal of Neurophysiology, 48,* 521–544.

Jones, E. G., & Porter, R. (1980). What is area 3a? *Brain Research Reviews, 2,* 1–43.

Jones, E. G., & Powell, T. P. S. (1970). An anatomical study of converging sensory pathways within the cerebral cortex of the monkey. *Brain, 93,* 793–820.

Juliano, S., Hand, P. J., & Whitsel, B. L. (1983). Patterns of metabolic activity in cytoarchitectural area SII and surrounding cortical fields of the monkey. *Journal of Neurophysiology, 50,* 961–980.

Kaas, J. H. (1983). What, if anything, is SI? Organization of first somatosensory area of cortex. *Physiological Reviews, 63,* 206–231.

Kaas, J. H., Nelson, R. J., Sur, M., Lin, C.-S., & Merzenich, M. M. (1979). Multiple representations of the body within the primary somatosensory cortex of primates. *Science, 204,* 521–523.

Kahneman, D. (1973). *Attention and effort.* Englewood Cliffs, NJ: Prentice-Hall.

Katz, D. (1989). *The world of touch.* (L. E. Krueger, Trans.) Hillsdale, NJ: Lawrence Erlbaum Associates.

Killackey, H. P., Koralek, K.-A., Chiaia, N. L., & Rhoades, R. W. (1989). Laminar and area differences in the origin of the subcortical projection neurons of the rat somatosensory cortex. *Journal of Comparative Neurology, 282,* 428–445.

Klatzky, R. L., Lederman, S., & Reed, C. (1987). There's more to touch than meets the eye: The salience of object attributes for haptics with and without vision. *Journal Experimental Psychology: General, 4,* 356–369.

Klatzky, R. L., Lederman, S. J., & Reed, C. (1989). Haptic integration of object properties: Texture, hardness, and planar contour. *Journal Experimental Psychology: Human Perception & Performance, 15,* 45–57.

Koch, K. W., & Fuster, J. M. (1989). Unit activity in monkey parietal cortex related to haptic perception and temporary memory. *Experimental Brain Research, 76,* 292–306.

Krubitzer, L., Clarey, J., Tweedale, R., Elston, G., & Calford, M. (1995). A redefinition of somatosensory areas in the lateral sulcus of macaque monkeys. *Journal of Neuroscience, 15,* 3821–3839.

Krubitzer, L. A., & Calford, M. B. (1992). Five topographically organized fields in the somatosensory cortex of the flying fox: Microelectrode maps, myeloarchitecture, and cortical modules. *Journal of Comparative Neurology, 317,* 1–30.

Krubitzer, L. A., Calford, M. B., & Schmid, L. M. (1993). Connections of somatosensory cortex in megachiropteran bats: The evolution of cortical fields in mammals. *Journal of Comparative Neurology, 327,* 473–506.

Krubitzer, L. A., & Kaas, J. H. (1992). The somatosensory thalamus of monkeys: Cortical connections and a redefinition of nuclei in marmosets. *Journal of Comparative Neurology, 319,* 123–140.

Lamb, G. D. (1983a). Tactile discrimination of textured surfaces: Peripheral neural coding in the monkey. *Journal of Physiology (London), 338,* 567–587.

Lamb, G. D. (1983b). Tactile discrimination of textured surfaces: Psychophysical performance measurements in humans. *Journal of Physiology (London), 338,* 551–565.

LaMotte, R. H., & Acuña, C. (1978). Defects in accuracy of reaching after removal of posterior parietal cortex in monkeys. *Brain Research, 139,* 309–326.

LaMotte, R. H., & Mountcastle, V. B. (1975). Capacities of human and monkeys to discriminate between vibratory stimuli of different frequency and amplitude: A correlation between neural events and psychophysical measurements. *Journal of Neurophysiology, 38,* 539–559.

LaMotte, R. H., & Mountcastle, V. B. (1979). Disorders in somesthesis following lesions of parietal lobe. *Journal of Neurophysiology, 42,* 400–419.

LaMotte, R. H., & Srinivasan, M. A. (1987a). Tactile discrimination of shape: Responses of slowly adapting mechanoreceptive afferents to a step stroked across the monkey fingerpad. *Journal of Neuroscience, 7,* 1655–1671.

LaMotte, R. H., & Srinivasan, M. A. (1987b). Tactile discrimination of shape: Responses of rapidly adapting mechanoreceptive afferents to a step stroke across the monkey fingerpad. *Journal of Neuroscience, 7,* 1672–1681.

LaMotte, R. H., & Srinivasan, M. A. (1987c). Tactile discrimination of shape: Responses of slowly and rapidly adapting mechanoreceptive afferents to a step indented into the monkey fingerpad. *Journal of Neuroscience, 7,* 1682–1697.

LaMotte, R. H., & Whitehouse, J. (1986). Tactile detection of a dot on a smooth surface: Peripheral neural events. *Journal of Neuroscience, 56,* 1109–1128.

Lederman, S. J. (1974). Tactile roughness of grooved surfaces: The touching process and effects of macro- and microsurface structure. *Perception and Psychophysics, 16,* 385–395.

Lederman, S. J. (1978). "Improving one's touch" . . . and more. *Perception and Psychophysics, 24,* 154–160.

Lederman, S. J., Browse, R. A., & Klatzky, R. L. (1988). Haptic processing of spatially distributed information. *Perception and Psychophysics, 44,* 222–232.

Lederman, S. J., & Klatzky, R. L. (1987). Hand movements: A window into haptic object recognition. *Cognitive Psychology, 19,* 342–368.

Lederman, S. J., & Klatzky, R. L. (1990). Haptic classification of common objects: Knowledge-driven exploration. *Cognitive Psychology, 22,* 421–459.

Lederman, S. J., Klatzky, R. L., & Reed, C. L. (1993). Constraints on haptic integration of spatially shared object dimensions. *Perception, 22,* 723–743.

Lederman, S. J., & Taylor, M. M. (1972). Fingertip force, surface geometry, and the perception of roughness by active touch. *Perception and Psychophysics, 12,* 401–408.

Lederman, S. J., Thorne, G., & Jones, B. (1986). Perception of texture by vision and touch: Multidimensionality and intersensory integration. *Journal of Experimental Psychology: Human Perception and Performance, 12,* 169–180.

Leinonen, L., Hyvärinen, J., Nyman, G., & Linnankoski, I. (1979). I. Functional properties of neurons in lateral part of associative area 7 in awake monkeys. *Experimental Brain Research, 34,* 299–320.

Leinonen, L., & Nyman, G. (1979). II. Functional properties of cells in anterolateral part of area 7 associative face area of awake monkeys. *Experimental Brain Research, 34,* 321–333.

Lin, C.-S., Merzenich, M. M., Sur, M., & Kaas, J. H. (1979). Connections of areas 3b and 1 of the parietal somatosensory strip with the ventroposterior nucleus in the owl monkey (*Aotus trivirgatus*). *Journal of Comparative Neurology, 185,* 355–372.

Loomis, J. M. (1982). Analysis of tactile and visual confusion matrices. *Perception and Psychophysics, 31,* 41–52.

Loomis, J. M., & Lederman, S. J. (1986). Tactual perception. In K. R. Boff, L. Kaufman & J. P. Thomas (Ed.), *Handbook of perception and human performance* (Vol. 2, pp. 1–41). New York: John Wiley.

Loomis, J. M., Lederman, S. J., Wake, H., & Fujita, N. (1993). Haptic identification of objects and their depictions. *Perception and Psychophysics, 54,* 170–178.

Lucier, G. E., Rüegg, D. C., & Wiesendanger, M. (1975). Responses of neurones in motor cortex and in area 3a to controlled stretches of forelimb muscles in cebus monkeys. *Journal of Physiology (London), 251,* 833–853.

Lund, J. S., Yoshioka, T., & Levitt, J. B. (1993). Comparison of intrinsic connectivity in different areas of macaque monkey cerebral cortex. *Cerebral Cortex, 3,* 148–162.

Manzoni, T., Conti, F., & Fabri, M. (1986). Callosal projections from area SII to SI in monkeys: Anatomical organization and comparison with association projections. *Journal of Comparative Neurology, 252,* 245–263.

Matelli, M., Camarda, R., Glickstein, M., & Rizzolatti, G. (1986). Afferent and efferent projections of the inferior area 6 in the macaque monkey. *Journal of Comparative Neurology, 251,* 281–298.

Melara, R. D., & Day, D. J. A. (1992). Primacy of dimensions in vibrotactile perception: An evaluation of early holistic models. *Perception and Psychophysics, 52,* 1–17.

Merzenich, M. M., Kaas, J. H., Sur, M., & Lin, C.-S. (1978). Double representation of the body surface within cytoarchitectonic areas 3b and 1 in "S1" in the owl monkey (*Aotus trivirgatus*). *Journal of Comparative Neurology, 181,* 41–74.

Merzenich, M. M., Nelson, R. J., Stryker, M. P., Cynader, M. S., Schoppmann, A., & Zook,

J. M. (1984). Somatosensory cortical map changes following digit amputation in adult monkeys. *Journal of Comparative Neurology, 224,* 591–605.

Mesulam, M. M., & Mufson, E. J. (1982). Insula of the old world monkey III: Efferent cortical output and comments on function. *Journal of Comparative Neurology, 212,* 38–52.

Millar, S. (1975). Effects of phonological and tactual similarity on serial object recall by blind and sighted children. *Cortex, 11,* 170–180.

Millar, S. (1978). Aspects of memory for information from touch and movement. In G. Gordon (Ed.), *Active touch: The mechanism of recognition of objects by manipulation* (pp. 215–227). New York: Pergamon Press.

Mishkin, M. (1979). Analogous neural models for tactual and visual learning. *Neuropsychologia, 17,* 139–151.

Moran, J., & Desimone, R. (1985). Selective attention gates visual processing in the extrastriate cortex. *Science, 229,* 782–784.

Moscovitch, M., & Behrmann, M. (1994). Coding of spatial information in the somatosensory system: Evidence from patients with neglect following parietal lobe damage. *Journal of Cognitive Neuroscience, 6,* 151–155.

Motter, B. C. (1993). Focal attention produces spatially selective processing in visual cortical areas V1, V2, and V4 in the presence of competing stimuli. *Journal of Neurophysiology, 70,* 909–919.

Motter, B. C. (1994a). Neural correlates of attentive selection for color or luminance in extrastriate area V4. *Journal of Neuroscience, 14,* 2178–2189.

Motter, B. C. (1994b). Neural correlates of feature selective memory and pop-out in extrastriate area V4. *Journal of Neuroscience, 14,* 2190–2199.

Mountcastle, V. B. (1984). Central nervous mechanisms in mechanoreceptive sensibility. In I. Darian-Smith (Ed.), *Handbook of physiology, Section I, the nervous system, Vol. III, sensory processes, Part 2* (pp. 789–878). Bethesda: American Physiological Society.

Mountcastle, V. B., LaMotte, R. H., & Carli, G. (1972). Detection thresholds for stimuli in humans and monkeys: Comparison with threshold events in mechanoreceptive afferent nerve fibers innervating the monkey hand. *Journal of Neurophysiology, 35,* 122–136.

Mountcastle, V. B., Lynch, J. C., Georgopoulos, A., Sakata, H., & Acuña, C. (1975). Posterior parietal association cortex of the monkey: Command functions for operations within extrapersonal space. *Journal of Neurophysiology, 38,* 871–908.

Mountcastle, V. B., Steinmetz, M. A., & Romo, R. (1990a). Frequency discrimination in the sense of flutter: Psychophysical measurements correlated with postcentral events in behaving monkeys. *Journal of Neurophysiology, 10,* 3032–3044.

Mountcastle, V. B., Steinmetz, M. A., & Romo, R. (1990b). Cortical neuronal periodicities and frequency discrimination in the sense of flutter. *Cold Spring Harbor Symposium on Quantitative Biology, 55,* 861–872.

Mountcastle, V. B., Talbot, W. H., Sakata, H., & Hyvärinen, J. (1969). Cortical neuronal mechanisms in flutter-vibration studied in unanesthetized monkeys. Neuronal periodicity and frequency discrimination. *Journal of Neurophysiology, 32,* 452–484.

Mufson, E. J., & Mesulam, M. M. (1982). Insula of the old world monkey. II. Afferent cortical input and comments on the claustrum. *Journal of Comparative Neurology, 212,* 23–37.

Murray, E. A., & Mishkin, M. (1984a). Relative contributions of SmII and area 5 to tactile discrimination of monkeys. *Behavioral Brain Research, 11,* 67–84.

Murray, E. A., & Mishkin, M. (1984b). Severe tactual as well as visual memory deficits follow combined removal of the amygdala and hippocampus in monkeys. *Journal of Neuroscience, 4,* 2565–2580.

Nelson, R. J., Smith, B. N., & Douglas, V. D. (1991). Relationships between sensory responsiveness and premovement activity of quickly adapting neurons in areas 3b and 1 of monkey primary somatosensory cortex. *Experimental Brain Research, 84,* 75–90.

Nelson, R. J., Sur, M., Felleman, D. J., & Kaas, J. H. (1980). Representations of the body surface in postcentral parietal cortex of *Macaca fascicularis*. *Journal of Comparative Neurology, 192,* 611–643.

Niebur, E., & Koch, C. (1994). A model for the neuronal implementation of selective visual attention based on temporal correlation among neurons. *Journal Computational Neuroscience, 1,* 141–158.

Olshausen, B. A., Anderson, C. H., & Van Essen, D. C. (1993). A neurobiological model of visual attention and invariant pattern recognition based on dynamic routing of information. *Journal of Neuroscience, 13,* 4700–4719.

Pandya, D. N., & Seltzer, B. (1982). Intrinsic connections and architectonics of posterior parietal cortex in the rhesus monkey. *Journal of Comparative Neurology, 204,* 196–210.

Paul, R. L., Merzenich, M., & Goodman, H. (1972). Representation of slowly and rapidly adapting cutaneous mechanoreceptors of the hand in Brodman's areas 3 and 1 of *Macaca mulatta*. *Brain Research, 36,* 229–249.

Penfield, W., & Boldrey, E. (1937). Somatic motor and sensory representation in the cerebral cortex of man as studied by electrical stimulation. *Brain, 60,* 389–443.

Penfield, W., & Rasmussen, T. (1950). *Secondary sensory and motor representation.* New York: Macmillan.

Phillips, C. B., Powell, T. P. S., & Wiesandanger, M. (1971). Projection from low threshold muscle afferents of hand and forearm to area 3a of baboon's cortex. *Journal of Physiology (London), 217,* 419–446.

Phillips, J. R., & Johnson, K. O. (1981a). Tactile spatial resolution. II. Neural representation of bars, edges, and gratings in monkey primary afferents. *Journal of Neurophysiology, 46,* 1192–1203.

Phillips, J. R., & Johnson, K. O. (1981b). Tactile spatial resolution. III. A continuum mechanics model of skin predicting mechanoreceptor responses to bars, edges, and gratings. *Journal of Neurophysiology, 46,* 1204–1225.

Phillips, J. R., Johnson, K. O., & Browne, H. M. (1983). A comparison of visual and two modes of tactual letter resolution. *Perception and Psychophysics, 34,* 243–249.

Phillips, J. R., Johnson, K. O., & Hsiao, S. S. (1988). Spatial pattern representation and transformation in monkey somatosensory cortex. *Proceedings of the National Academy of Sciences (USA), 85,* 1317–1321.

Pons, T. P., Garraghty, P. E., Cusick, C. G., & Kaas, J. H. (1985a). A sequential representation of the occiput, arm, forearm and hand across the rostrocaudal dimension of areas 1, 2 and 5 in macaque monkeys. *Brain Research, 335,* 350–353.

Pons, T. P., Garraghty, P. E., Cusick, C. G., & Kaas, J. H. (1985b). The somatotopic organization of area 2 in macaque monkeys. *Journal of Comparative Neurology, 241,* 445–466.

Pons, T. P., Garraghty, P. E., Friedman, D. P., & Mishkin, M. (1987). Physiological evidence for serial processing in somatosensory cortex. *Science, 237,* 417–420.

Pons, T. P., Garraghty, P. E., & Mishkin, M. (1992). Serial and parallel processing of tactual information in somatosensory cortex of rhesus monkeys. *Journal of Neurophysiology, 68,* 518–527.

Pons, T. P., & Kaas, J. H. (1985). Connections of area 2 of somatosensory cortex with the anterior pulvinar and subdivisions of the ventroposterior complex in Macaque monkeys. *Journal of Comparative Neurology, 240,* 16–36.

Pons, T. P., & Kaas, J. H. (1986). Corticocortical connections of area 2 of somatosensory cortex in Macaque monkeys: A correlative anatomical and electrophysiological study. *Journal of Comparative Neurology, 248,* 313–335.

Pons, T. P., Wall, J. T., Garraghty, P. E., Cusick, C. G., & Kaas, J. H. (1987). Consistent features of the representation of the hand in area 3b of macaque monkeys. *Somatosensory Research, 4,* 309–331.

Poranen, A., & Hyvärinen, J. (1982). Effects of attention on multiunit responses to vibration in the somatosensory regions of the monkey brain. *Electroencephalography and Clinical Neurophysiology, 53,* 525–537.

Posner, M. I. (1978). *Chronometric explorations of mind.* Hillsdale, NJ: Lawrence Erlbaum Associates.

Posner, M. I., & Dehaene, S. (1994). Attentional networks. [Review]. *Trends in Neuroscience, 17,* 75–9.

Posner, M. I., & Petersen, S. E. (1990). The attention system of the human brain. *Annual Review of Neuroscience, 13,* 25–42.

Posner, M. I., & Presti, D. E. (1987). Selective attention and cognitive control. *Trends in Neuroscience, 10,* 13–17.

Posner, M. I., Walker, J. A., Friedrech, F. J., & Rafal, R. D. (1984). Effects of parietal injury on covert orienting of attention. *Journal of Neuroscience, 4,* 1863–1874.

Post, L. J., & Chapman, C. E. (1991). The effects of cross-modal manipulations of attention on the detection of vibrotactile stimuli in humans. *Somatosensory and Motor Research, 8,* 149–57.

Post, L. J., Zompa, I. C., & Chapman, C. E. (1994). Perception of vibrotactile stimuli during motor activity in human subjects. *Experimental Brain Research, 100,* 107–20.

Powell, T. P. S., & Mountcastle, V. B. (1959). Some aspects of the functional organization of the cortex of the postcentral gyrus of the monkey: A correlation of findings obtained in a single unit analysis with cytoarchitecture. *Bulletin Johns Hopkins Hospital, 105,* 133–162.

Preuss, T. M., & Goldman-Rakic, P. S. (1989). Connections of the ventral granular frontal cortex of macaques with perisylvian premotor and somatosensory areas: Anatomical evidence for somatic representation in primate frontal association cortex. *Journal of Comparative Neurology, 282,* 293–316.

Preuss, T. M., & Goldman-Rakic, P. S. (1991). Architectonics of the parietal and temporal association cortex in the strepsirhine primate *Galago* compared to the anthropoid primate *Macaca. Journal of Comparative Neurology, 310,* 475–506.

Ramachandran, V. S. (1993). Behavioral and magnetoencephalographic correlates of plasticity in the adult human brain. *Proceedings of the National Academy of Sciences (USA), 90,* 10413–10420.

Randolph, M., & Semmes, J. (1974). Behavioral consequences of selective subtotal ablations in the postcentral gyrus of *Macaca mulatta. Brain Research, 70,* 55–70.

Rausell, E., & Jones, E. G. (1995). Extent of intracortical arborization of thalamocortical axons as a determinant of representational plasticity in monkey somatic sensory cortex. *Journal of Neuroscience, 15,* 4270–88.

Ray, R. H., Mallach, L. E., & Kruger, L. (1985). The response of single guard and down hair-mechanoreceptors to moving air-jet stimulation. *Brain Research, 346,* 333–347.

Reed, C. L., Lederman, S. J., & Klatzky, R. L. (1990). Haptic integration of planar size with hardness, texture, and planar contour. *Canadian Journal Psychology, 44,* 522–545.

Ridley, R. M., & Ettlinger, G. (1976). Impaired tactile learning and retention after removals of the second somatic sensory projection cortex (SII) in the monkey. *Brain Research, 109,* 656–660.

Ridley, R. M., & Ettlinger, G. (1978). Further evidence of impaired tactile learning after removals of the second somatic sensory projection cortex (SII) in the monkey. *Experimental Brain Research, 31,* 465–488.

Rizzolatti, G., Gentilucci, M., Fogassi, L., Luppino, G., Matelli, M., & Ponzoni-Maggi, S. (1987). Neurons related to goal-directed motor acts in inferior area 6 of the macaque monkey. *Experimental Brain Research, 67,* 220–224.

Rizzolatti, G., Gentilucci, M., & Matelli, M. (1985). Selective spatial attention: One center, one circuit, or many circuits? In M. I. Posner & O. S. Marin (Ed.), *Attention and performance XI* (pp. 251–265). Hillsdale, NJ: Lawrence Erlbaum Associates.

Roberts, T. S., & Akert, K. (1963). Insular and opercular cortex and its thalamic projection in *Macaca mulatta. Schweizer Archiv für Neurologie, Neurochirurgie und Psychiatrie, 92,* 1–45.

Robinson, C. J., & Burton, H. (1980a). Somatotopographic organization in the second somatosensory area of M. fascicularis. *Journal of Comparative Neurology, 192,* 43–67.

Robinson, C. J., & Burton, H. (1980b). Organization of somatosensory receptive fields in cortical areas 7b, retroinsula, postauditory and granular insula of M. fascicularis. *Journal of Comparative Neurology, 192,* 69–92.

Robinson, C. J., & Burton, H. (1980c). Somatic submodality distribution within the second somatosensory (SII), 7b, retroinsular, postauditory, and granular insular cortical areas of *M. fascicularis. Journal of Comparative Neurology, 192,* 93–108.

Robinson, D. L., McClurkin, J. W., Kertzman, C., & Petersen, S. E. (1991). Visual responses of pulvinar and collicular neurons during eye movements of awake, trained macaques. *Journal of Neurophysiology, 66,* 485–496.

Ruiz, S., Crespo, P., & Romo, R. (1995). Representation of moving tactile stimuli in the somatic sensory cortex of awake monkeys. *Journal of Neurophysiology, 73,* 525–537.

Sakata, H., Takaoka, Y., Kawarasaki, A., & Shibutani, H. (1973). Somatosensory properties of neurons in the superior parietal cortex (area 5) of the rhesus monkey. *Brain Research, 64,* 85–102.

Sanides, F., & Krishnamurti, A. (1967). Cytoarchitectonic subdivisions of sensorimotor and prefrontal regions and of bordering insular and limbic fields in slow loris (*Nycticebus coucang coucang*). *Journal für Hirnforschung, 9,* 226–252.

Sathian, K., & Burton, H. (1991). The role of spatially selective attention in the tactile perception of texture. *Perception and Psychophysics, 50,* 237–248.

Sathian, K., Goodwin, A. W., John, K. T., & Darian-Smith, I. (1989). Perceived roughness of a grating: Correlation with responses of mechanoreceptive afferents innervating the monkeys fingerpad. *Journal of Neuroscience, 9,* 1273–1279.

Schmidt, R. J., Schady, W. J. L., & Torebjörk, H. E. (1990). Gating of tactile input from the hand. I. Effects of finger movement. *Experimental Brain Research, 79,* 97–102.

Schneider, R. J., Friedman, D. P., & Mishkin, M. (1993). A modality-specific somatosensory area within the insula of the rhesus monkey. *Brain Research, 621,* 116–120.

Shiffrin, R. M., Craig, J. C., & Cohen, E. (1973). On the degree of attention and capacity limitation in tactile processing. *Perception and Psychophysics, 13,* 328–336.

Sinclair, R., & Burton, H. (1990). Psychometric and SI neurometric functions during active touch of gratings in man and monkey. *Society of Neuroscience Abstracts, 16,* 1080.

Sinclair, R. J., & Burton, H. (1991a). Neuronal activity in the primary somatosensory cortex in monkeys (*Macaca mulatta*) during active touch of textured surface gratings: Responses to groove width, applied force and velocity of motion. *Journal of Neurophysiology, 66,* 153–169.

Sinclair, R. J., & Burton, H. (1991b). Tactile discrimination of gratings: Psychophysical and neural correlates in human and monkey. *Somatosensory and Motor Research, 8,* 241–248.

Sinclair, R. J., & Burton, H. (1993a). Neuronal activity in the second somatosensory cortex of monkeys (*Macaca mulatta*) during active touch of gratings. *Journal of Neurophysiology, 70,* 331–350.

Sinclair, R. J., & Burton, H. (1993b). Responses in monkey primary somatosensory cortex during controlled passive tactile stimulation with gratings and bars. *Society of Neuroscience Abstracts, 19,* 765.

Sinclair, R. J., & Burton, H. (in press). Discrimination of vibrotactile frequencies in a delayed pair comparison task. *Perception and Psychophysics.*

Sinclair, R. J., Sathian, K., & Burton, H. (1991). Neuronal responses in ventroposterolateral nucleus of thalamus in monkeys (*Macaca mulatta*) during active touch of gratings. *Somatosensory and Motor Research, 4,* 293–300.

Srinivasan, M. A., & LaMotte, R. H. (1987). Tactile discrimination of shape: Responses of slowly and rapidly adapting mechanoreceptive afferents to a step indented into the monkey fingerpad. *Journal of Neuroscience, 7,* 1682–1697.

Stepniewska, I., Preuss, T. M., & Kaas, J. H. (1993). Architectonics, somatotopic organization, and ipsilateral cortical connections of the primary motor area (M1) of owl monkeys. *Journal of Comparative Neurology, 330,* 238–271.

Stevens, R. T., London, S. M., & Apkarian, A. V. (1993). Spinothalamocortical projections to the secondary somatosensory cortex (SII) in squirrel monkey. *Brain Research, 631,* 241–246.

Sullivan, E. V., & Turvey, M. T. (1974). On the short-term retention of serial, tactile stimuli. *Memory and Cognition, 2,* 600–606.

Sur, M., Wall, J. T., & Kaas, J. H. (1984). Modular distribution of neurons with slowly adapting and rapidly adapting responses in area 3b of somatosensory cortex in monkeys. *Journal of Neurophysiology, 51,* 724–744.

Talbot, W. H., Darian-Smith, I., Kornhuber, H. H., & Mountcastle, V. B. (1968). The sense of flutter-vibration: Comparison of the human capacity with response patterns mechanoreceptive afferents from the monkey hand. *Journal of Neurophysiology, 31,* 301–334.

Taylor, M. M., & Lederman, S. J. (1975). Tactile roughness of grooved surfaces: A model and the effect of friction. *Perception and Psychophysics, 17,* 23–36.

Taylor, M. M., Lederman, S. J., & Gibson, R. H. (1973). Tactual perception of texture. In E. C. Carterette & M. P. Friedman (Eds.), *Handbook of perception: Biology of perceptual systems, Vol. III* (pp. 251–272). New York: Academic Press.

Torebjörk, H. E., Schady, W., & Ochoa, J. (1984). Sensory correlates of somatic afferent fibre activation. *Human Neurobiology, 3,* 15–20.

Treisman, A., & Gormican, S. (1988). Feature analysis in early vision: Evidence from search asymmetries. *Psychology Reviews, 95,* 15–48.

Treisman, A., & Souther, J. (1985). Search asymmetry: A diagnostic for preattentive processing of separable features. *Journal Experimental Psychology, 114,* 285–310.

Valbo, A. B., & Johansson, R. S. (1984). Properties of cutaneous mechanoreceptors in the human hand related to touch sensation. *Human Neurobiology, 3,* 3–14.

Verrillo, R. T. (1962). Investigation of some parameters of the cutaneous threshold for vibration. *The Journal of the Acoustical Society of America, 34,* 1768–1772.

Verrillo, R. T. (1970). Subjective magnitude functions for vibrotaction. *IEEE Transactions on Man-Machine Systems, 11,* 19–24.

Verrillo, R. T. (1971). Vibrotactile thresholds measured at the finger. *Perception and Psychophysics, 9,* 329–330.

Vierck, C. J. (1974). Tactile movement detection and discrimination following dorsal column lesions in monkeys. *Experimental Brain Research, 20,* 331–346.

Vogt, C., & Vogt, O. (1919). Ergebnisse unseren hirnforschung vierte mitteilung: Die pysiologische bedeutung der architektonischen rindenreizungen. *Journal of Psychology & Neurology, 25,* 279–461.

Warren, S., Hamalainen, H. A., & Gardner, E. P. (1986a). Objective classification of motion- and direction-sensitive neurons in primary somatosensory cortex of awake monkeys. *Journal of Neurophysiology, 56,* 598–622.

Werner, G., & Whitsel, B. L. (1968). Topology of the body representation in somatosensory area I of primates. *Journal of Neurophysiology, 31,* 856–869.

Whang, K. C., Burton, H., & Shulman, G. L. (1991). Selective attention in vibrotactile tasks: Detecting the presence and absence of amplitude change. *Perception and Psychophysics, 50,* 157–165.

Whitsel, B. L., Franzen, O., Dreyer, D. A., Hollins, M., Young, M., Essick, G. K., & Wong, C. (1986). Dependence of subjective traverse length on velocity of moving tactile stimuli. *Somatosensory Research, 3,* 185–96.

Whitsel, B. L., Petrucelli, L. M., & Werner, G. (1969). Symmetry and connectivity in the map of the body surface in somatosensory area II of primates. *Journal of Neurophysiology, 32,* 170–183.

Whitsel, B. L., Roppolo, J. R., & Werner, G. (1972). Cortical information processing of stimulus motion on primate skin. *Journal of Neurophysiology, 35,* 691–717.

Wiesendanger, M. (1973). Input from muscle and cutaneous nerves of the hand and forearm to neurones of the precentral gyrus of baboons and monkeys. *Journal of Physiology (London), 228,* 203–219.

Woolsey, C. N. (1943). "Second" somatic receiving areas in the cerebral cortex of cat, dog and monkey. *Federation Proceedings, 2,* 55.

Woolsey, C. N. (1958). Organization of somatic sensory and motor areas of the cerebral cortex. In H. P. Harlow & C. N. Woolsey (Eds.), *Biological and biochemical basis of behavior* (pp. 63–82). Madison: University of Wisconsin Press.

Woolsey, C. N., & Fairman, D. (1946). Contralateral, ipsilateral, and bilateral representation of cutaneous receptors in somatic areas I and II of the cerebral cortex of the pig, sheep, and other mammals. *Surgery, 19,* 684–702.

Woolsey, C. N., Marshall, W., & Bard, P. (1942). Representation of cutaneous tactile sensibility in the cerebral cortex of the monkey as indicated by evoked potentials. *Bulletin Johns Hopkins Hospitals, 70,* 399–441.

Woolsey, C. N., & Walzl, E. M. (1981). Cortical auditory area of *Macaca mulatta* and its relation to the second somatic sensory area (Sm II): Determination by electrical excitation of auditory nerve fibers in the spiral osseous lamina and by click stimulation. In C. N. Woolsey (Ed.), *Cortical sensory organization. Volume 3: Multiple auditory areas.* (pp. 232–266). Clifton, NJ: Humana Press.

Woolsey, C. N., & Wang, G.-H. (1945). Somatic sensory areas I and II of the cerebral cortex of the rabbit. *Federation Proceedings, 4,* 79.

Young, M. P. (1992). Objective analysis of the topological organization of the primate cortical visual system. *Nature, 358,* 152–155.

Zhou, Y.-D., & Fuster, J. M. (1992). Unit discharge in monkey's parietal cortex during perception and mnemonic retention of tactile features. *Society of Neuroscience Abstracts, 18,* 706.

CHAPTER 4

Nociception and Pain: Evolution of Concepts and Observations

Edward R. Perl
Lawrence Kruger

Throughout history, the relationship of pain to other sensory experiences from the body has presented a major conceptual hurdle in defining its underlying mechanisms. Under most circumstances, pain is associated with a subjective component, suffering. Suffering or discomfort is, of course, not unique to pain, although pain is more consistently linked to suffering than other sensory experiences. The language of pain has been confounded by the use of the same words to describe both the perceptual recognition of the experience and the unpleasantness or emotion associated with it. For example, Aristotle placed pain at one pole of an emotional continuum opposite to pleasure and, thus, did not categorize pain as a sensation equivalent to touch, vision, or hearing. He and Greek philosophers influenced by him also did not attribute perception and other mental activities to the brain, designating the heart as their source. It was not until several centuries later that physician–anatomists of Alexandria, especially Galen in the second century A.D., proposed the vital part played by the nervous system in perception and cognition. The history of thinking and investigations about pain during antiquity is beyond this brief chapter; several useful sources provide detail (Boring, 1942; Castiglioni, 1947; Dallenbach, 1939; Garrison, 1969; Keele, 1957; Kruger & Kroin, 1978; Perl, 1984b; Procacci & Maresca, 1984; Sigerist, 1951; Soury, 1899; Walker, 1951). With few exceptions, (cf.

Pain and Touch

Kruger & Kroin, 1978; Perl, 1984b), past literature does not clearly distinguish between the discharge of sense organs or that of central neurons they directly influence (now termed *nociception*) and the ensuing, complex perceptual, emotional and motor reactions. This chapter focuses on the 19th- and 20th-century development of and obstacles to the concept that afferent organs and central systems exist that function selectively in signaling noxious events. Nonetheless, we emphasize that the legacy of old ideas such as pain representing an "emotion" pervades even modern thinking about pain mechanisms.

I. THE BELL–MAGENDIE LAW

Possibly the major impetus to modern thinking about the relation between the nervous system and afferent messages came about as the result of Charles Bell's (1811) recognition of differences in function between the dorsal and ventral roots of the spinal cord and Francois Magendie's (1822) simple, but clear, experimental evidence for the sensory nature of the dorsal roots. The concept that dorsal roots are sensory and the ventral roots are motor has come to be called the Bell–Magendie Law. In spite of minor challenges, it has provided a basic principle for investigation of sensory and motor function since the early 19th century. Modern workers should be mindful of possible exceptions to this rule. Sherrington (1894) noted that after ventral root section a few proximal myelinated axons degenerate and a few distal axons remain intact, suggesting that not all ventral root fibers originate from output neurons of the spinal cord. More recently, a substantial number of unmyelinated fibers in ventral roots have been proposed to be sensory because they are derived from dorsal root ganglion neurons (Clifton, Coggeshall, Vance, & Willis, 1976; Coggeshall, Coulter & Willis, 1974; Coggeshall & Ito, 1977). It was later observed that ventral root afferent fibers undergo sharp turns to loop back and enter the spinal cord via dorsal roots (Risling & Hildebrand, 1982). Nonetheless, there is evidence that a few afferent fibers may enter the spinal cord directly through ventral roots (Coggeshall, 1985; Light & Metz, 1978).

The afferent function of fibers entering the spinal cord via the dorsal roots appears firmly established, but this does not mean that some dorsal root fibers do not have effector functions as well. Bayliss (1901) showed that antidromic excitation of afferent fibers elicits a pronounced vasodilatation in the region of those fibers' peripheral distribution, a phenomenon to which he gave the name, "axon reflex." The efferent effects mediated by afferent fibers include plasma extravasation resulting in tissue swelling as well as vasodilatation (Lewis, 1942); effects mediated by ATP and vasoactive neuropeptides (Holton & Holton, 1954; Lembeck, 1983). The vascular and other efferent effects from activity in sensory fibers appear to be principally the result of peptidergic unmyelinated fibers excited by noxious stimuli (Lembeck, 1983). Nociceptive sensory elements releasing neuropeptides have peripheral trophic functions, which led to the suggestion that they could be labeled "noceffectors" to distinguish them from the substantial population of nociceptive primary afferent neurons devoid of such agents (Kruger & Halata, 1996; Kruger, Silverman, Mantyh, Sternini, & Brecha, 1989).

The basic principle of the separation of function by the spinal dorsal and ventral roots set the stage for exploring the nature of afferent nerves. One result was the codification of the doctrine of "Specific Nerve Energies" in the influential treatise by Johannes Müller (1842). Müller argued that "the nerve of each sense seems to be capable of one kind of sensation only, and not those belonging to other sense organs; thus, one sensory nerve cannot take the place and perform the function of the nerve of another sense" (pp. 183). This concept was applied by Müller to most of the senses with the notable exception of pain, which he relegated to common sensibility (*gemeingefühl*). Müller's argument for specific function by sensory nerves came at a propitious time because the experimental groundwork for understanding nerve function was being laid in the same period. Matteucci demonstrated the presence of an electric current produced by injury of tissue in 1841, and two years later DuBois-Reymond (1843) showed the electrical polarization of tissues. Subsequently, Helmholtz ingeniously made the first measurements of the conduction velocity of nerve, and in 1866 Bernstein (cf. Bernstein, 1868), a student of both DuBois-Reymond and Helmholtz, showed that the conduction of activity by a nerve consisted of the passage of a wave of electrical negativity.

II. SENSORY DISSOCIATION

In the latter half of the 19th century, observations on the behavioral effects of nervous system lesions produced experimentally in animals or by disease in humans provided the impetus for ideas about pain. The effect of lesions of particular sectors of the spinal cord in young dogs was found by Schiff (1858) to dissociate touch and pain deficits and led him to propose that pain is a specific sensation. Similar dissociations in people and animals were reported by Brown-Séquard (1860), including alterations in the reaction to light touch compared to pin prick in advanced cases of syphilis. A problem at the time and continuing to the present is the categorization of somatic sensory modalities by a wide range of descriptors that included touch, pressure, tingling, stinging, pain, itch, cold, and warmth. Evidence to allow assignment of specific nerve fibers to each of these categories of experience was missing, and "common sensibility" became for some "the residue of the unsorted sensations and feelings" (Dallenbach, 1939, p. 336). This included pain, which some authors did not regard as specific but related to the strength of the stimulus and direct excitation of nerve trunks (Weber, 1846), a concept that Erb (1874) explicitly formulated as the "intensive theory."

By the late 19th century the debate about the sensory status of pain was divided into three main views: the classic one, pain as an emotional state; a more recent idea, pain as a concomitant of all sensations when sufficiently intense; and a concept with roots back to the great Moslem physician, Avicenna (980–1037 A.D.) of pain as a specific sensation derived from particular afferent organs. Residues of all three of these interpretations persist to recent times in the somewhat less than orderly classification of knowledge about this subject.

III. EARLY CUTANEOUS PSYCHOPHYSICS AND THE STRUCTURE OF AFFERENT TERMINALS

The rigorous approach by the physicist–physician, Helmholtz, in the earlier part of the 19th century undoubtedly influenced the emergence of a psychophysics of cutaneous sensation. Late in the century this new field led to the discovery by Blix (1884) that the sensibility of the skin was not uniformly distributed across the surface but rather consisted of punctate regions of heightened sensibility to contact pressure; an observation soon confirmed by Goldscheider (1885). Max von Frey (1894, 1897) took up this question using careful quantitative methods and described a mosaic of distinct tactile, cold, warm, and pain spots distributed across the skin with distinctive regional variations. The regional differences led von Frey to deduce morphological correlates for the separate varieties of punctate sensibility by comparing the relative concentrations of the different kinds of spots in different body areas with reported frequency of occurrence of particular structurally distinct forms of neural terminations. The histologists of the late 19th century described numerous encapsulated endings of the skin, and these von Frey attributed to pressure, cold, and warmth spots. He assigned the large category of thin fibers with the unencapsulated "free" endings to the pain spots (von Frey, 1894, 1897). At the turn of the century, Charles Sherrington, the premier neurophysiologist of the period, in an important textbook chapter on cutaneous sensibility (Sherrington, 1900) supported von Frey's deductions about histology. Sherrington argued from his own observations on tooth pulp, cornea, and the inner surface of the eardrum that those regions where stimulation elicited only reports of pain were innervated by unmyelinated or very thin axons whose terminals lacked encapsulation. Perhaps more importantly, in his seminal volume, *Integrative Action of the Nervous System,* Sherrington (1906) overcame the dilemma of the various stimuli effective in producing pain. He suggested that the common denominator was damage of tissue or the threat of such damage and labeled these stimuli, noxious. This was the origin of the concept that the reception of noxious, normally painful stimuli constitutes nociception and afferent neurons signaling such events as *nocireceptors* (nociceptors).

Von Frey's deductions about the relationship of specific nerve terminal structures to the various punctate regions of cutaneous sensibility and even the idea of stable cutaneous receptive loci were seriously challenged (Dallenbach, 1929; Nafe, 1934). Nonetheless, the view that pain was initiated by "free" nerve endings from unmyelinated fibers became a textbook dictum until relatively recently (Perl, 1995; Sinclair, 1967), although evidence had surfaced that thinly myelinated fibers innervated the cornea and the tooth pulp from which only pain was alleged to be reported. Graham Weddell and colleagues (Weddell, Palmer, & Pallie, 1955) championed a widely held view during the mid-20th century that free nerve endings could subserve different modalities, including pain, touch, cold, and warmth and argued that there was "inadequate evidence of the existence of specific nociceptors" (Sinclair, 1967). As we shall see, only in the latter third of the 20th century did the concept of specific nociceptors become accepted, and attempts to characterize distinctive morphological features for different classes of nociception have developed even more recently (Kruger & Halata, 1996).

IV. AFFERENT FIBER DIAMETER AND SENSORY FUNCTION

Sherrington's reflex studies (1906) led him to consider the "flexion reflex," a protective reaction because it caused withdrawal of a limb from the locus of a noxious stimulus to bring it closer to the body and thereby become less vulnerable. He believed nocireceptive terminals to be the free nerve endings seen histologically in most epithelial structures; such terminations originated from fine nerve fibers, thereby relating this group to pain. Head, Rivers, and Sherren (1905) described the poorly differentiated sensory experiences appearing in the return of function to a denervated region as *protopathic*, a category that included pain. About a decade later, Ranson succeeded in visualizing the very large number of unmyelinated fibers in peripheral nerve and dorsal roots by modifying the silver impregnation method using pyridine and inferentially related them to pain and other protopathic sensations (Ranson, 1914, 1915). Subsequently, Ranson and Billinglsey (1916) found that interruption of the lateral part of the dorsal root, which Ranson (1914) had confirmed to contain most of the thinner fibers, eliminated the blood pressure elevation evoked by a noxious stimulus. The separation of dorsal roots into medial and lateral divisions was an idea considerably debated subsequently and apparently represents a concept misunderstood by some observers. The actual separation of the two divisions in some species occurs just at the junction with, or actually within, the spinal cord; the early observations by Lissauer (1885), Bechterew (1887), and Edinger (1892) as elaborated by Ranson have been confirmed in recent times (see Perl, 1984b).

A new wave of information derived from the work in St. Louis by Erlanger and Gasser, who developed the amplified cathode ray oscilloscope and succeeded in recording the compound action potential of nerve (Gasser & Erlanger, 1922; Gasser, Erlanger, & Bishop, 1924), a feat that had escaped earlier investigators due to the limited high-frequency response of recording devices previously available (e.g., string or capillary galvanometers and moving iron oscillograph). This was followed by evidence of the relationship between the diameter of fibers in a nerve trunk and the compound action potential (Gasser & Erlanger, 1927) and then the demonstration that pressure and local anesthetics differentially block components of the compound action potential (Gasser & Erlanger, 1929). The initial work on the compound action potential of peripheral nerve dealt with the activity generated by myelinated fibers, but soon the most slowly conducting or C deflection was uncovered and related to activity generated by unmyelinated fibers of sympathetic and afferent origin (Erlanger & Gasser 1930; Heinbecker, 1928/29; Heinbecker, Bishop, & O'Leary, 1933a). The studies on the compound action potential provided one stratagem for correlating reflexes or behavioral reactions as well as human sensory reports with fiber diameter. The larger fibers possessed lower electrical thresholds, rapid conduction velocity, and a greater sensitivity to pressure block. The thinner myelinated fibers and unmyelinated fibers required stronger electrical stimuli for activation and were more susceptible to local anesthetic block than the larger fibers. This enabled the differential excitation or blocking of one group relative to the other. Clues concerning conduction velocity and fiber size soon led to conclusions that the larger, more rapidly conducting population was associated with tactile or proprioceptive experiences, whereas the thinner, more slowly conducting category was related to pain and temperature sense and the reactions associated with them (Clark, Hughes, & Gasser, 1935; Gasser, 1943; Lewis, Pickering, & Rothschild, 1931; Lewis & Pochin, 1937; Zotterman, 1933). These simple correlations between conduction velocity of afferent fibers and the fiber's involvement in particular sensory experiences were widely accepted; however, objections to extensions of this notion to presumptions of discrete sets of specific receptors and the diameters of their axons, asphyxic susceptibility, and action of local anesthetics emerged in time (see reviews by Melzack & Wall, 1962; Perl, 1995; Sinclair, 1955; Sweet, 1959; Tower, 1943).

V. EARLY REPORTS OF SIGNALS FROM PRIMARY AFFERENT FIBERS

In parallel with the deciphering of the compound action potential, Adrian and Zotterman in Cambridge, England, showed that it was possible to record electrical activity from single peripheral nerve fibers in fine nerve bundles, either those occurring naturally or split from larger nerves (Adrian, 1926, 1928; Adrian & Zotterman, 1926a,b). Initial success with this technique was largely limited to tactile evoked discharges from relatively larger diameter afferent myelinated fibers or the equivalently large-diameter axons from motor neurons. These early studies did not uncover unique signals related to pain-producing stimuli in the mammal, although in the frog Adrian (1930) noted distinctive differences in the nature of action potentials in the dorsal cutaneous nerves evoked by tactile stimulation and that provoked by acid; the latter matched in size and configuration of activity recorded from postsynaptic sympathetic efferent fibers.

There were technical problems in obtaining unequivocal recordings from the most slowly conducting single fibers. In 1936 Zotterman published records from the mammalian lingual nerve, some of which he interpreted, on the basis of amplitude of the action potentials, to be generated by unmyelinated fibers. Certain of these small amplitude potentials were responsive to innocuous thermal stimuli, whereas others required tissue-damaging, heat stimuli. In the same period, Sarah Tower (1935) described activity of afferent fibers innervating the cornea, which on the basis of anatomical studies were unmyelinated, and differentiated them from those innervating the sclera, iris, and lens of the eye. In a study that proved important for the development of ideas about pain mechanisms, in 1939 Zotterman correctly showed that some discharges with characteristics of those generated by unmyelinated afferent fibers were excited by the most gentle of mechanical stimulation of the skin of cat; other similar impulses were initiated by thermal stimuli or strong mechanical stimulation. By the mid-1940s uncertainty appeared about the peripheral source of pain, with Tower (1940, 1943) expressing doubt about the existence of selective fibers for pain based upon her studies of the cornea. It should be noted that at the same time, others were convinced that unmyelinated fibers carried messages essential for pain and pain-associated reactions (Gasser, 1943). The cornea is an exceptional model for studying the integument because stimuli elsewhere judged innocuous can be injurious and painful.

VI. THE QUESTION OF TWO PAINS

That myelinated fibers may be involved in initiating pain was pointedly addressed in a human experiment by Heinbecker, Bishop, and O'Leary in 1933; they stimulated electrically the sural nerve in a conscious subject who was about to have a gangrenous leg amputated and found that electrical stimuli activating only the more rapidly conducting myelinated fibers (over 30 m/sec) produced an ill-defined tactile experience. When the stimulus strength was increased to activate fibers conducting around 25 m/sec, the slower myelinated range, without including the C (unmyelinated) component, the subject reported pain. Early nerve block experiments also provided evidence that pain was evocable by activity in thin, myelinated fibers as well as the unmyelinated population (Zotterman, 1933). The concept of a single stimulus evoking two distinctive pain experiences, related to different sets of afferent fibers, was an outcome of Lewis and Pochin's (1937) observation that the interval between such double experiences altered systematically at different distances of the stimulated site from the spinal cord. Not all workers accepted the notion of a "first" and "second" pain, as is evident from the controversy between Landau and Bishop (1953) and Jones (1956). Landau and Bishop (1953; Bishop & Landau, 1958) argued that first pain was aroused by activity in myelinated fibers and second pain derived from discharges in unmyelinated fibers; however, Jones (1956) was unable to reproduce two temporally distinct types of pain.

VII. THE ERA OF "SINGLE-UNIT" ANALYSES

The world-wide conflict in 1939 through 1945 slowed investigation on sensory systems of the body, but its end heralded a new level of sophistication in electronics and optics. This improved technology eventually paid dividends in the search for the peripheral sensory systems related to pain. The pre-war recordings from individual primary afferent fibers were accomplished largely through selection of thin peripheral cutaneous nerves, relatively simple teasing of such nerves, and the use of selective stimulation of restricted regions from which an individual fiber could be excited, the "receptive field." In a true departure, Tasaki's laboratory, in work begun before Japan's entry into the conflict, utilized a compound microscope and special chambers to dissect peripheral nerves down to a visible single fiber in toad and cat (Maruhashi, Mizuguchi, & Tasaki, 1952). They described recordings from some myelinated fibers that were most effectively excited by strong, presumably noxious mechanical stimuli or heat and a few unmyelinated fibers that were difficult to excite mechanically but that were activated by thermal or strong mechanical stimuli. Support for distinctive painful experiences from activity in different groups of thin primary afferent fibers came from a repeat of the classic human experiment by Heinbecker, Bishop, and O'Leary (1933b) using electrical stimulation of intact cutaneous nerves. At intensities exciting thin myelinated fibers (Aδ) a sharp localized pain was reported. Increase of the stimulus strength to activate also the unmyelinated (C-fibers) component altered the characteristics of the pain and made it intolerable; however, repetitive stimuli were required for eliciting a sensory report (Collins, Nulsen, & Randt, 1960). More recently, it has been shown that small doses of opiates do not alter perception of short latency pain induced by electrical stimulation of the skin at intensities activating myelinated fiber but significantly suppress the delayed pain evoked by C-fiber intensities of stimulation (Cooper, Veirck, & Yeomans, 1986; Yeomans, Cooper, & Vierck, 1996).

An ingenious electrophysiological approach in the 1950s further confused the story. Douglas and Ritchie (1957a,b) showed collision of impulses, and thereby refractoriness of fibers, generated by an electrical stimulus to a nerve with discharges evoked by "natural" stimuli applied to the skin to suppress the components of the compound potential. In skin nerves of the cat, they found light touch to profoundly diminish both the A-(myelinated fiber) and the C-(unmyelinated fiber) deflections, demonstrating that in this species a substantial proportion of the slowly conducting fibers must be excited by gentle tactile events. Furthermore, both touch and cooling stimuli markedly reduced most of the first part of the C-deflection, whereas only the later part of the C-deflection was suppressed by heat stimuli. These observations were interpreted as indicating that the C-(unmyelinated) fibers were not a homogeneous group specially subserving pain-related reactions, and even led to question the contribution of C-fiber activity to sensation (Douglas & Ritchie, 1959). At the time of these studies, there was substantial interest among workers in England in the lack of selectivity in responsiveness of primary afferent fibers. This concept was furthered by studies interpreted as suggesting that the experiences of pain, cold, warmth, and touch could be initiated from the human cornea (Lele & Weddell, 1956). The cornea is innervated by thin, almost exclusively unmyelinated fibers. These results may have been influenced by the nature of the stimuli employed and possible stimulus spread to adjacent tissues. Nonetheless, there was correspondence in time between rejection of the notion of specific "pain" afferent fibers and the reemergence of an intensity theory of pain (Sinclair, 1955; Weddell et al., 1955).

At about this time, a major advance was made by Ainsley Iggo in Edinburgh, who improved techniques for splitting peripheral nerves and obtained convincing records from single afferent fibers conducting at C-velocity in viscera and skin. His studies described a variety of types of primary afferent units in the C population of the cat (Iggo, 1957a,b; 1958, 1959, 1960). Some cutaneous units with C-afferent fibers were excited by noxious levels of heat and strong mechanical stimuli (Iggo, 1959), and his survey of 58 C-fiber recordings emphasized the existence of a continuum of mechanical responsiveness, ranging from minimal contact to noxious levels, rejecting the idea that C-fibers were exclusively nociceptive (Iggo, 1960). Although the significance of afferent fibers responding solely to strong mechanical or intense thermal stimuli was not widely appreciated then, other evidence for modality-specific C-fibers appeared (Iriuchijima & Zotterman, 1960). Recordings of muscle afferent fibers conducting at the slower myelinated range as well were published in the early 1960s, including some labeled as "pressure–pain receptors." The slowly conducting myelinated population proved to be mixed; the same range of fiber diameters also contained low-threshold mechanoreceptive elements (Bessou & LaPorte, 1961; Paintal, 1960). This period produced the most extensive survey to date of cutaneous sensory units with myelinated fibers, a sample in which only 7 of over 400 units exhibited high thresholds for mechanical stimulation (Hunt & McIntyre, 1960).

By the early 1960s there were apparent inconsistencies in the electrophysiological evidence on the peripheral sensory units responsible for pain. On the one hand, work on the compound action potential suggested that activity in the slower myelinated fiber group (Aδ) and unmyelinated fibers evoked pain-related reactions or were essential for them. On the other hand, it seemed clear that these categories of thin afferent fibers included many if not a predominance of sensory elements excited by innocuous stimuli. At the same time, reports of sensory units in the myelinated and the unmyelinated groups that required intense or noxious stimuli for activation were few in number and had not been shown to exhibit a set of coherent characteristics typical of sensory units related to tactile or proprioceptive sense.

VIII. HEAD'S PROTOPATHIC CONCEPT

Nineteenth-century clinical observations on the loss of contralateral pain and temperature sensation with retention of ipsilateral tactile and proprioceptive sensibilities along with anatomical studies defined a crossed pathway for pain and temperature in the anterolateral sector of the spinal cord, which came to be known as Gower's tract (Gower, 1886). By early in the 20th century, surgical chordotomy for the relief of intractable pain became a practical procedure (see reviews by Walker, 1951; White & Sweet, 1955). In contrast to the crossed anterolateral pathway, tactile and proprioceptive reactions were found to depend upon an ipsilateral dorsal spinal pathway, presumed to be the dorsal columns. As mentioned earlier, in the first part of the 20th century, Henry Head, a distinguished British neurologist, from observations on the sensory sequellae of peripheral nerve division and these differences in afferent pathways in the spinal cord, divided somatic sensation into two forms: epicritic and protopathic (Head, 1920; Head et al., 1905;). He presumed the epicritic pathway to underlie fine discriminative capacity, subserved by the dorsal column and medial lemniscus pathway. Protopathic sensation was assumed to serve a protective or guardian role and was conceived as primitive, nonspecific, and diffuse, originating from nonspecialized, peripheral free nerve endings. The latter's activity was conducted by thin axons into the spinal cord and transmitted centrally via the contralateral anterolateral funiculus.

The failure in the first half of the 20th century to identify specific nociceptive neurons peripherally and centrally led to a resuscitation of Head's notions of a genetically older, diffuse system (Herrick & Bishop,

1958; Rose & Mountcastle, 1959). These concepts were bolstered by reports of thalamic neurons responsive to noxious stimuli with broadly distributed, bilateral, and variable receptive fields (Poggio & Mountcastle, 1960) and of neurons in medial thalamic regions responsive to strong stimuli (Albe-Fessard & Kruger, 1962; Perl & Whitlock, 1961). In this setting, Lundberg and Oscarsson (1961) described differences in what they called the "dynamic range" of central neurons in ascending somatosensory pathways; some neurons receiving input from both large-diameter myelinated fibers and smaller diameter muscle afferent fibers had a more extensive range of response frequencies than those excited by tactile stimuli alone. Wall and Taub (1962) noted similar multireceptive neurons in the trigeminal system, where interestingly they were present in the part of the lemniscal projection system not normally identified with pain mechanisms. Mendell (1966) subsequently showed that some neurons of the spinal dorsal horn receiving both A- and C-fiber inputs exhibited a graduated range of discharge frequency parallel to recruitment of progressively smaller diameter fibers and again related this to "dynamic" properties of the neurons. This eventually gave birth to the term *wide-dynamic-range* (WDR) *neuron,* signifying a neuron with convergent inputs from a wide spectrum of primary afferent fibers and the receptors they serve. Unfortunately, this term is an oxymoron; many neurons of the central nervous system exhibit a broad range of discharge frequencies according to the synaptic efficacy of their inputs and yet do not necessarily exhibit the multireceptive characteristics implied by its usage for pain mechanisms. The essential point, however, is that multireceptive neurons exhibiting progressively higher frequencies of discharge for progressively stronger stimulation or recruitment of progressively finer fibers in afferent volleys became in the minds of some a keystone for afferent projections in association with pain. This terminology and concept (e.g., Mayer, Price, & Becker, 1975; Price & Mayer, 1975) found a receptive environment of contemporary interest in diffuse and nonspecific systems in the reticular core of the brain (Magoun, 1959; Morruzzi & Magoun, 1949). In the same period, descending control of ascending somatosensory activity was uncovered (Hagbarth & Kerr, 1954), and the part that such mechanisms can play in switching or gating spinal withdrawal reflexes was demonstrated (Eccles & Lundberg, 1959; Kuno & Perl, 1960). In addition, this era saw description and evidence for presynaptic inhibition and its use to explain interaction between afferent fibers in the dorsal horn of the spinal cord (Mendell & Wall, 1964).

IX. SPINAL GATES AND A THEORY

The mixture of evidence over the first 40 years of electrophysiological studies on sensory nerve fibers and the arguments from the Oxford commentators (Sinclair, 1955; Weddell, 1955) provided the background for an explicit challenge to the existence of specific primary afferent neurons for cutaneous sense in a review by Melzack and Wall (1962). Melzack and Wall interpreted the existing evidence as demonstrating a continuous distribution of primary afferent thresholds rather than discrete classes of sense organs, denying the existence of specific nociceptors. In 1965 Melzack and Wall culminated this argument with a theory about pain based upon the notions of (a) a continuous distribution of thresholds and adaptation rates by primary afferent fibers to various stimuli, and (b) a gating mechanism in the spinal cord that shifted the output of central afferent neurons according to the balance of activity between thin- and thick-diameter primary afferent fibers. The latter notion found support in Noordenbos's (1959) clinical observations that aberrant pain could be accounted for by excessive activity in small-diameter fibers, particularly when coupled with a reduction in signals from larger myelinated fibers. Descending control was invoked to explain variations in pain under different conditions in the original "gate control" hypothesis; its central idea, based on interactions between the inputs of different sensory modalities, had preoccupied the late 19th-century psychophysicists (cf., von Frey). Indeed, masking suppression of pinprick pain by vibration was one of the many sensory "illusions" demonstrated in that era. There also were several subsequent attempts to vitalize theories suggesting interactions between large- and small-diameter fibers to account for varieties of pain (e.g., Bishop, 1946; Kruger & Michel, 1962) and failure up to this time to detect separate neural systems serving detection of noxious stimuli.

In the Melzack and Wall proposal of 1965, neurons in the dorsal horn of the spinal cord transmitting activity related to pain centrally were excitable by innocuous as well as noxious stimuli. The activity of these neurons was controlled by descending as well as segmental mechanisms. Neurons of the substantia gelatinosa of the spinal dorsal horn were presumed to modulate presynaptically the primary afferent projections upon these transmission neurons, suppressing activity from larger fibers when the small-fiber activity reached a certain level or outlasted that of the thicker fibers. The theory was provocative in combining the terminology of digital electronics and the concept of presynaptic inhibition.

The gate theory created substantial ferment over the decade following its introduction. However, from the beginning there were a number of pieces of conflicting evidence. First, there were incorrect assumptions concerning both resting and evoked activity in C-afferent fibers. Soon after the introduction of the theory, detailed documentation of the existence of specific kinds of nociceptors emerged along with evidence that nociceptors evoke selective excitation of neurons in the superficial spinal cord (see below). Both of these findings were directly contradictory to basic assumptions of the theory. On the other hand, descending control of ascending sensory messages has been borne out to apply to nocireceptive central projections. Melzack and Wall (1970) subsequently modified the gate control theory to encompass the mounting evidence for specific nociceptors; but, they retained the central transmission neuron as a multireceptive or WDR type. There is now ample evidence for the existence of selective nocireceptive neurons along with the nociceptive multireceptive (WDR) types. How the two work together in the transmission of activity related to pain remains a matter to be clarified (Perl, 1984a,b, 1993; Willis, 1993). Regardless of the formidable factual obstacles encountered by the gate control theory, it proved valuable in attracting many clinicians as well as basic scientists into exploration of mechanisms underlying pain in normal creatures and under pathological circumstances.

X. DOCUMENTATION OF SPECIFIC NOCICEPTORS

The gate control theory represented sentiment against the concept of "labeled" lines, that is, particular receptors connected to a more or less dedicated projection system as an underlying mechanism for somatic sensation, including pain. At the time of the theory's proposal, serious concern existed about the nature of the sampling of the thinner end of the afferent fiber spectrum. The paucity of examples of sensory units with slowly conducting afferent fibers whose characteristics fit Sherrington's idea of nocireception in the eyes of some observers did not prove the absence of such sensory neurons. Until that time, the technique of searching for activity from a single nerve fiber consisted of tedious microdissection of a nerve bundle with successive division until the discharge of an individual fiber could be identified. This almost never involved dissection to a single fiber itself but rather a combination of splitting the nerve bundle into fine filaments and selective stimulation of the receptive field until an identifiable unitary discharge was evoked. By necessity, the procedure made it likely that afferent neurons readily excited by innocuous stimulation of the skin or other tissue would be selected for study; their response was easiest to evoke, and the tedious process of microdissection to isolate the unitary action potential would not be in vain. A novel approach to an unbiased sampling of peripheral nerve fibers was introduced by Burgess and Perl (1967). They employed fine glass microelectrodes inserted into the nerve and an electrical search stimulus to the whole nerve to identify a recording from a single fiber before examination of the peripheral receptive field. This approach yielded a population of cutaneous afferent fibers conducting in the slowest range of myelinated fibers (<30 m/sec, Aδ) that had notably high mechanical thresholds. These high-threshold units were (a) excitable from a series of spots surrounded by unresponsive zones, (b) not activated by strong, blunt compression, and (c) unresponsive promptly to even noxious heat. In cat hairy skin such myelinated high-threshold receptive units made up a substantial proportion of the slowly conducting myelinated population (6–30 m/sec); however, the majority of this conduction velocity range proved to have distinctly different characteristics. The latter were excitable by the weakest of mechanical disturbances in quite differently organized receptive fields. Following Sherrington's proposal, Burgess and Perl (1967) labeled the highest threshold mechanoreceptors as *mechanical nociceptors*. Essentially the same class of high-threshold mechanoreceptors (i.e., mechanical nociceptors) were demonstrated in monkey cutaneous nerves shortly thereafter; in monkey they made up approximately 10% of the myelinated fibers encountered using the microelectrode search technique (Perl, 1968). In contrast to the success with myelinated fibers, glass micropipette electrodes in peripheral nerve or dorsal roots did not yield stable recordings from C-fibers. This prompted Bessou and Perl (1969) to return to the classic teased filament preparation; in a year they characterized over 100 cutaneous C-fibers using the selection technique of a unitary response to electrical stimulation of the whole nerve employed for the myelinated fiber studies. Consistent with earlier evidence from cat hairy skin, about one-half of their sample was excited by the most gentle of mechanical disturbances of the skin or hairs. On the other hand, roughly 40% of their population was classified as *polymodal nociceptors* because these units were effectively excited by strong mechanical stimuli, noxious heat, or irritant chemicals, but did not respond to innocuous cooling and gentle mechanical manipulations. A smaller proportion (~10%) of the C-fiber sample was activated only by strong mechanical stimuli and not excited by noxious heating or cooling. A few of the C-fiber sensory units (~2%) responded to innocuous thermal stimuli with little or no response to even strong mechanical stimuli, constituting the specific thermal receptors known from their early characterization by Zotterman (1936) and Hensel, Iggo, and Witt (1960). Additional analyses of the cutaneous C-fiber population were performed on coccygeal segments supplying the tail, which permitted microelectrode recordings from the cell bodies of the dorsal root ganglia; this approach permitted comparison of a large sample of C- and A-afferent fiber units (Bessou, Burgess, Perl, & Taylor, 1971). The latter work established that the low-threshold C-fiber mechanoreceptors do not generate patterns or frequencies of discharge that distinguish innocuous contact from tissue-damaging stimuli, thereby providing evidence for the now widely held belief that low-threshold units do not transmit information indicative of noxious stimulation. Comparable findings for primate C-fibers also were soon provided (Kumazawa & Perl, 1977). It should be noted that the large proportion of C-fiber low-threshold mechanoreceptors noted in early studies may reflect a peculiarity of the afferent population makeup in hairy skin of species such as the cat or of particular body regions. Primate nerves supplying distal limb regions contain a smaller proportion of low-threshold C-fiber units than appear in cat or in nerves supplying more proximal body parts (as also seems to be the case in regions of glabrous skin) (Kumazawa & Perl, 1977; Lynn & Carpenter, 1982; Ochoa & Torebjörk, 1989). There now is evidence that innocuous C-mechanoreceptors are present in human hairy skin (Vallbo, Olausson, Wessberg, & Norrsell, 1993).

The studies identifying the C-polymodal nociceptors revealed that these units, unlike the low-threshold C-fiber mechanoreceptors, display a remarkable enhanced responsiveness following initial skin damage; that is, they "sensitize" (Bessou & Perl, 1969; Kumazawa & Perl, 1977). Sensitization had been observed previously in amphibian (Habgood, 1950) and in mammalian cutaneous units (Witt, 1962; Witt & Griffin, 1962). These earlier reports of sensitization were largely ignored (see Perl, 1995), and, in fact, the variability and lack of specificity they appeared to display encouraged some critics of the concept of nociceptors to dismiss the importance of such afferent units for a significant role in pain perception. Sensitization was recognized initially in units of the polymodal nociceptors types; however, it has been shown for myelinated high-threshold mechanoreceptors as well (Fitzgerald & Lynn, 1977). Sensitization of nociceptors is now firmly established and is a much studied phenomenon. It is an important factor in explaining the basis of pain arising from inflamed tissues (see Lynn and Perl, Chap. 5, this volume).

XI. HUMAN AFFERENT FIBER STUDIES

The observations relating components of the compound action potential to perceived sensory experiences already considered above revealed the differences in the pain experienced from activation of the slower myelinated fibers and that provoked by the presence of activity in C-fibers. These differences represent only part of the variety of experiences that are labeled "pain." They emphasize the necessity for more distinctive semantics to distinguish between afferent impulses (i.e., nociception) and the integrative activity in the central nervous system (CNS) that gives rise to the sensation of "pain." Recording sensory activity in human subjects has helped close the gap between these two processes. In 1967 Hagbarth and Vallbo introduced the technique of recording electrical activity from individual nerve fibers in awake persons by the insertion of a fine metal electrode percutaneously into nerve bundles. The metallic needle electrodes, insulated except close in the tip, are large compared to fine glass pipettes, but because of their relatively low noise, in the immediate vicinity of nerve fibers they permit recording of small extracellular currents associated with the passage of impulses. This procedure enabled recording from both single C- and myelinated fibers (Torebjörk & Hallin, 1970; von Hees & Gybels, 1972). The early findings further secured the correlation between C-afferent impulse activity and human reports of a delayed, dull pain. Subsequent work suggested that pathology in human peripheral nerves is associated with abnormal "spontaneous" activity (Ochoa, Torebjörk, Culp & Schady, 1982; Torebjörk, Ochoa, & McCann, 1979). Eventually, human intraneural recordings with such metal microelectrodes correlated receptive field, receptor modality characteristics, and the reported effects of electrical stimulation through the recording electrode for both myelinated and unmyelinated afferent units (Konietzny, Perl, Trevino, Light, and Hensel, 1981; Torebjörk & Ochoa, 1980; Torebjörk, Vallbo & Ochoa, 1987). Observations from different laboratories established that stimulation through the electrode (from which a recording was obtained) of a putative single C-fiber with characteristics of a polymodal nociceptor evokes a dull pain-like sensation projected accurately to the receptive field (Konietzny, Perl, Trevino, Light, & Hensel, 1981; Torebjörk & Ochoa, 1980). Stimulation at the recording site for a myelinated cutaneous mechanical nociceptor also was found to produce a pain-like sensation projected to the fibers' receptive field. Moreover, only a few impulses were necessary to establish the sensory experience (Konietzny et al., 1981). Stimulation at intraneural recording sites for some putative C-fiber polymodal nociceptors is reported to elicit "itch" rather than pain; modulation of the frequency of stimulation would not convert an itch or pain experience into the other, suggesting the existence of quite specific projections for sensory qualities from particular C-fiber afferent units (Ochoa, 1984; Torebjörk & Ochoa, 1981). The results of the intraneural stimulation have been challenged by questioning the possibility of recording the electrical activity from a single unmyelinated fiber using a metallic electrode whose tip diameter is larger than the cross-sectional diameter of the recorded fiber (Wall & MacMahon, 1985). One explanation for this apparent paradox comes from electron microscopic evidence that in human cutaneous nerves unmyelinated axons are often individually isolated within separate Schwann sheaths, a situation that could explain the ability to obtain recordings from single C-fibers (Ochoa, 1984). The anatomical situation in human nerve differs from that in many experimental animal nerves in which several unmyelinated fibers are enclosed in bundles surrounded by a Schwann cell sheath. In another answer to this challenge, Torebjörk et al. (1987) presented convincing data that intraneural microstimulation in human subjects can activate single primary afferent fibers in isolation; in many trials such stimulation was found to evoke a unique cutaneous sensory experience referred to the cutaneous receptive field of the recorded unitary action potential.

The observations on the effects of intraneural stimulation in human beings have profound importance for theories on the relationship between primary afferent neurons and perceptual experience. At this writing, selectivity of function through signaling by primary afferent neurons seems unquestioned; however, the issue of the extent to which a pattern of activity and interaction between different kinds of sensory units participate in the eventual perceptual experience is still to be settled. Some commentators still resist notions of selectivity in central somatosensory function in relation to pain, and raise combative objection to arguments favoring specific relationships between afferent unit characteristics and sensation. Resolution of opinions on the role of different afferent units in integrative experiences will require more evidence and time.

XII. PERIPHERAL NEUROPATHY AND PAIN

The dissociation of pain from other bodily sensory experiences as evidence for the sensory nature of pain has already been mentioned. Clinically, such situations are reasonably common due to diseases that affect development or function of peripheral nerves. They were not readily interpretable until the relatively recent advent of percutaneous recording of compound action potentials and the resolution capabilities of electron microscopy that enable reasonable quantitative sampling of the finest of nerve fibers in dorsal roots or peripheral nerves. Cases of apparent congenital lack of reactivity to painful stimuli have been known since the early 1930s (Dearborn, 1932). Some 30 years after this initial description, Swanson, Buchan, and Alvord (1965), in a case of congenital insensitivity to pain, documented a dramatic reduction in thin dorsal root fibers and small dorsal root ganglion neurons along with a concomitant reduction in the size of Lissauer's tract and the spinal trigeminal tract. A similar absence of thinly myelinated and unmyelinated fibers was reported for a patient with congenital insensitivity to pain as part of the Riley-Day syndrome (Aguayo, Cherunada, & Bray, 1971). Subsequently, in several neuropathies, a pattern of sensory loss, changes in the compound action potential, and alterations in the morphometry of fiber size established strong correlations between selective loss of pain sensibility and a decreased number of thinly myelinated and unmyelinated fibers in cutaneous nerves (e.g., Dyck, 1993; Dyck & Lambert, 1969; Dyck, Lambert, & O'Brien, 1976; Thomas, 1974). On the other hand, such studies have failed to provide an anatomical substratum for painful neuropathies and hyperalgesia (Thomas, 1979).

Although pathophysiology of abnormal pain is beyond the scope of this overview, one recently emerged idea deserves comment. For many years, a syndrome with aberrant pain as a dominant symptom has been recognized, on occasion, to follow upon partial injury of a peripheral nerve or the innervation to tissue. S. Weir Mitchell, the "father" of American neurology, described and named the syndrome *causalgia* over a century ago (Mitchell, 1872). Causalgia has been related to sympathetic efferent activity (Leriche,

1916) and often, at least temporarily, can be ameliorated by block or removal of the sympathetic supply to the affected region (White & Sweet, 1955). Recently, partial injury of mixed nerves has been shown to induce an adrenergic excitation of some nociceptors, suggesting that an alteration in the spectrum of responsiveness by nociceptors may underlie this pathological pain state (Perl, 1994; Sato & Perl, 1991). On the other hand, this and other varieties of "abnormal" pain are unlikely to be explicable on a strictly peripheral basis, even though peripheral events may serve as initial triggers for ensuing central alterations and dysfunction. Trigeminal neuralgia in which excruciating paroxysmal pain can be elicited by ordinarily innocuous tactile stimuli (called allodynia in current usage) may be another example of complex interactions between peripheral and central processes. Allodynia, pain elicited by normally nonpainful stimuli, presents a knotty problem for conceptualization of a mechanism independent of nociceptor input. This implies a CNS abnormality. There is also the experimental phenomenon of *secondary hyperalgesia* in which pain is produced by normally innocuous stimuli in regions near or surrounding injured tissue. It has been suggested for many years that secondary hyperalgesia is of central origin even though it may follow a noxious peripheral event (Hardy, Wolff, & Goodell, 1952). The suggested underlying mechanisms generally invoke enhanced excitability of central neurons consequent to persisting changes induced by specific synaptic mediators. The multireceptive or WDR neuron has been argued to play a crucial part (see Treede & Magerl, 1995; Willis, 1993). With central nervous lesions, there is the possibility that inhibitory mechanisms have been suppressed or inactivated due to lesions of selective nociceptive pathways. Another possibility associated with either CNS or peripheral nerve lesions is a reorganization of central terminals of primary afferent fibers to include new targets or a new balance of targets for low-threshold sensory units producing potent excitatory contacts on projection neurons the input of which is normally dominated by nociceptors (cf. Bullitt, Stofer, Vierck, & Perl, 1988).

XIII. SPINAL NOCIRECEPTIVE NEURONS

Factors underlying genesis of the gate control theory included the paucity of evidence for both peripheral sensory receptors selectively responsive to pain-provoking stimuli and a similar dearth of evidence concerning central neurons with selective nocireceptive features. The interest aroused by the gate theory caused substantial ferment in the decade after its proposal, and the demonstration of the existence of specific nociceptors shortly after the theory's introduction has already been touched upon. Disagreements about the interpretation of dorsal root potentials and the indirect nature of the evidence that was used to indicate presynaptic inhibitory processes soon followed its proposal; experimental observations revealed features that seemed incompatible with the theory's assumptions and predictions. For example, central terminal excitability changes proposed by the theory proved to be contrary to those observed (e.g., Whitehorn & Burgess, 1973) as well as assumptions regarding the nature of resting and evoked activity in C-afferent fibers (Bessou & Perl, 1969). Another discrepancy challenging the theory came from electrophysiological identification of neurons shown by dye deposition to be in the most superficial part of the spinal dorsal horn (principally lamina I) that were uniquely activated by input from thin afferent fibers and by noxious or thermal stimuli (Christensen & Perl, 1970). These selectively nocireceptive neurons had relatively circumscribed excitatory receptive fields and received inhibitory input from body regions outside the excitatory zone. Perl and co-workers established that some of the selectively nociceptive neurons could be antidromically activated from the lateral funiculus of the opposite side of the spinal cord, thereby providing evidence of central nociceptive and thermoreceptive-specific neurons projecting into a contralateral ascending system (Kumazawa, Perl, Burgess, & Whitehorn, 1975). A part of this population appeared to represent the marginal zone neurons shown in human studies to undergo retrograde atrophy in appropriate segments following contralateral spinal cord lesions that produced profound defects in pain and temperature sensibility (Foerster & Gagel, 1932; Kuru, 1949). Subsequently, it was shown that the superficial part of the spinal cord with its dominant input from thinly myelinated and unmyelinated afferent fibers from the periphery had a complex functional organization containing both neurons selectively excited by noxious, thermal, and also by specific forms of tactile stimuli (Kumazawa & Perl, 1978; Light, Trevino, & Perl, 1979; Rèthelyi, Light, & Perl, 1989). Retrograde and anterograde anatomical tracing studies have demonstrated that the marginal zone of the spinal dorsal horn contributes a large proportion (20 to over 60%) of neurons with fibers that reach thalamic regions via a crossed pathway from spinal lumbar and cervical regions (Craig, Linington & Kniffki, 1989; Jones, Apkarian, Stevens, & Hodge, 1987; Stevens, Hodge, and Apkarian, 1989; Willis, Kenshalo, & Leonard, 1979). Existence of nocireceptive and thermoreceptive neurons in the *pericornual* (most superficial) layer of the spinal trigeminal nucleus homolog of the spinal dorsal horn has been established as well (Mosso & Kruger, 1972). At the same time, the new retrograde labeling techniques have shown a substantial number of neurons in deeper layers, including some in the cat ventral horn, to project over contralateral ascending tracts (Trevino & Carstens, 1975; Trevino, Coulter, & Willis, 1973).

As a counter to the evidence for selective noci-receptive neurons, a number of analyses have described spinal neurons in deeper portions of the spinal dorsal horn that can be retrogradely activated from thalamic regions (see review by Willis & Coggeshall, 1978). These projection neurons of the deeper layers of the spinal cord that contribute importantly to the spinothalamic system are mostly relatively large with distinctive multireceptive characteristics; typically they are activated by movement of hairs or light pressure to the skin but exhibit increased frequency of discharge in response to pinching and noxious heating. Such WDR neurons appear to project at least partially in parallel with the selective nocireceptive or thermoreceptive neurons. Parenthetically, it might be noted that multireceptive neurons contribute to lemniscal and spinal cervical ascending projection systems as well. Such nonselective projections have been suggested to be involved in altering states of consciousness and awareness and to modify the sensitivity of higher centers (Perl, 1984a). Thus, an important consideration in understanding the information coding for nociception and the sensation of pain is that the crucial pathway for ascending nociceptive information contains at least two classes of neurons. One class is excited largely, if not exclusively, by noxious stimuli, and another is activated by a variety of different somatosensory events and the afferent units detecting them. The latter include some neurons that respond to noxious stimuli by their highest frequency and most persistent discharge. How these two distinctive forms of information transfer combine to produce sensation remains a mystery.

XIV. THE BRAIN AND PAIN

More than a century after the crossed lateral–anterolateral pathway was defined and first studied, little doubt remains about its crucial role in the transmission of activity leading to pain. On the other hand, ideas about the termination of this pathway have been under constant revision. The fact that this crossed spinal pathway projects to the thalamus is not disputed (Perl, 1984b; White & Sweet, 1955); the controversial issues concern its infrathalamic terminations and which thalamic structures receive input. The ascending anterolateral spinal tract's contribution to several brain stem regions caudal to the thalamus is deserving of note. In addition to terminations in the medullary and midbrain reticular region, there are notable projections to the parabrachial region and to the periaqueductal gray. The parabrachial termination stems, to a large extent, from nociceptive-specific spinal lamina I (marginal zone) neurons (see Light, 1992). It is possible that some of the lower and midbrain stem projections of the anterolateral tract are related to autonomic, respiratory, or altering effects of noxious input.

The existence of a relationship between the thalamus and the experience of pain derives, in part, from analyses by Head and Holmes (1911) of the *thalamic syndrome*. In this syndrome, first described by Dejerine and Roussy (1906), aberrant pain plays an important part. The thalamus is a complex structure; its evolutionary development parallels that of the cerebral cortex in many ways. The connections between the various nuclear groups making up the thalamus among themselves and to the cerebral cortex are complicated and their description often confounded by problems of nomenclature. To some extent this explains difficulty in reconciling the various observations concerning the anatomical termination of the ascending ventrolateral spinal tracts in thalamic regions, physiological or pathophysiological observations on electrophysiological recordings from the thalamus, and the effects of electrical stimulation of thalamic loci. Mehler, Feferman, and Nauta (1960) indicated that there were at least three major areas of termination of ascending ventrolateral tracts in the primate thalamus; these included a medial projection, a ventrolateral one, and posterior medial zone. All three of these general regions have been reported to contain neurons with nocireceptive characteristics, although the features in different areas vary considerably (Albe-Fessard & Kruger, 1962; Casey, 1966; Perl & Whitlock, 1961; Poggio & Mountcastle, 1960). Both multireceptive neurons excitable from large portions of the body, and neurons selectively excited by noxious stimuli from relatively small receptive fields in a somatotopically arranged pattern have been reported (Honda, Mense & Perl, 1983; Kenshalo, Giesler, Leonard, & Willis, 1980). Recording and stimulation studies in human patients undergoing sterotaxic exploration of the rostral brain stem have supported the idea that parts of the thalamus are associated with pain perception or expression and are suggestive of the conclusion that a fairly localized region in the ventral posterior portion of the thalamus are related to a topographic arrangement associated with pain (Halliday & Logue, 1972; Hassler, 1958, 1960; Lenz et al., 1993). A common factor in several of these studies is the identification of a relatively circumscribed thalamic nuclear group ventral to a major somatotopically organized tactile region in which stimulation evokes painful sensations referred to particular contralateral loci. The recent report by Craig, Busnell, Zhang, and Blomquist (1994) of a localized nucleus region in the ventral medial thalamus (VMpo) in monkey describes specific cold, specific nocireceptive, and mixed noci- and thermoreceptive neurons with distinctive topographical inputs and immunocytochemical properties that on the basis of cytology and immunocytochemical markers might be related to an homologous region in the human thalamus.

The focus on the importance of the thalamus for pain perception has led to the belief that pain is not dependent upon cerebral cortical mechanisms. Evidence to the contrary has existed for a long time. The paucity of observations showing that localized stimulation of the cortex in human beings evokes pain has contributed to underestimating the importance of cortical mechanisms for pain (Penfield & Boldrey, 1937) as has the rarity of disturbances of pain perception as a consequence of cerebral cortical damage. On the other hand, several clinical studies have demonstrated that loss of the capacity to recognize painful stimuli on the contralateral side of the body does occur in a topographic fashion after small lesions of the immediate postcentral "primary" somatosensory cortex (Lewin & Phillips, 1952; Marshall, 1951; Russell, 1945). The loss of sensation due to cortical lesions can persist for years and may or may not be related to disturbances of thermal sensation in the same regions (Marshall, 1951). The portion of the primary somatosensory cortical region involved in pain perception appears to lie deep in the central sulcus (area 3), thus rarely stimulated in explorations of the human cortex. Consonant with descriptions of the loss of the ability to recognize painful stimuli after focal lesions of the postcentral gyrus are observations in monkeys that both selectively nocireceptive and WDR-type neurons are present in the immediate postcentral region (Kenshalo & Isensee, 1983).

On the other hand, the concept that a particular cortical area is solely responsible for the experience of pain or the recognition of noxious stimuli is inconsistent with present information. Large parietal lobe lesions involving the postcentral cortex do not result in loss of recognition of painful stimulation (Russell, 1945). Furthermore, there is evidence that the secondary somatosensory region has importance for pain recognition (Biemond, 1956; Craig et al., 1994). Recent positron emission tomography (PET) studies have demonstrated that both regions equivalent to the primary somatosensory region and the second somatosensory area in conscious human subjects exhibit activity particularly related to noxious stimuli (Bushnell, Craig, Reiman, Yun, & Evans, 1995; Coghill et al., 1994). The emerging evidence that a particular nucleus or cerebral locus does not underlie a complex function by the awake brain deserves emphasis in thinking about thalamic or cerebral cortical involvement in the process of detecting or distinguishing painful stimuli from other events and in generating the experience of pain. It is more likely that multiple portions of the brain process information in concert to produce perceptual or behavioral actions (e.g., Goldman-Rakic, 1984).

The substantial variability in the magnitude of pain or the tolerance to noxious stimuli exhibited even by trained experimental subjects indicates that the influence of context and learning on central processing should be taken into account in considering a response to noxious input. Thus, although scaling of the intensity of painful stimuli can follow the basic features of intensity scaling observed for other sensory systems, grading discomfort/suffering/tolerability may vary substantially in different individuals (see Gracely & Wolskee, 1983; Gracely & Naliboff, chap. 6, this volume). These differences indicate a greater complexity for the type of response that might be classified as suffering or discomfort than for nociception. For example, deprivation of early exposure to noxious stimuli can result in a dramatic failure of appropriate protective or avoidance responses (Nissen, Chow, & Semmes, 1951; Melzack & Scott, 1957). The relative lack of such studies until the latter part of the 20th century undoubtedly reflects that past research has concentrated on the consequence of transient injurious events and their physiological sequelae. Many contemporary investigations of pain place greater emphasis on differences between the immediate effects of an afferent nociceptive barrage and the long-term effects of injury to peripheral tissue or innervation more closely related to pain abnormalities seen clinically.

One cannot touch upon the central nervous mechanisms associated with pain without at least mentioning the problem of "central pain," that is, pain of central origin. The devastating phenomenon of central pain is recognized ordinarily as a sequel to a CNS lesion. Cassinari and Pagni (1969) argued that a common feature in many cases of pain of central origin is interruption of the crossed "spinothalamic" (anterolateral) pathway. Injury of the dorsal root entry zone and the superficial dorsal horn are associated with a more regionally referred variety of central pain. One speculative explanation for pain of central origin is that it represents loss of inhibitory connections to crucial central neurons. These might originate from other central regions or be part of the anterolateral projection from either selective nocireceptive neurons (e.g., lamina I) or the nonselective projection (WDR). The situation is not simple though, because central pain is not a concomitant of every central nervous lesion. A frugal explanation hypothesizes that the activity of certain central neurons (possibly in several locations) represents a signal to the process underlying the conscious appreciation of pain. This activity normally is generated by ascending impulses conducted over the anterolateral spinal tracts. With pathological loss of some suppressive inhibitory connections, ongoing activity is conceivably enhanced in these same neurons and leads to the experience of pain independent of input from the periphery (Livingston, 1943).

XV. PAIN CONTROL AND TREATMENT

Therapy is usually ignored in histories of research on pain, but, the methods of dealing with clinical pain after the mid-19th century have had considerable conceptual impact. The relief of pain by the introduction of chemical anesthetics was of profound importance. In spite of controversies over priority in the discovery of ether and chloroform anesthesia in the mid-19th century (Davison, 1965), the production of loss of consciousness and all sensory experience by inhaling these gases had significant impact on ideas about pain mechanisms because of a peculiar dissociation. Snow (1847) noted that a critical level of ether inhalation resulted in retention of consciousness while eliminating the discomfort of pain. A noxious stimulus could be recognized but it was tolerable, a phenomenon that came to be known as "ether analgesia." This separation between perception and affect stresses the point made earlier about the differences among phenomena labeled by the inclusive word *pain*.

Another problematic example in the semantics of pain can be found in the action of opiates. Morphine and other opiates are commonly called analgesics, yet, although they may ameliorate clinical pain, they do not necessarily alter the threshold for detectability of noxious stimuli or even the recognition that a stimulus is painful (Beecher, 1959). Thus, tolerance for pain or for suprathreshold noxious stimuli may be achieved without eliminating nociception. Another dissociation between nociception and pain as a recognized experience was popularized by demonstration of methods known as *mesmerism,* now called hypnotism, by Franz Mesmer in the early 19th century (cf., Hilgard & Hilgard, 1975). Scientific exploration of hypnotism has not advanced significantly, however, it is apparent that *hypnotic analgesia* is not the same as *analgesia* produced by opiates. The lack of conceptual appreciation of the mechanisms linking nociception, tolerance, and suffering are certainly part of the reason that a parsimonious explanation is lacking for the range of therapeutic modalities known to be efficacious in pain control. Until there is better understanding of the neural basis of consciousness, perception, and emotion, the relation of the brain to pain will remain elusive. At present, we are limited to a modest understanding of nociception and glimpses into regions of the nervous systems that might be responsible for its transformation into the sensory experience we call *pain.*

Acknowledgments

Writing of this chapter was partially supported by grants from the National Institutes of Health NS-10321 and NS-5685.

References

Adrian, E. D. (1926). The impulses produced by sensory nerve endings. Part I. *Journal of Physiology* (London), *61,* 4–72.

Adrian, E. D. (1928). The discharge of impulses in motor nerve fibres. Part I. *Journal of Physiology* (London), *66,* 81–101.

Adrian, E. D. (1930). Impulses in sympathetic fibres and in slow afferent fibres. (Abstract). *Journal of Physiology* (London), *70,* xx–xxi.

Adrian, E. D., & Zotterman, Y. (1926a). The impulses produced by sensory nerve endings. Part III. Impulses set up by touch and pressure. *Journal of Physiology* (London), *61,* 465–483.

Adrian, E. D., & Zotterman, Y. (1926b). The impulses produced by sensory nerve-endings. Part 2. The response of a single end-organ. *Journal of Physiology* (London), *61.* 11–171.

Aguayo, A. J., and Cherunada, P. V. N., & Bray, G. M. (1971). Peripheral nerve abnormalities in the Riley-Day syndrome. *Archives of Neurology, 24,* 106–116.

Albe-Fessard, D., & Kruger, L. (1962). Duality of unit discharges from cat centrum medianum in response to natural and electrical stimulation. *Journal of Neurophysiology, 25,* 3–20.

Bayliss, W. M. (1901). On the origin from the spinal cord of the vasodilator fibres of the hind limb, and the nature of these fibres. *Journal of Physiology* (London), *26,* 173–209.

Bechterew, W. (1887). Ueber die hinteren Nervenwurzeln, ihre Endigung in der grauen Substanz des Rückenmarkes und ihre centrale Fortsetzung im letzteren. [On the posterior nerve roots their ending in the gray matter of the spinal cord and their central course in the latter]. *Archives of Anatomical Physiology and Anatomy Abstracts, 11,* 126–136.

Beecher, H. K. (1959). *Measurement of subjective responses: Quantitative effects of drugs.* London: Oxford University Press.

Bell, C. (1811). *Idea of a new anatomy of the brain submitted for the observations of his friends.* London: Strahan and Preston.

Bernstein, J. (1868). Uber den zeitlichen Verlauf der negativen Schwan Kung des Nervenstroms. [On the time course of negative variation of nerve currents.] *Archives für die gesamte physioloige des menschen und der tiere, 1,* 173.

Bessou, P., Burgess, P. R., Perl, E. R., & Taylor, C. B. (1971). Dynamic properties of mechanoreceptors with unmyelinated (C) fibers. *Journal of Neurophysiology, 34,* 116–131.

Bessou, P., & LaPorte, Y. (1961). Étude des recepteurs musculaires innervés par les fibres afférentes du groupe III (fibres myelinisées fines), chez le chat. [Study of the muscular receptors innervated by Group III afferent fibres (fine myelinated fibers) in the cat]. *Archives of Italian Biology, 99,* 293–321.

Bessou, P., & Perl, E. R. (1969). Response of cutaneous sensory units with unmyelinated fibers to noxious stimuli. *Journal of Neurophysiology, 32,* 1025–1043.

Biemond, A. (1956). The conduction of pain above the level of the thalamus opticus. *Archives of Neurology and Psychiatry, 75,* 231–244.

Bishop, G. H. (1946). Neural mechanisms of cutaneous sense. *Physiological Reviews, 26,* 77–102.

Bishop, G. H., & Landau, W. M. (1958). Evidence for a double peripheral pathway for pain. *Science, 128,* 712–713.

Blix, M. (1884). Experimentelle Beiträge zur Lösing der Frage über die specifische Energie der Hautnerven. [Experimental contribution to the solution of the question on the specific energies of cutaneous nerves]. *Zeitschrift für Biologie, 20,* 141–160.

Boring, E. G. (1942). *Sensation and perception in the history of experimental psychology.* New York: Appleton.

Brown-Séquard, C. E. (1860). *Course of lectures on the physiology and pathology of the central nervous system.* Philadelphia: Collins.

Bullitt, E., Stofer, W. D., Vierck, C. J., & Perl, E. R. (1988). Reorganization of primary afferent nerve terminals in the spinal dorsal horn of the primate caudal to anterolateral chordotomy. *Journal of Comparative Neurology, 270,* 549–558.

Burgess, P. R., & Perl, E. R. (1967). Myelinated afferent fibers responding specifically to noxious stimulation of the skin. *Journal of Physiology* (London), *190,* 541–562.

Bushnell, M. C., Craig, A. D., Reiman, E. M., Yun, L.-S., & Evans, A. (1995). Cerebral activation in the human brain by pain, temperature and an illusion of pain. *Society of Neuroscience Abstracts, 21,* 1637.

Casey, K. L. (1966). Unit analysis of nociceptive mechanisms in the thalamus of the awake squirrel monkey. *Journal of Neurophysiology, 29,* 727–750.

Cassinari, V., & Pagni, C. (1969). *Central pain. A neurosurgical survey.* Cambridge, MA: Harvard University Press.

Castiglioni, A. (1947). *A history of medicine.* New York: Knopf.

Christensen, B. N., & Perl, E. R. (1970). Spinal neurons specifically excited by noxious or thermal stimuli: Marginal zone of the dorsal horn. *Journal of Neurophysiology, 33,* 293–307.

Clark, D., Hughes, J., & Gasser, H. S. (1935). Afferent function in the group of nerve fibers of slowest conduction velocity. *American Journal of Physiology, 114,* 69–76.

Clifton, C. L., Coggeshall, R. E., Vance, W. H., & Willis, W. D. (1976). Receptive fields of unmyelinated ventral root afferent fibers in the cat. *Journal of Physiology* (London), *256,* 573–600.

Coggeshall, R. E. (1985). An overview of dorsal root axon branching and ventral root afferent fibers. *Neurology and Neurobiology, 14,* 105–110.

Coggeshall, R. E., Coulter, J. D., & Willis, W. D. (1974). Unmyelinated axons in the ventral roots of the cat lumbosacral enlargement. *Journal of Comparative Neurology, 153,* 39–58.

Coggeshall, R. E., & Ito, H. (1977). Sensory fibers in ventral roots L7 and S1 in the cat. *Journal of Physiology* (London), *267,* 215–235.

Coghill, R. C., Talbot, J. D., Evans, A. C., Meyer, E., Gjedde, A., Bushnell, M. C., & Duncan, G. H. (1994). Distributed processing of pain and vibration by the human brain. *Journal of Neuroscience, 14,* 4095–4108.

Collins, W. F., Nulsen, F. E., & Randt, C. T. (1960). Relation of peripheral nerve fiber size and sensation in man. *Archives of Neurology, 3,* 381–397.

Cooper, B. Y., Vierck, C. J., & Yeomans, D. C. (1986). Selective reduction of second pain sensations by system morphine in humans. *Pain, 24,* 93–116.

Craig, Jr., A. D., Linington, A. J., & Kniffki, K.-D. (1989). Cells of origin of spinothalamic tract projections to the medial and lateral thalamus in the cat. *Journal of Comparative Neurology, 289,* 568–585.

Craig, Jr., A. D., Bushnell, M. C., Zhang, E.-T., & Blomqvist, A. (1994). A thalamic nucleus specific for pain and temperature sensation. *Nature, 372,* 770–773.

Dallenbach, K. M. (1929). A bibliography of the attempts to identify the functional end-organs of cold and warmth. *American Journal of Psychology, 41,* 344.

Dallenbach, K. M. (1939). Pain: History and present status. *American Journal of Psychology, 52,* 331–347.

Davison, M. H. A. (1965). *The evolution of anesthesia.* Altrincham, UK: John Sheratt and Son.

Dearborn, G. P. N. (1932). A case of congenital general pure analgesia. *Journal of Nervous and Mental Diseases, 75,* 612–615.

Dejerine, J., & Roussy, G. (1906). Memoires originaux. [Original memories]. *Revue de Neurologie (Paris), 14,* 521–532.

Douglas, W. W., & Ritchie, J. M. (1957a). A technique for recording functional activity in specific groups of medullated and nonmedullated fibres in whole nerve trunks. *Journal of Physiology* (London), *138,* 19–30.

Douglas, W. W., & Ritchie, J. M. (1957b). On excitation of non-medullated afferent fibres in the vagus and aortic nerves by pharmacological agents. *Journal of Physiology* (London), *138,* 31–43.

Douglas, W. W., & Ritchie, J. M. (1959). The sensory functions of the non-myelinated afferent nerve fibers from the skin. In G. E. W. Wolstenholme & M. O'Conner (Eds.), *Pain and itch—Nervous mechanisms* (Ciba Foundation Study Group No. 1). (pp. 26–40). London: Churchill.

DuBois-Reymond, E. (1843). Vorläufiger Abriss einer Untersuchung über sogenannten Froschstrom und über die elektromotorischen Fische. [Preliminary report on the examination of the so-called frog current and the electric organ of fishes]. *Annalen der Physik und Physikalischen Chemie, 134,* 1–30.

Dyck, P. J. (1993). Neuronal atrophy and degeneration predominantly affecting peripheral sensory and autonomic neurons. In P. J. Dyck, P. K. Thomas, J. W. Griffin, P. A. Low, & J. F. Podulso (Eds.), *Peripheral neuropathy* (3rd ed.) (pp. 1065–1093). Philadelphia: W. B. Saunders Company.

Dyck, P. J., & Lambert, E. H. (1969). Dissociated sensation in amyloidosis. *Archives of Neurology, 20,* 490–507.

Dyck, P. J., Lambert, E. H., & O'Brien, P. (1976). Pain in peripheral neuropathy related to rate and kind of fiber degeneration. *Neurology, 28,* 466–471.

Eccles, R. M., & Lundberg, A. (1959). Supraspinal control of interneurons mediating spinal reflexes. *Journal of Physiology* (London), *147,* 565–584.

Edinger, L. (1892). *Zwolf Vorlesungen über den Bau der nervosen Centralorgane 3. Auflage.* [Twelve lectures on the structure of the central nervous organ. (3rd ed.)]. Leipzig: F. C. W. Vogel.

Erb, W. H. (1874). *Handbuch der Krankheiten des Nervensystems II.* [Handbook of diseases of the nervous system]. Leipzig: Davis.

Erlanger, J., & Gasser, H. S. (1930). The action potential in fibers of slow conduction in spinal roots and somatic nerves. *American Journal of Physiology, 92,* 43–82.

Fitzgerald, M., & Lynn, B. (1977). The sensitization of high threshold mechanoreceptors with myelinated axons by repeated heating. *Journal of Physiology (London), 365,* 549–563.

Foerster, O., & Gagel, O. (1932). Die Vorderseitenstrangdurchschneidung beim menschen. [Anterior cordotomy in man]. *Zentralblatt für die Gesamte Neurologie and Psychiatrie, 138,* 1–92.

Garrison, F. H. (1969). *History of neurology* (Revised and enlarged by C. McHenry, Jr.). Springfield, IL: Thomas.

Gasser, H. S. (1943). Pain-producing impulses in peripheral nerves. *Proceedings of the Association for Research in Nervous Mental Diseases, 23,* 44–62.

Gasser, H. S., & Erlanger, J. (1922). A study of the action currents of nerve with the cathode ray oscillograph. *American Journal of Physiology, 62,* 496–524.

Gasser, H. S., & Erlanger, J. (1927). The role played by the sizes of the constituent fibers of a nerve trunk in determining the form of its action potential wave. *American Journal of Physiology, 80,* 522–547.

Gasser, H. S., & Erlanger, J. (1929). The role of fiber size in the establishment of a nerve block by pressure or cocaine. *American Journal of Physiology, 88,* 581–591.

Gasser, H. S., Erlanger, J., & Bishop, G. H. (1924). The compound nature of the action current of nerve disclosed by the cathode ray oscilloscope. *American Journal of Physiology, 70,* 624–666.

Goldscheider, A. (1885). Neue Tatsachen über die Hautsinnesnerven. [New facts on the cutaneous sensory nerves]. *Archives of Anatomy and Physiology (Physiol. Suppl. Vol.),* 1–110.

Goldman-Rakic, P. S. (1984). Modular organization of prefrontal cortex. *Trends in Neuroscience, 7,* 419–429.

Gowers, W. R. (1886). *Manual of diseases of the nervous system* (1st ed.). London: Churchill.

Gracely, R. H., & Wolskee, P. J. (1983). Semantic functional measurement of pain: Integrating perception and language. *Pain, 15,* 389–398.

Habgood, J. S. (1950). Sensitization of sensory receptors in the frog's skin. *Journal of Physiology (London), 111,* 195–213.

Hagbarth, K.-E., & Kerr, D. I. B. (1954). Central influences on spinal afferent conduction. *Journal of Neurophysiology, 17,* 295–307.

Hagbarth, K.-E., & Vallbo, Å. B. (1967). Mechanoreceptor activity recorded percutaneously with semi-microelectrodes in human peripheral nerves. *Acta Physiologica Scandinavia, 69,* 121–122.

Halliday, A. M., & Logue, V. (1972). Painful sensations evoked by electrical stimulation in the thalamus. In G. G. Somjen (Ed.), *Neurophysiology studied in man* (pp. 221–230). Amsterdam: Excerpta Medica Foundation.

Hardy, J. D., Wolff, H. G., & Goodell, H. (1952). *Pain sensations and reactions.* New York: Haffner Publishing.

Hassler, R. (1958). Functional anatomy of the thalamus. *Annual Volume of Physiology and Experimental Medicine Science, India, 1,* 56–91.

Hassler, R. (1960). Die zentralen Systeme des Schmerzes. [The central system of pain]. *Acta Neurochirgirie, 8,* 353–423.

Head, H. (1920). *Studies in neurology.* London: Frowde, Hodder & Stoughton.

Head, H., & Holmes, G. (1911). Sensory disturbances from cerebral lesions. *Brain, 34,* 102–254.

Head, H., Rivers, W. H. R., & Sherren, J. (1905). The afferent nervous system from a new aspect. *Brain, 28,* 99–115.

Heinbecker, P. (1928/29). Properties of unmyelinated fibers of nerve. *Proceedings of the Society for Experimental Biology and Medicine, 26,* 349–351.

Heinbecker, P., Bishop, G. H., & O'Leary, J. (1933a). Nature and source of fibers contributing to the saphenous nerve of the cat. *American Journal of Physiology, 104,* 23–35.

Heinbecker, P., Bishop, G. H., & O'Leary, J. (1933b). Pain and touch fibers in peripheral nerves. *Archives of Neurology and Psychiatry, 29,* 771–789.

Hensel, H., Iggo, A., & Witt, I. (1960). A quantitative study of sensitive cutaneous thermo-receptors with C afferent fibres. *Journal of Physiology (London), 153,* 113–126.

Herrick, C. J., & Bishop, G. H. (1958). A comparative survey of the spinal lemniscus systems. In H. H. Jasper et al. (Eds.), *Reticular formation of the brain.* Henry Ford Hospital International Symposium (pp. 353–364). Boston: Little Brown.

Hilgard, E. R., & Hilgard, J. R. (1975). *Hypnosis in the relief of pain.* Los Altos, CA: William Kaufman.

Holton, F. A., & Holton, P. (1954). The capillary dilator substances in dry powders of spinal roots: A possible role of adenosine triphosphate in chemical transmission from nerve endings. *Journal of Physiology (London), 126,* 124–140.

Honda, C. N., Mense, S., & Perl, E. R. (1983). Neurons in ventrobasal region of cat thalamus selectively responsive to noxious mechanical stimulation. *Journal of Neurophysiology, 49,* 662–673.

Hunt, C. C., & McIntyre, A. K. (1960). An analysis of fiber diameter and receptor characteristics of myelinated cutaneous afferent fibers in cat. *Journal of Physiology* (London), *153,* 99–112.

Iggo, A. (1957a). Gastro-intestinal tension receptors with unmyelinated afferent fibres in the vagus of the cat. *Quarterly Journal of Experimental Physiology, 42,* 130–143.

Iggo, A. (1957b). Gastric mucosal chemoreceptors with vagal afferent fibres in the cat. *Quarterly Journal of Experimental Physiology, 42,* 398–409.

Iggo, A. (1958). The electrophysiological identification of single nerve fibres, with particular reference to the slowest-conducting vagal afferent fibres in the cat. *Journal of Physiology, 142,* 110–126.

Iggo, A. (1959). A single unit analysis of cutaneous receptors with C afferent fibers. In G. E. W. Wolstenholme & M. O'Conner (Eds.), *Pain and itch—Nervous mechanisms* (pp. 41–56). (Ciba Foundation Study Group No. 1). London: Churchill.

Iggo, A. (1960). Cutaneous mechanoreceptors with afferent C fibres. *Journal of Physiology (London), 152,* 337–353.

Iriuchijima, J., & Zotterman, Y. (1960). The specificity of afferent cutaneous C fibres in mammals. *Acta Physiologica Scandinavia, 49,* 267–278.

Jones, M. H. (1956). Second pain: Fact or artifact? *Science, 124,* 442–443.

Jones, M. W., Apkarian, A. V., Stevens, R. T., & Hodge, Jr., C. J. (1987). The spinothalamic tract: An examination of the cells of origin of the dorsolateral and ventral spinothalamic pathways in cats. *Journal of Comparative Neurology, 260,* 349–361.

Keele, K. D. (1957). *Anatomies of pain.* Oxford: Blackwell.

Kenshalo, Jr., D. R., Giesler, Jr., G. J., Leonard, R. B., & Willis, W. D. (1980). Responses of neurons in primate ventral posterior lateral nucleus to noxious stimuli. *Journal of Neurophysiology, 43,* 1594–1614.

Kenshalo, Jr., D. R., & Isensee, O. (1983). Effects of noxious stimuli on primate SI cortical neurons. In J. J. Bonica, U. Lindblom, & A. Iggo, (Eds.), *Advances in pain research and therapy* (Vol. 5, pp. 139–145). New York: Raven.

Konietzny, F., Perl, E. R., Trevino, D., Light, A., & Hensel, H. (1981). Sensory experiences in man evoked by intraneural electrical stimulation of intact cutaneous afferent fibers. *Experimental Brain Research, 42,* 219–222.

Kruger, L., & Halata, Z. (in press). Structure of nociceptive 'endings.' In F. Cervero & C. Belmonte (Eds.), *Neurobiology of nociceptors.* Oxford: Oxford University Press.

Kruger, L., & Kroin, S. A. (1978). A brief historical survey of concepts in pain research. In E. C. Carterette & M. P. Friedman (Eds.), *Handbook of perception* (Vol. VIB, pp. 159–179). New York: Academic Press.

Kruger, L., & Michel, F. (1962). Reinterpretation of the representation of pain based on physiological excitation of single neurons in the trigeminal sensory complex. *Experimental Neurology, 5*, 157–178.

Kruger, L., Silverman, J. D., Mantyh, P. W., Sternini, C., & Brecha, N. C. (1989). Peripheral patterns of calcitonin gene-related peptide (CGRP) general somatic sensory innervation: Cutaneous and deep terminations. *Journal of Comparative Neurology, 280*, 291–302.

Kumazawa, T., & Perl, E. R. (1977). Primate cutaneous sensory units with unmyelinated (C) afferent fibers. *Journal of Neurophysiology, 40*, 1325–1338.

Kumazawa, T., & Perl, E. R. (1978). Excitation of marginal and substantia gelatinosa neurons in the primate spinal cord: Indications of their place in dorsal horn functional organization. *Journal of Comparative Neurology, 177*, 417–434.

Kumazawa, T., Perl, E. R., Burgess, P. R., & Whitehorn, P. (1975). Ascending projections from marginal zone (lamina I) neurons of the spinal dorsal horn. *Journal of Comparative Neurology, 162*, 1–12.

Kuno, M., & Perl, E. R. (1960). Alteration of spinal reflexes by interaction with suprasegmental and dorsal root activity. *Journal of Physiology* (London), *151*, 103–122.

Kuru, M. (1949). *Sensory paths in the spinal cord and brain stem of man.* Tokyo: Sogensya.

Landau, W., & Bishop, G. H. (1953). Pain from dermal, periosteal, and fascial endings and from inflammation. *Archives of Neurology and Psychology, 69*, 490–504.

Lele, P. P., & Weddell, G. (1956). The relationship between neurohistology and corneal sensibility. *Brain, 79*, 119–154.

Lembeck, F. (1983). Sir Thomas Lewis's nocifensor system, histamine and substance P-containing primary afferent nerves. *Trends in Neuroscience, 6*, 106–108.

Lenz, F. A., Seike, M., Richardson, R. T., Lin, Y. C., Baker, F. H., Khoja, I., Yeager, C. J., & Gracely, R. H. (1993). Thermal and pain sensations evoked by microstimulation in the area of human ventrocaudal nucleus (Vc). *Journal of Neurophysiology, 70*, 200–212.

Leriche, R. (1916). De la causalgie. [On causalgia]. *La Presse Medicale, 23*, 177–180.

Lewin, W., & Phillips, C. G. (1952). Observations on partial removal of the post-central gyrus for pain. *Journal of Neurology, Neurosurgery, and Psychiatry, 15*, 143–147.

Lewis, T. (1942). *Pain.* New York: Macmillan.

Lewis, T., & Pochin, E. E. (1937). The double pain response of the human skin to single stimulus. *Clinical Science, 3*, 67–76.

Lewis, T., Pickering, G. W., & Rothschild, P. (1931). Centripetal paralysis arising out of arrested bloodflow to the limb, including notes on a form of tingling. *Heart, 16*, 1–32.

Light, A. R. (1992). *The initial processing of pain and its descending control: Spinal and trigeminal systems.* Basel, Switzerland: S. Karger.

Light, A. R., & Metz, C. B. (1978). The morphology of the spinal cord efferent and afferent neurons contributing to the ventral roots of the cat. *Journal of Comparative Neurology, 179*, 501–516.

Light, A. R., Trevino, D. L., & Perl, E. R. (1979). Morphological features of functionally defined neurons in the marginal zone and substantia gelatinosa of the spinal dorsal horn. *Journal of Comparative Neurology, 186*, 151–171.

Lissaur, H. (1885). Beitrag zur pathologischen Anatomie der Tabes dorsalis und zum Faserverlauf im menschlichen Rückenmark. [Contribution to the pathological anatomy of tabes dorsalis and on the fiber course in the human spinal cord.] *Neurologisches Zentralblatt, 4*, 245.

Livingston, W. K. (1943). *Pain mechanisms. A physiologic interpretation of causalgia and its related states.* New York: Macmillan

Lundberg, A., & Oscarsson, O. (1961). Three ascending spinal pathways in the dorsal part of the lateral funiculus. *Acta Physiologica Scandinavia, 51*, 1–16.

Lynn, B., & Carpenter, S. E. (1982). Primary afferent units from the hairy skin of the rat hind limb. *Brain Research, 238*, 29–43.

Magendie, F. (1822). Expériences sur les fonctions des racines des nerfs rachidiens. [Experiments on the function of the spinal nerve roots.] *Journal of Physiology and Experimental Pathology, 2,* 276–279, 366–71.

Magoun, H. W. (1959). *The working brain.* Springfield, IL: Thomas.

Marshall, J. (1951). Sensory disturbances in cortical wounds with special reference to pain. *Journal of Neurology, Neurosurgery and Psychiatrist, 14,* 187–204.

Maruhashi, J., Mizuguchi, K., & Tasaki, I. (1952). Action currents in single afferent nerve fibres elicited by stimulation of the skin of the toad and the cat. *Journal of Physiology, 117,* 129–151.

Mayer, D. J., Price, D. D., & Becker, D. P. (1975). Neurophysiological characterization of the anterolateral spinal cord neurons contributing to pain perception in man. *Pain, 1,* 51–58.

Mehler, W. R., Feferman, M. E., & Nauta, W. J. H. (1960). Ascending axon degeneration following anterolateral cordotomy. An experimental study in the monkey. *Brain, 83,* 718–750.

Melzack, R., & Scott, T. H. (1957). The effects of early experience on the response to pain. *Journal of Comparative Physiology and Psychology, 50,* 155–161.

Melzack, R., & Wall, P. D. (1962). On the nature of cutaneous sensory mechanisms. *Brain, 85,* 331–356.

Melzack, R., & Wall, P. D. (1965). Pain mechanisms: A new theory. *Science, 150,* 971–979.

Melzack, R., & Wall, P. D. (1970). Psychophysiology of pain. *International Anesthesiology Clinics, 8,* 3–34.

Mendell, L. M. (1966). Physiological properties of unmyelinated fiber projection to the spinal cord. *Experimental Neurology, 16,* 316–332.

Mendell, L. M., & Wall, P. D. (1964). Presynaptic hyperpolarization: A role for fine afferent fibers. *Journal of Physiology, 172,* 274–294.

Mitchell, S. W. (1872). *Injuries of nerves and their consequences.* Philadelphia: Lipincott.

Morruzzi, G., & Magoun, H. W. (1949). Brain stem reticular formation and activation of the EEG. *Electroencephalography and Clinical Neurophysiology, 1,* 455–473.

Mosso, J. A., & Kruger, L. (1972). Spinal trigeminal neurons excited by noxious and thermal stimuli. *Brain Research, 38,* 206–210.

Müller, J. (1842). *Elements of physiology* (W. Baly, Trans.). London: Taylor & Walton.

Nafe, J. P. (1934). The pressure, pain, and temperature senses. In C. Murchison (Ed.), *Handbook of general experimental psychology.* Worchester, PA: Clark Univ. Press.

Nissen, H. W., Chow, K. L., & Semmes, J. (1951). Effects of restricted opportunity for tactile, kinesthetic, and manipulative experience on the behavior of a chimpanzee. *American Journal of Physiology, 164,* 485–507.

Noordenbos, W. (1959). *Pain.* Amsterdam: Elsevier.

Ochoa, J. (1984). Peripheral unmyelinated units in men: Structure, function, disorder, and role in sensation. In L. Kruger & J. C. Liebeskind (Eds.), *Advances in Pain Research and Therapy,* (Vol. 6, pp. 53–68). New York: Raven Press.

Ochoa, J. L., & Torebjörk, E. (1989). Sensations evoked by intraneural microstimulation of C nociceptor fibers in human skin nerves. *Journal of Physiology* (London), *415,* 583–599.

Paintal, A. S. (1960). Functional analysis of Group III afferent fibres of mammalian muscles. *Journal of Physiology* (London), *152,* 250–270.

Ochoa, J., Torebjörk, H. E., Culp, W. J., & Schady, W. (1982). Abnormal spontaneous activity in single sensory nerve fibers in humans. *Muscle Nerve, 5,* S74–S77.

Penfield, W., & Boldrey, E. (1937). Somatic motor and sensory representation in the cerebral cortex of man as studied by electrical stimulation. *Brain, 60,* 389–443.

Perl, E. R. (1968). Myelinated afferent fibers innervating the primate skin and their response to noxious stimuli. *Journal of Physiology* (London), *197,* 593–615.

Perl, E. R. (1984a). Why are selectively responsive and ultireceptive neurons both present in somatosensory pathways? In C. von Euler, O. Franzén, U. Lindblom, & D. Ottoson (Eds.), *Somatosensory mechanisms* (Vol. 41, pp. 141–161). Wenner-Gren International Symposium Series. London: Plenum Press.

Perl, E. R. (1984b). Pain and nociception. In J. Darian-Smith (Ed.), *Handbook of physiology. The nervous system* (Vol. 3, pp. 915–975). Bethesda, MD: American Physiological Society.

Perl, E. R. (1993). Multireceptive neurons and mechanical allodynia. *American Pain Society Journal, 2*, 37–41.

Perl, E. R. (1994). Causalgia and reflex sympathetic dystrophy revisited. In J. Boivie, P. Hansson, & U. Lindblom (Eds.), *Touch, temperature, and pain health and disease: Mechanisms and assessments, progress in pain research management* (Vol. 3, pp. 231–248). Seattle: IASP Press.

Perl, E. R. (in press). Pain and the discovery of nociceptors. In F. Cervero & C. Belmonte (Eds.), *Neurobiology of nociceptors*. Oxford: Oxford University Press.

Perl, E. R., & Whitlock, D. G. (1961). Somatic stimuli exciting spinothalamic projections to thalamic neurons in cat and monkey. *Experimental Neurology, 3*, 256–296.

Poggio, G. F., & Mountcastle, V. B. (1960). A study of the functional contributions of the lemniscal and spinothalamic systems to somatic sensibility: Central nervous mechanisms in pain. *Bulletin of Johns Hopkins Hospital, 106*, 266–316.

Price, D. D., & Mayer, D. J. (1975). Neurophysiological characterization of the anterolateral quadrant neurons subserving pain in *M. mulatta*. *Pain, 1*, 59–72.

Procacci, P., & Maresca, M. (1984). Pain concept in western civilization: A historical review. *Advances in Pain Research and Therapy, 7*, 1–11.

Ranson, S. W. (1914). The tract of Lissauer and the substantia gelatinosa Rolandi. *American Journal of Anatomy, 16*, 97–126.

Ranson, S. W. (1915). Unmyelinated nerve-fibers as conductors of protopathic sensation. *Brain, 38*, 381–389.

Ranson, S. W., & Billinglsey, P. E. (1916). The conduction of painful afferent impulses in the spinal nerves. Studies in vasomotor reflex arcs. II. *American Journal of Physiology, 40*, 571–584.

Réthelyi, M., Light, A. R., & Perl, E. R. (1989). Synaptic contacts on physiologically defined neurons of the superficial dorsal horn. *Advances in Pain Research and Therapy, 9*, 139–147.

Risling, M., & Hildebrand, C. (1982). Occurrence of unmyelinated axon profiles at distal, middle and proximal levels in the ventral root L7 of cats and kittens. *Journal of Neurological Science, 56*, 219–231.

Rose, J. E., & Mountcastle, V. B. (1959). Touch and kinesthesis. In J. Field, H. Magoun, & V. Hall (Eds.), *Handbook of physiology: Section I. Neurophysiology* (Vol. 1, pp. 387–429). Washington, DC: American Physiology Society.

Russell, W. R. (1945). Transient disturbances following gunshot wounds of the head. *Brain, 68*, 79–97.

Sato, J., & Perl, E. R. (1991). Adrenergic excitation of cutaneous pain receptors induced by peripheral nerve injury. *Science, 251*, 1608–1610.

Schiff, M. (1858). *Lehrbuch der Physiologie des Menschen*. [Textbook of human physiology]. Lahr: Schauenburg.

Sherrington, C. S. (1894). On the anatomical constitution of nerves. *Journal of Physiology* (London), *17*, 211–258.

Sherrington, C. S. (1900). Cutaneous sensations. In E. A. Sharpey-Schäfer (Ed.), *Textbook of physiology* (Vol. II: pp. 920–1001). Edinburgh: Y. J. Pentland.

Sherrington, C. S. (1906). *The integrative action of the nervous system*. New Haven, CT: Yale University Press.

Sigerist, H. E. (1951). *A history of medicine: Primitive and archaic medicine* (Vol. 1). London: Oxford University Press.

Sinclair, D. C. (1955). Cutaneous sensation and the doctrine of specific energy. *Brain, 78*, 584–614.

Sinclair, D. C. (1967). *Cutaneous sensation*. London: Oxford University Press.

Snow, J. (1847). *On the inhalation of the vapour of ether*. London: Churchill.

Soury, J. (1899). *Le système nerveux centrale. Structure et fonctions: histoire critique des theories et des doctrines*. [The central nervous system: Structure and functions; a historical critique of theories and doctrines]. Paris: Carré et Naud.

Stevens, R. T., Hodge, Jr., C. J., & Apkarian, A. V. (1989). Medial, intralaminar, and lateral terminations of lumbar spinothalamic tract neurons: A fluorescent double-label study. *Somatosensory and Motor Research, 6*, 285–308.

Swanson, A. G., Buchan, G., & Alvord, E. C. (1965). Anatomic changes in congenital insensitivity to pain. *Archives of Neurology, 12*, 12–18.

Sweet, W. H. (1959). Pain. In J. Field, H. Magoun, & V. Hall (Eds.), *Handbook of physiology: Section I. Neurophysiology* (Vol. 1, pp. 459–506). Washington, DC: American Physiological Society.

Thomas, P. K. (1974). The anatomical substratum of pain. Evidence derived from morphometric studies on peripheral nerve. *Canadian Journal of Neurological Science, 1*, 92–97.

Thomas, P. K. (1979). Painful neuropathies. *Advances in Pain Research and Therapy, 3*, 103–110.

Torebjörk, H. E., & Hallin, R. G. (1970). C-fibre units recorded from human sensory nerve fascicles *in situ. Acta Society Medicus Upsalensis, 75*, 81–84.

Torebjörk, H. E., & Ochoa, J. (1980). Specific sensations evoked by activity in single identified sensory units in man. *Acta Physiologica Scandinavia, 110*, 445–447.

Torebjörk, H. E., Ochoa, J., & McCann, F. V. (1979). Paresthesiae: Abnormal impulse generation in sensory nerve fibres in man. *Acta Physiologica Scandinavia, 105*, 518–520.

Torebjörk, E., Vallbo, Å. B., & Ochoa, J. L. (1987). Intraneural microstimulation in man. Its relation to specificity to tactile sensations. *Brain, 110*, 1509–1529.

Tower, S. S. (1935). Nerve impulses from receptors in the cornea. *Proceedings of the Society for Experimental Biology (N.Y.), 32*, 590–592.

Tower, S. S. (1940). Units for sensory reception in cornea; with notes on nerve impulses from sclera, iris and lens. *Journal of Neurophysiology, 3*, 486–500.

Tower, S. S. (1943). Pain: Definition and properties of the unit for sensory reception. *Proceedings of the Association for Research in Nervous Mental Diseases, 23*, 16–43.

Treede, R.-D., & Magerl, W. (1995). Modern concepts of pain and hyperalgesia: Beyond the polymodal C-nociceptor. *News in Physiological Sciences, 10*, 216–228.

Trevino, D. L., & Carstens, E. (1975). Confirmation of the location of spinothalamic neurons in the cat and monkey by the retrograde transport of horseradish peroxidase. *Brain Research, 98*, 177–182.

Trevino, D. L., Coulter, J. D., & Willis, W. D. (1973). Location of cells of origin of spinothalamic tract in lumbar enlargement of the monkey. *Journal of Neurophysiology, 36*, 750–761.

Vallbo, Å., Olausson, H., Wessberg, J., & Norrsell, U. (1993). A system of unmyelinated afferents for innocuous mechanoreception in the human skin. *Brain Research, 628*, 301–304.

von Frey, M. (1894). Beiträge zur Physiologie des Schmerzsinns. [Contribution to the physiology of pain sensibility]. *Berichte über die Verhandlung der Königlich Sächsischen Gessellschaft der Wissenschaften zu Leipzig. Mathematisch-Physische Classe, 46*, 185–196.

von Frey, M. (1897). Untersuchengen über die sinnesfunctionen der menschlichen haut. Erste Abhandlung: Druckempfindung und Schmerz. [Investigations on the sensory function of human skin. Part I. Pressure sensation and pain]. *Abhandlungen der Mathematische-Physischen Classe der Koniglich Sächsischen Gesellschaft der Wissenschaften, 48*, 175–266.

von Hees, J., & Gybels, J. M. (1972). Pain related to single afferent C fibres from human skin. *Brain, Research, 48,* 397–400.

Walker, A. E. (1951). *A history of neurological surgery.* Baltimore: Williams & Wilkins.

Wall, P. D., & MacMahon, A. (1985). Microneuronography and its relation to perceived sensation: A critical review. *Pain, 21,* 209–229.

Wall, P. D., & Taub, A. (1962). four aspects of trigeminal nucleus and a paradox. *Journal of Neurophysiology, 25,* 110–126.

Weber, E. H. (1846). Der Tastsinn und das Gemeingefühl. [The sense of touch and the sense of the body]. In R. Wagner (Ed.), *Handwortesbuch der Physiologie,* Vol. III. Braunschiveig.

Weddell, G. (1955). Somesthesis and the chemical senses. *Annual Review of Psychology, 6,* 119–136.

Weddell, G., Palmer, E., & Pallic, W. (1955). Nerve endings in mammalian skin. *Biological Reviews, 30,* 159–195.

White, J. C., & Sweet, W. H. (1955). *Pain: Its mechanisms and neurosurgical control.* Springfield, IL: Thomas.

Whitehorn, D., & Burgess, P. R. (1973). Changes in polarization of central branches of myelinated mechanoreceptor and nociceptor fibers during noxious and innocuous stimulation of the skin. *Journal of Neurophysiology, 36,* 226–237.

Willis, Jr., W. D. (1993). Mechanical allodynia. A role for sensitized nociceptive tract cells with convergent input from mechanoreceptors and nociceptors? *American Pain Society Journal, 2,* 23–33.

Willis, W. D., & Coggeshall, R. E. (1978). *Sensory mechanisms of the spinal cord.* New York: Plenum Press.

Willis, W. D., Kenshalo, D. R., & Leonard, R. B. (1979). The cells of origin of the primate spinothalamic tract. *Journal of Comparative Neurology, 188,* 543–574.

Witt, I. (1962). Aktivität einzelner C-Fascrn bei schmerzhaften und nicht schmerzhaften Hautreizen. [Activation of individual C-fibers by means of painful and non-painful skin stimulations]. *Acta Neurovegetativa (Wien), 24,* 208–219.

Witt, I., & Griffin, J. P. (1962). Afferent cutaneous C-fibre reactivity to repeated thermal stimuli. *Nature, 194,* 776.

Yeomans, D. C., Cooper, B. Y., and Vierck, C. J., Jr. (in press). Effects of systemic morphine on responses to first or second pain sensations of primates.

Zotterman, Y. (1933). Studies in the peripheral nervous mechanism of pain. *Acta Medicus Scandinavia, 80,* 185–242.

Zotterman, Y. (1936). Specific action potential sin the lingual nerve of cat. *Scandinavian Archives, 75,* 105–120.

Zotterman, Y. (1939). Touch, pain and tickling: an electriphysiological investigation on cutaneous sensory nerves. *Journal of Physiology* (London), *95,* 1–28.

CHAPTER 5

Afferent Mechanisms of Pain

Bruce Lynn
Edward R. Perl

I. INTRODUCTION

Most pain of peripheral origin is associated with strong stimulation, with tissue injury, or with inflammation. This chapter surveys what is known about the afferent mechanisms underlying such pain. The basic thesis is that specialized nociceptors are not only the principal, but in fact for most tissues, the sole primary afferent neurones that normally encode stimuli causing pain. This general finding does not, of course, mean that activation of nociceptors always evokes pain. In fact, the evidence is that very low-frequency activation of the most common cutaneous nociceptors is largely ignored under normal circumstances. As in all sensory pathways, signals from the primary afferent neurones are subject to extensive processing before reaching higher centers and before conscious sensations occur. Even simple segmental reflexes such as the flexion reflex are under strong central control. The final pain sensation is the product of both afferent signaling and central processing. We consider here the peripheral afferent part of the pain story. Several recent reviews deal extensively with central processing related to nociception and pain: Craig (1994), Guilbaud, Bernard, and Besson (1994), Light (1992), Woolf (1994). Chapter 1 of this volume provides an overview of concepts on central mechanisms related to pain.

II. FIBER SIZE, NERVE BLOCK, AND PAIN

Even before extensive recordings of the activity from single primary afferent fibers were available, there was a body of data indicating that peripheral signaling relative to pain was largely the preserve of small-calliber afferent fibers. These data came from human and animal experiments using selective stimulation and/or differential blocking of nerve trunks. It was known that weak electrical stimulation of nerve trunks in animals activated only large myelinated, Aαβ fibers. Similar stimulation in humans did not evoke pain and in animals did not initiate aversive reactions. On the other hand, stimulation strong enough to excite small myelinated (Aδ) and unmyelinated (C) fibers caused pain in humans and resulted in aversive reactions in animals. During surgical operations, a small number of direct, simultaneous comparisons were made between sensations produced by electrical stimulation of human nerve trunks and nerve fiber groups excited (determined by recording of compound action potentials). The results confirmed the deduction from the earlier studies that Aαβ fibers were not involved in signaling about pain but Aδ and C-fibers were (Collins et al., 1960).

The development of microneurography, in which unitary signals are recorded from nerves in awake human subjects, has allowed more extensive studies comparing the sensations evoked by stimulation of different fiber groups. It has proved possible to collect reports of sensations that accompany stimulation of successively more slowly conducting fibers while monitoring nerve conduction with a sensitivity that can detect unitary activity. The results provide further confirmation that Aαβ fibers are not involved in nociception, whereas Aδ and C-fibers are. Even high-frequency stimulation of fibers conducting at >30 m/sec does not produce the experience of pain, whereas single stimuli to fibers with conduction velocities <20 m/sec evoke pain.

A second line of evidence comes from studies using differential nerve block. As reviewed by Raymond and Gissen (1987), many investigations have shown that local anesthetics produce an early block of the thin fibers (C and Aδ) followed by a later block of large-diameter (Aαβ) fibers. Compression of nerve gives the opposite sequence with conduction in large-diameter fibers blocking first and thinner fibers later. Human studies using similar methods have found that local anesthetics blocked pain before touch, whereas compression blocked touch before pain. This result is entirely consistent with the conclusion from graded nerve stimulation that small-diameter fibers are involved in signaling pain, but large fibers are not. Microneurographic recordings have also enabled direct comparisons to be made of blockage of particular fiber groups and sensory loss in humans. The results again confirm earlier conclusions that in order to prevent pain sensations, both Aδ and C-fibers need to be blocked. Microneurography studies also show that when only C-fibers are conducting, noxious stimuli, whether mechanical (pinprick), thermal (heat), or chemical (capsaicin), are still painful (Hallin & Torebjörk & Hallin, 1973; Torebjörk, Lundberg, and LaMotte, 1992).

III. SINGLE-UNIT STUDIES

A. Criteria for Classification of Nociceptors

1. Nociceptors versus Receptors for Innocuous Stimuli

The key criterion defining a nociceptor is clearly the ability to encode preferentially information about noxious stimuli. It is not enough for a sensory unit to respond to noxious events, the responses must be distinct from those evoked by innocuous stimuli. Thus, many low-threshold cutaneous mechanoreceptors fire to strong, noxious pressure, but do so in a way identical to their responses to light, innocuous pressure (e.g., Bessou, Burgess, Perl, & Taylor, 1971; Perl, 1968). The difference between the input–output relation of typical mechanoreceptors and nociceptors is shown in the upper part of Figure 1. In somatic tissues (e.g., skin, muscle) a clear distinction is possible between afferents that are mechanoreceptive and others that are obviously nociceptive. It should be noted that many nociceptors do have thresholds towards the upper end of the innocuous range, but nevertheless encode mostly in noxious range.

In visceral tissues the situation is not as clear. There can be a problem in deciding the nature of an appropriate test stimulus for visceral nociceptors. Many tissue-damaging stimuli to visceral structures are not reported as painful by human subjects (Lewis, 1942). Therefore, the link between nociception and pain in viscera may be less direct than for skin and other somatic tissues. Another difficult question is the status of afferent units whose input–output relation appears to span a significant part both of the innocuous and the noxious intensity ranges (Fig. 1, lower part). There may also be legitimate questions about the fitness of the stimuli used. Although it is argued that some visceral afferent fibers have the possibility of both contributing to physiological regulation within the innocuous range and to nociceptive reactions, including pain (Cervero & Jänig, 1992), more information is needed about the natural and pathological circumstances under which such afferent elements are activated. This matter is considered again in the section on visceral nociceptors.

Low resolution is one criterion that clearly appears inappropriate for the classification of nociceptors. The use of this property derives from the idea that nociceptors are "primitive." In fact, nociceptors turn out to be just as specialized as other afferent units, and their coding when tested by stimuli

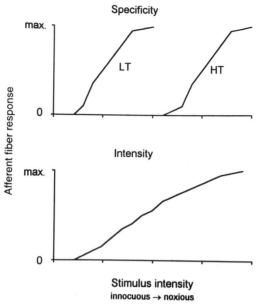

FIGURE 1   Schematic stimulus–response functions portraying two classical theories for the encoding of noxious stimuli by peripheral afferent fibers. The specificity theory postulates the existence of two different classes of primary afferent fibers: low threshold (LT) and high threshold (HT) responsible for the encoding of innocuous and noxious events respectively. The intensity theory proposes that all stimuli are encoded by a single homogeneous population of afferent fibers with a wide-dynamic range covering innocuous and noxious stimulus levels. (From Cervero & Jänig, 1992. Reprinted with permission of Elsevier Trends Journals.)

adequate for them can be as precise as by mechanoreceptors or thermo-receptors. Note also that several kinds of nociceptors can show high sensitivity to certain endogenous chemical agents (e.g., bradykinin).

3. Classes of Nociceptor

Many different forms of stimulation are painful, notably high mechanical stresses, extremes of heat or cold, and the presence of certain chemical agents. Few nociceptors have been tested with all these stimuli, and of those that have been tested, rather few were excited by all. However, one large subset of units do respond to heating, to strong pressure, and to some irritant chemicals, and these have been designated as polymodal nociceptors (Bessou & Perl, 1969). Another commonly observed class respond to strong mechanical stimulation, but not (or only very poorly) to heat, cold, or chemical irritants and are labeled mechanical nociceptors (also sometimes called high-threshold mechanoreceptors or HTM). Other types of nociceptors have also been reported, and will be mentioned in the following sections. In evaluating classifications of nociceptors it is important to remember that numbers in subclasses often depend on range of stimuli used for testing. For example, the relative numbers of mechanical (only) nociceptors and of mechanical-plus-heat-responsive units will vary with the size of heat test stimulus. In addition, the past history of stimulation may affect the classification, because, for example, heat sensitization can convert mechanical nociceptors into mechano-heat-responsive units (Fitzgerald & Lynn, 1977).

Currently there is substantial interest in the question of whether nociceptors exist that only become responsive after activation by chemical agents (McMahon & Koltzenburg, 1990). Physical stimuli such as heating or pressure fail to excite some dorsal root fibers, and the proportions of such "silent" or "sleeping" afferents can be quite high in deep tissues (e.g., urinary bladder: Habler, Jänig, & Koltzenburg, 1990; joints: Schaible & Schmidt, 1985). In searching for specifically chemosensitive nociceptors it is difficult to deliver chemical stimuli repeatedly and reliably. For skin, an attempt has been made to get around this problem by using electrical search stimuli. A small conducting probe pressed on the skin surface proves to be an efficient method for exciting all classes of cutaneous sensory fibers. Using this approach, the proportion of C-fiber afferents reported in current work to have negligible sensitivity to skin heating, cooling, and pressure is 10–20% in humans, monkey, and rat (Heppelmann, Messlinger, Neiss, & Schmidt, 1995; Kress, Koltzenburg, Reeh, & Handwerker, 1992; Meyer, Cohen, Davis, Treede, & Campbell, 1991). In interpreting these figures it is important to remember that nerves primarily innervating skin also contain small but significant numbers of fibers with terminals in subcutaneous structures (e.g., veins: Michaelis, Goder, Habler, & Jänig, 1994) as well as postganglionic sympathetic fibers.

B. Cutaneous Nociceptors

1. Myelinated Fiber Nociceptors

a. Mechanical Nociceptors

Cutaneous receptors responding preferentially to strong or overtly noxious pressure have been established to exist in all mammalian species and all skin areas so far examined (humans: Adriaensen, Gybels, & Handwerker, 1983; cat: Burgess & Perl, 1967; rabbit: Fitzgerald & Lynn, 1977; rat: Lynn & Carpenter, 1982; monkey: Perl, 1968) as well as in cold-blooded vertebrates (Liang & Terashima, 1993). Their discharge is relatively slowly adapting, and in hairy skin on the limbs the large receptive fields consist of many distinct points (Fig. 2). On the glabrous skin of hands or feet (Lynn &

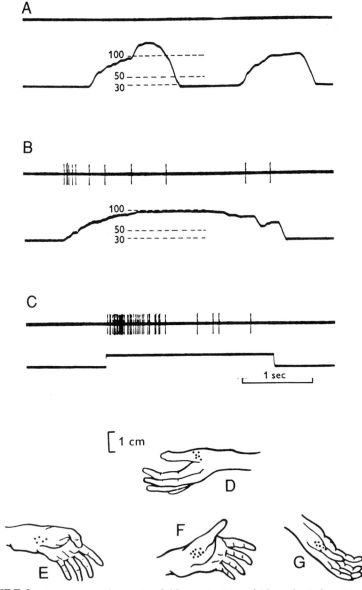

FIGURE 2 Responses and receptive field organizations of Aδ mechanical nociceptors of monkey glabrous skin. (A) Pressure with blunt probe (2.2-mm diameter) to receptive area on glabrous skin. Measured force is shown in grams. (B) Pressure using needle tip probe. (C) Pinch of receptive field (lower beam) with toothed forceps. (D–G): Receptive fields of individual Aδ mechanical nociceptors on monkey hand. Each dot represents a point from which responses could be evoked. Equivalent stimuli between the marked points were ineffectual. (Adapted with permission from Figs. 1 and 3 in Perl, 1968.)

Shakhanbeh, 1988) and on the skin of the face (Dubner, Sumino, & Stark-man, 1973; Hu & Sessle, 1988), receptive fields may be small and comprise just one zone. Conduction velocities cover most of the myelinated fiber range. Values as low as 2–3 m/sec, almost in the C-fiber range, have been reported (e.g., Adriaensen et al., 1983; Handwerker, Anton, & Reeh, 1987), whereas the highest conduction velocity reported is 65 m/sec (Burgess, Petit, & Warren, 1968). The majority of mechanical nociceptors, however, conduct within the Aδ range (i.e., at 5–25 m/sec). Pressure sensitivity also varies widely across this population, with pressure thresholds using small calibrated bristles varying from 0.1 g to over 50 g. Although subgroupings have been designated on the basis of pressure thresholds (Burgess & Perl, 1967), these appear to represent ranges within a single population rather than distinctive subclasses. An unusual feature of up to 20% of A-mechanical nociceptors in both cat and monkey is the presence of a subcutaneous and a cutaneous receptive field (Light & Perl, 1983, Mense, Light, & Perl, 1981).

A-fiber mechanical nociceptors are the only cutaneous nociceptor units for which information is available about the morphology of the terminals. Experiments marking receptive points for cat hairy skin units by Kruger, Perl, and Sedivic (1981) found the consistent presence of intraepidermal nerve profiles in which the terminals remained enclosed by a Schwann cell process and its basal lamina and sometimes contained clusters of vesicles.

b. Heat or Chemically Sensitive A-Fiber Nociceptors

Many afferent C-fibers are excited by skin heating. On the other hand, reaction times to noxious heat are too fast, for example in humans 0.45 s from the hand, to involve C-fiber conduction (Campbell & LaMotte, 1983; Lewis & Pochin, 1937; Price, Hu, Dubner, & Gracely, 1977). Some A-mechanical nociceptors do give delayed responses to heating, but thresholds are high and the latencies long in undamaged skin (although they may become lower with sensitization, see later). In primates there does appear to be a distinct population of A-fiber nociceptors that respond vigorously to skin heating, as well as to noxious pressure (Dubner, Price, Beitel, & Hu, 1977). Adriaensen et al. (1983) also describe mechano-heat nociceptors with Aδ axons innervating human skin. Aδ units in primate skin with a short latency response to heating to 40–50°C have been designated Type II A-MH (A-mechano-heat) units by Meyer and Campbell (1986). Other Aδ units excited by heat after a long latency and usually with thresholds >50°C have been labeled Type I A-MH. Type II A-MH units are plentiful in hairy skin, but are reported to be missing in glabrous skin (Campbell & Meyer, 1986). Occasional reports of mechano-heat A-fiber nociceptors in subprimates have also appeared (e.g. cat: Beck, Handwerker, & Zimmerman, 1974; Roberts & Elardo, 1985; rat: Szolcsányi, Anton, Reeh, & Handwerker,

1988), although other studies have not uncovered such units (e.g., cat: Burgess & Perl, 1967; rat: Lynn & Shakhanbeh, 1988) or noted only a very few (cat, rabbit: Fitzgerald & Lynn, 1977).

A number of reports indicate that at least some A-fiber nociceptors are excited by irritant chemicals. This is not an effect on A-mechanical nociceptor units of hairy skin of rat or rabbit (Szolcsányi, 1987; Szolcsányi et al., 1988). On the other hand, some A-mechano-heat units also respond to algesic chemicals (i.e., show polymodal responsiveness more typical of C-fiber nociceptors) (see below). For example, many A-fiber nociceptors from the cat paw fire to intra-arterial 5HT or bradykinin (Beck & Handwerker, 1974). In human nerves, a majority of A-nociceptors were excited by the cutaneous application of methylene chloride (Adriaensen et al., 1983). Whether this result indicates widespread chemosensitivity among human A-nociceptors or is the consequence of direct tissue injury will only become clear when studies with other, less severe, chemical stimuli are carried out.

2. C-Fiber Nociceptors

a. Polymodal Nociceptors

Polymodal nociceptors comprise the major class of C-fiber afferents from mammalian skin. They respond to firm pressure, to heating, and to irritant chemicals (Fig. 3). Such units have been described in humans, monkeys, and many subprimate species. They are found in hairy and glabrous skin (e.g., Beitel & Dubner, 1976; Bessou & Perl, 1969; Lynn & Carpenter, 1982; Torebjörk, 1974, van Hees & Gybels, 1972). Quantitative properties vary markedly from unit to unit and between species and skin areas. Typically, conduction velocities average 0.8–0.9 m/sec in larger mammals, punctate pressure thresholds are reported to be 0.5–5 g, and heat thresholds 40–55°C. No obvious subclasses appear for these measures (Lynn & Carpenter, 1982). A subgroup of polymodal nociceptors have been reported to discharge in response to extreme cold (e.g., Georgopoulos, 1976); however, many of the C-polymodal units develop an excitatory response to cooling after a prior noxious heat exposure (Bessou & Perl, 1969). Receptive fields increase with body size, but usually consist of a single zone with clear borders. In this respect they differ markedly from the A-mechanical nociceptors that have multipoint fields. Innervation densities are high; for example, on the top of the foot of the rat, it has been calculated that there are 8–11/mm^2 (Lynn, 1984; Lynn & Baranowski, 1987). Responses have been reported to a range of irritants including dilute acid, histamine, bradykinin, capsaicin, and itch powder (e.g., Beck & Handwerker, 1974; Bessou & Perl, 1969; Szolcsányi, 1987; Tuckett & Wei, 1987).

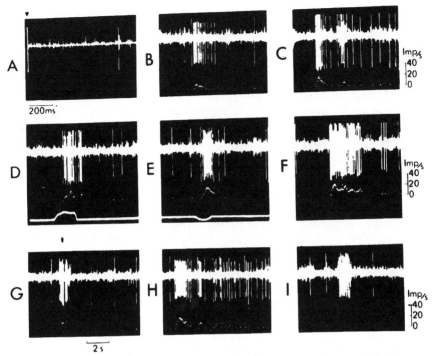

FIGURE 3   Action potentials recorded from a polymodal nociceptor in the human peroneal nerve by microneurography. Responses to various stimuli applied to the receptive field
on the big toe. Dots below action potential traces show instantaneous frequency of firing. (A)
Responses to electrical stimulation at long latency indicating a conduction velocity in the
C-fiber range (0.64 m/sec). (B) Sustained pressure with a 2-g von Frey bristle. (C) Repeated
firm stroking with a small wooden stick, evoking a sensation of slight aching. (D) Blunt
pressure of approximately 15 g with a 1-mm diameter probe with time course indicated by the
strain gauge signal. (E) Pointed stimulator (diameter 0.1 mm), 5 g force, indicated by down-
ward-going deflection of trace. (F) Needle penetration through skin causing pricking and
delayed pain. (G) After application of itch powder, subject felt burning itch. (H) After penetration of skin with a nettle leaf evoking pain followed by itch. (I) Touching skin briefly with
glowing match. (From Torebjörk, 1974. Reprinted with permission.)

b. C-Mechanical Nociceptors

A small number of C-fibers have pressure thresholds in the nociceptor
range, but no responses to heat (Bessou & Perl, 1969). Some of the reported
examples might have responded to heating if higher stimuli had been tested,
but others do seem to be heat insensitive. In a recent survey of monkey
hairy skin units, only 3 out of 62 C-fibers with high-threshold pressure receptive fields did not fire to heating (Baumann, Simone, Shain, & LaMotte,

1991). This is a similar proportion to that found in the rat, in which only four units in a sample that included 80 polymodal nociceptors were classified as mechanical nociceptors (Lynn & Carpenter, 1982).

c. Heat Nociceptors

C-afferent units firing preferentially to noxious heating, but only weakly or not at all to pressure, have been reported in small numbers in several species. Baumann et al. (1991) found 5/72 (7%) C-fibers from monkey hairy skin to respond to noxious heating, but not pressure or cold, and a similar proportion (6%) of such units were reported in microneurography studies using an electrical search stimulus in the peroneal nerve supplying hairy skin on the leg in humans (Schmidt et al., 1995). Georgopoulos (1976) reported 8/191 (4%) such units in monkey glabrous skin, three of which also fired to cooling. In hairy skin of the pig, 7/37 (19%) C-fibers were of this type (Lynn, Faulstroh, & Pierau, 1995). However, in rat and cat, only a very small number of such units have been reported. It is of interest that all five of this kind of heat nociceptor tested in monkeys were vigorously excited by the irritant chemical capsaicin (Baumann et al., 1991) and $0.1 \mu M$ bradykinin activated four pig units tried (Lynn, Schütterle, & Pierau, 1996).

d. Cold Nociceptors

Nociceptors that only fire to severe cooling, although reported, are rare (Baumann et al., 1991). A few C-fiber units of the polymodal and mechanical nociceptor classes however also fire to strong cooling, and these may be important for signaling cold pain (see above). The insensitive thermoreceptor units described by LaMotte and Thalhammer (1982) were not considered likely to be nociceptive by those authors. A significant population of nociceptors with responses to both strong pressure and to cooling, termed mechanical-cold nociceptors, has been described in the skin of the guinea pig and rabbit (Shea & Perl, 1985a; Sugiura, Hosoya, Ito, & Kohno, 1988; Sugiura, Lee, & Perl, 1986; Sugiura, et al., 1993).

e. Nociceptors Firing Only to Irritant Chemicals

Use of electrical stimuli to search for afferent units to be tested has revealed the existence of C-afferent fibers that only respond, at least in undamaged skin, to irritant chemical stimulation. For example, in recordings in humans 7/22 (32%) of C-fibers with no detected heat or pressure receptive field were excited by topical mustard oil, a painful chemical irritant (Schmidt et al., 1995). In addition, both units that responded to the irritants and those that did not developed sensitivity to pressure and/or heat following the chemical stimulation. Overall, however, the number of nociceptors in mammalian skin excited by irritants, but not by physical stimuli, appears to be small.

C. Skeletal Muscle and Joints

Skeletal muscle and joints, like skin, are innervated by specialized small myelinated or unmyelinated nociceptive afferent fibers. In these tissues the fibers are often designated group III and IV rather than Aδ and C, a difference in nomenclature that goes back to the earlier days of muscle afferent studies (Lloyd & Chang, 1948). The size spectrum of somatic subcutaneous afferent fibers does differ slightly from that associated with the cutaneous innervation. Again, like skin, there are also specialized low-threshold mechanoreceptors with similar-sized fibers (Bessou & Laporte, 1961), and muscle has, in addition, C-fiber thermoreceptors.

Skeletal muscle nociceptors can be subdivided into mechanical and polymodal types, another similarity to cutaneous nociceptors. A receptive field, often quite extensive in the muscle, yields responses to strong local pressure; activity is also evoked by extreme stretch and maximal tetanic contractions (Mense, 1993). A concentration of mechanical nociceptor fields appears at muscle–tendon junctions (Abrahams, Lynn, & Richmond, 1984; Mense & Meyer, 1985), and equivalent units also have fields in tendons themselves (Mense & Meyer, 1985). Essentially similar units have been reported in other species including humans (Simone, Caputi, Marchettini, & Ochoa, 1991).

Polymodal units respond to strong mechanical stimuli and also to close-arterial injections of irritant chemicals, notably bradykinin. An early report described group III receptors in cat muscle that were excited by strong pressure and by local injection of hypertonic NaCl (Paintal, 1960). Most polymodal units also are excited by noxious heating with thresholds of 38–48°C on the muscle surface (Kumazawa & Mizumura, 1977). As shown in Figure 4, some polymodal units fire strongly during ischmic contractions; this is one of the few properties that appears restricted to C-afferent fibers (Mense & Stahnke, 1983). In other properties, the same range is seen in Aδ and C-muscle afferent fibers. An interesting difference from skin nociceptors is that around 25% of muscle nociceptors display low-frequency background firing in the absence of any stimulation (Mense, 1993); the extent to which the ongoing discharge may be related to the trauma associated with preparation of the tissue or past stimulation is not certain.

Many Aδ and C-receptors in joint tissue are excited only by strong mechanical stimuli and by irritant chemicals, qualifying them as nociceptors. Typically, such units respond to extreme joint rotation and twisting, have a receptive field for strong pressure on part of the joint tissue, and respond to close-arterial injections of bradykinin, capsaicin, or other inflammatory mediators or irritants (see review by Schaible & Grubb, 1993). A population of mechanical nociceptors without chemosensitivity has not been described. In tissue associated with joints, as for skeletal muscle, there

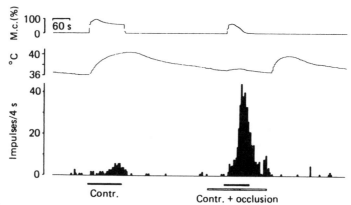

FIGURE 4   Responses of a muscle C-fiber nociceptor to contraction (Contr.) of the muscle and occlusion of the arterial supply. Effect of contraction produced by maximal stimulation of the muscle nerve (left) and to occlusion of arterial supply plus contraction (right). Solid horizontal bars mark time of electrical stimulation of muscle nerve. Occlusion of the arterial supply marked by open horizontal bar. Upper trace shows force developed by the muscle as a percentage of maximal contraction. Middle trace indicates the intramuscular temperature; note rise with contraction and after release of arterial occlusion. Lower trace shows histogram of impulses (4 sec bins). This afferent fiber was not activated by muscle stretch. (Adapted with permission from Mense & Stahnke, 1983.)

are few reported differences between the properties of Aδ and C units. Morphologically, joint nociceptors are thought to correspond to simple unencapsulated terminals, some of which (including elements in the synovial layer) contain sensory neuropeptides (Calcinonin gene-related peptide or CGRP, Substance P). Ultrastructure of nerve terminals in joints shows them to be of variable diameter, giving a "string of beads" appearance and that the Schwann cell covering is incomplete with some patches of axon membrane exposed directly to the interstitial space (Heppelmann, Messlinger, Neiss, & Schmidt, 1990). Terminals with myelinated stem fibers maintain a neurofilament core within the terminal region, whereas those with unmyelinated stem axons do not (Heppelmann et al., 1990).

A considerable proportion (around half) of small afferent fibers from the joints cannot be excited by mechanical search stimuli or physiological joint movements (Schaible & Schmidt, 1983). These receptive fibers often become active in the presence of acute inflammation (Schaible & Schmidt, 1985). It was the study of these joint receptors that gave rise to the concept of silent or "sleeping" nociceptors (see Schmidt et al., 1994). This idea will be discussed further in the section on modulation of nociceptor sensitivity by inflammation.

D. Cornea

All forms of stimulation of the cornea can give rise to pain, so its innervation is of considerable interest. In taking this into account, it must be kept in mind that the cornea is easily damaged and even stimuli as gentle as brushing with a soft material can be injurious. Information on corneal receptors largely comes from studies on anesthetized felines. A considerable controversy about corneal afferent fibers occurred in the mid-1950s (see chap. 1, this volume, by Stevens & Green). Cornea is innervated by Aδ and C-fibers and most of these respond to gentle brushing, to heating, and to irritant chemical stimuli, so the majority are polymodal in characteristic and probably act as nociceptors (Belmonte & Giraldez, 1981; Gallar, Pozo, Tuckett, & Belmonte, 1993). Some Aδ units respond only to mechanical stimuli and so behave like mechanical nociceptors (Belmonte, Gallar, Pozo, & Rebollo, 1991), whereas some C-units do not respond to gentle mechanical stimuli, but do respond to moderate cooling (Gallar et al., 1993). Because cooling the cornea is painful, this latter group may have a role of specialized cold nociceptors. Overall, the properties of corneal nociceptors appear somewhat different from those of skin nociceptors in the same species.

The cornea represents a valuable preparation for studying transduction of noxious stimuli to electrical signals because applied substances can penetrate to the nerve endings more easily than in most other tissues and because it is possible to visualize terminals while recording from fiber bundles (MacIver & Tanelian, 1993). A particularly interesting dissociation has been established using the corneal innervation; calcium channel blockers were found to reduce corneal nociceptor sensitivity to heat and acid while leaving sensitivity to mechanical stimuli unchanged (Pozo, Gallego, Gallar, & Belmonte, 1992).

E. Teeth

As for the cornea, stimulation of the tooth pulp and dentine gives rise predominantly to pain. The dental nerves contain Aαβ, Aδ, and C fibers that often exhibit a marked increase in conduction velocity from tooth root to trigeminal ganglion (Cadden, Lisney, & Matthews, 1983). Those with nociceptive response profiles include Aαβ units conducting at up to 58 m/sec in cats (Lisney, 1978). Dental nociceptors usually respond to a range of stimuli: heating, scratching the exposed dentinal surface, or applying chemicals to it, so they are polymodal in character. There are, however, clear differences between A and C units. A-fiber units respond well to all stimuli that affect the hydraulic state of the dentinal tubules, including solutions of high-osmotic pressure or temperature extremes (Narhi,

Yamamoto, Ngassapa, & Hirvonen, 1994). C-fiber units are readily excited by stimuli that act on the pulp itself and respond less well to stimuli that only affect dentine. Most pulp C-fiber units are activated by heat (thresholds around 42°C) and by local probing (Jyvasjarvi, Kniffki, & Mengel, 1988). Many also respond to cold, but at longer latency and with a more prolonged time course than dentine Aδ units (Mengel, Stiefenhofer, Jyvasjarvi, & Kniffki, 1993). In addition, many dental C-fiber afferent units are excited by chemical agents applied to the pulp, such as KCl or bradykinin (Jyvasjarvi & Kniffki, 1989; Jyvasjarvi et al., 1988; Narhi et al., 1994).

The tooth socket is also innervated by specialized nociceptors. Strong pressure on the intact tooth excites these afferent fibers, but stimuli localized to the dentine or the pulp do not affect them. Both Aδ and C units have nociceptive pressure thresholds, a slowly adapting response and slight directional selectivity (Mengel, Jyvasjarvi, & Kniffki, 1992; Mengel, Jyvasjarvi,

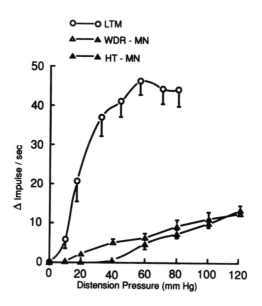

FIGURE 5   The stimulus–response functions of three classes of afferent fiber responding to distension of the esophagus in the anesthetized opossum. LTM, low-threshold vagal mechanoreceptors responding to small pressure changes and showing no significant increase in firing with pressure in the noxious range. Two subclasses of nociceptor are recorded from sympathetic nerves to the esophagus: HT-MN, high-threshold mechanical nociceptor, and WDR-MN, wide-dynamic range MN. Although designated MN, both classes also fired to the irritant chemical bradykinin (Sengupta et al., 1992). HT-MN firing is only in the noxious range. WDR-MN firing is over both innocuous and noxious pressure ranges. Note the similarity between the two presumed nociceptor classes and the marked difference between them and the vagal mechanoreceptors. It is instructive to compare the real functions shown here with the schematic functions in Figure 1. (Adapted with permission from Sengupta et al., 1990. Reprinted with permission).

& Kniffki, 1993). The high-threshold units differ distinctively from the periodontal mechanoreceptors that have low thresholds and a marked directional selectivity (see Cash & Linden, 1982). Some Aδ periodontal nociceptors respond solely to pressure. However, many discharge to cold and a few are excited by heat and exhibit weak responses to chemical stimuli (Mengel, Jyvasjarvi, & Kniffki, 1993). Part of the population of periodontal C-units are activated by pressure, heat, and cold. Others discharge only to thermal stimuli (either heat or cold or both) plus strong pressure. Approximately half of periodontal C-units are strongly activated by chemical stimulation (Mengel et al., 1992).

F. Visceral Sensory Units

Specific nociceptive afferent fibers have been reported for all visceral organs for which sensory innervation has been systematically studied, although there are opinions to the contrary (see Cervero & Jänig, 1992). Visceral nociceptive fibers usually pass centrally in the sympathetic nerves. Common characteristics are high mechanical thresholds, responses to irritant chemicals, and when tested, responses to heat. Organs with documented nociceptors are the heart (Baker, Coleridge, Coleridge, & Nerdrum, 1980; Uchida, Kamisaka, Murao, & Ueda, 1974), the gall bladder (Cervero, 1982), the urinary bladder (Habler et al., 1990; Sengupta & Gebhart, 1994), the testis (Kumazawa & Mizumura, 1980a,b), the uterus (Berkley, Robbins, & Sato, 1988, 1993), and the esophagus (Sengupta et al., 1990). In general, the nociceptive afferent innervation contains both Aδ and C-fibers with few marked differences between the properties of the two categories reported so far. Ischemia excites some visceral nociceptors, for example in the heart (Uchida & Murao, 1974a). In some tissues the number of nociceptors excited by mechanical stimuli is increased when inflammation is present or there is prior chemical stimulation (Habler et al., 1990). In this regard, the afferent innervation of viscera bear a similarity to that of joints.

Visceral afferents are by no means all nociceptive. Most viscera also contain specialized mechanoreceptors, and these have fibers predominantly passing through parasympathetic nerves (vagus, pelvic nerves), although they appear in other nerves as well (e.g., Bessou & Perl, 1966). In addition, there are afferent units with response ranges spanning both innocuous and noxious levels of stimulation. Now that it is established that viscera are innervated with specialized nociceptors, at first consideration these spanning fibers seem redundant. In fact, at least some of the "wide-range" units signal mainly in the nociceptive range. For example, Figure 5 shows average input–output functions for three classes of esophageal afferents: mechanoreceptors, nociceptors, and wide-range units. The wide-range units closely resemble nociceptors for all but near-threshold stimuli. The wide-range fibers also are similar to the frankly nociceptive ones by being 100% responsive to bradykinin (Sengupta et al., 1992). The ability of wide-range units to encode information in the innocuous range (up to around 40 mm Hg) of hollow organs is very poor in comparison to the vagal low-threshold mechanoreceptors. In retrospect, it seems that too much attention has been paid to thresholds of visceral afferent units, and not enough to their suprathreshold response ranges. In the skin many clearly nociceptive afferents do have thresholds at innocuous mechanical stimulus levels. As will be discussed below, the low-frequency firing evoked by innocuous stimuli to sensory units may be filtered in the central nervous system (CNS). For visceral afferent elements such as the esophageal wide-range units, a similar mechanism may be involved, and the encoding of stimuli in the noxious range may be their sole function. Further studies of suprathreshold responses of wide-range units from other viscera are needed to establish the extent to which these also encode information primarily in the noxious range.

IV. QUANTITATIVE COMPARISONS OF AFFERENT FIRING AND PAIN SENSATIONS

There should be good agreement between the levels of stimuli that excite nociceptors and those that evoke pain because the definition of units as nociceptors takes into account the relation of firing thresholds to human pain thresholds. Initially, such comparisons were made across species. Responses of sensory units from experimental animals were compared to human reports of sensation to equivalent stimuli (see review by Perl, 1984). Direct comparisons of nociceptor firing and sensory responses became possible with the development of microneurography, the recording of single-unit discharges from nerves in conscious human subjects using percutaneously placed tungsten microelectrodes. Before reviewing some of the main results from such studies, the effects of stimulating through microneurography electrodes and intraneural microstimulation (INMS) will be addressed (see also chap. 1, this volume).

Two approaches with this technique have been employed. In one, the microelectrode is slowly manipulated in a cutaneous nerve until a unitary action potential is recorded, either spontaneously active or evoked by stimulating peripherally in the innervation region of the nerve. The responsive characteristics of the unit are determined by applying various stimuli to the receptive region, and then brief electrical pulses are applied through the recording microelectrode while the latter remains in place. INMS at loci from which Aδ and C-fiber nociceptor discharges were recorded have been reported to initiate pain-like sensations referred to the unit's receptive field (Konietzny, Perl, Trevino, Light, & Hensel, 1981). The alternative approach involves stimulating while slowly moving a microneurography electrode through a cutaneous nerve. The latter reveals some sites where pain sensations are felt. With threshold stimulation the sensation may be quite simple in character and have a limited peripheral projection. Again, keeping the electrode positioned at such a site and switching to recording mode often reveals unitary action potentials from high-threshold receptive units. Sites from which C-polymodal activity can be recorded give rise on stimulation to burning pain (or occasionally to itch) if the sensation is projected to hairy skin, and to a dull pain without thermal character in glabrous skin (Marchettini, Cline, & Ochoa, 1990; Ochoa & Torebjörk, 1989; Torebjörk, LaMotte, & Robinson, 1984a; Torebjörk & Ochoa, 1980, 1981). Sites from which Aδ activity can be recorded, on the other hand, typically give rise to sharp, pricking pain (Marchettini et al., 1990; Torebjörk, LaMotte, & Robinson, 1984). In either search technique, spatially restricted sensations of single quality are regularly felt with near-threshold INMS. As stimulus strength is increased, the sensations become more complex in quality and location, presumably as more fibers are recruited. If stimulus strength is kept low and frequency is increased, the pain magnitude will rise, but with no reported change in quality or localization (Konietzny et al., 1981; Lundberg, Jorum, Holm, & Torebjörk, 1992; Ochoa & Torebjörk, 1989; Torebjörk & Ochoa, 1981). The results of INMS have therefore indicated that excitation of small numbers, possibly individual, nociceptors can cause pain sensations in normal subjects. In addition, it appears that the character of the pain varies between Aδ and C-nociceptors, and also with skin type.

The validity of the conclusions from INMS have been questioned because at first sight it appears unlikely that a microneurography electrode would be able to excite single, or at most very few, C-fibers. Block of conduction in fibers excited by INMS also has been suggested (Wall & McMahon, 1985, 1986). Convincing evidence for the functional continuity of nerves undergoing INMS and that at least for A-fibers, genuine single-fiber stimulation is possible has been presented (Torebjörk, Vallbo, & Ochoa, 1987). Investigators using INMS have generally considered that stimulation of small groups of C-fibers, rather than single units, could be occurring in most cases. In practice, the all-or-none behavior of the sensation in many instances and the close correspondence of unitary receptive fields and referred sensation argues strongly that single unmyelinated units can be excited. Given carefully graded stimuli, even if an electrode is close to many fibers the relationship to each will differ and so at threshold only one will be excited to fire. After all, in recordings from large nerve bundles in animal experiments, it is often possible to excite just one out of many C-fibers with a liminal stimulus. Theoretically and practically, a small microneurography electrode should be capable of doing the same.

Microneurography recording has allowed simultaneous comparisons of nociceptor discharges and subjective sensations. This is a less direct way of assessing the actions of single fibers because, although only one is recorded, even the most spatially restricted stimulus is unlikely to excite just that single fiber. Nevertheless, some convincing correlations between nociceptor firing and pain sensation have appeared. It should be noted that most microneurography data on pain relates to C-polymodal nociceptor observations. With noxious heating of hairy skin, C-polymodal units were found to have thresholds for initial discharges of 41–43°C, whereas subjective pain thresholds were 41–49°C (Torebjörk, Schady, & Ochoa, 1984b). For brief stimuli (5 sec) up to 53°C, ratings of pain sensation intensity were linearly related to C-polymodal firing frequencies (Torebjörk, LaMotte, & Robinson, 1984). The small mismatch, with pain thresholds being slightly higher than nociceptor thresholds, has been seen in other studies and appears to indicate that a certain level of activity is needed, estimated as >0.4 Hz, for pain to be reported (van Hees & Gybels, 1981). Mechanical stimulation reveals a larger discrepancy with units firing at >1 Hz before any pain is felt (van Hees & Gybels, 1981; Koltzenburg & Handwerker, 1994). The mechanical stimuli used in these tests were always more spatially restricted than the heat stimuli, and so provided less scope for spatial summation. That difference could explain higher relative sensory in comparison to neural thresholds. An interesting alternative idea is that the simultaneous low-threshold mechanoreceptor activation when mechanical stimuli are used leads to inhibitory gating at the spinal cord level (Adriaensen et al., 1983; see Handwerker & Kobal, 1993). However, INMS frequencies of <1 Hz generally do not give rise to pain sensation, and for such situations there is little simultaneous input in large A-fiber mechanoreceptors (Ochoa & Torebjörk, 1989), arguing against the mechanoreceptor gating proposal.

For repetitive or long duration heat or pressure stimuli (more than 1–2 min), pain usually increases but nociceptor discharge does not, or may even fall off. However, during the relatively low-frequency firing after painful chemical stimulation, there is a reasonable match between nociceptor firing and pain for 5–8 min (Handwerker et al., 1991). With relatively stronger physical stimuli it seems likely that the slow increases in pain sensation reflect the onset of central mechanisms enhancing or prolonging nociceptor-induced activity.

V. NOCICEPTOR SENSITIZATION

A. Inflammation and Nociceptors

The neural systems involved in nociception and the sensation of pain are modifiable. Potent mechanisms exist for both increasing and decreasing their sensitivity. This section considers mechanisms that operate at the level of the afferent terminals (nociceptor sensitization). It is essential to keep in mind that in parallel with the alterations in behavior of primary afferent units, changes also can and do occur at the first synaptic area in the nociceptive pathway in the spinal cord and trigeminal nucleus (McMahon, Lewin, & Wall, 1993; Woolf, 1994). The overall effects of, for example, tissue inflammation, will therefore also reflect consequential second- and higher-order changes at synaptic regions in nociceptive pathways.

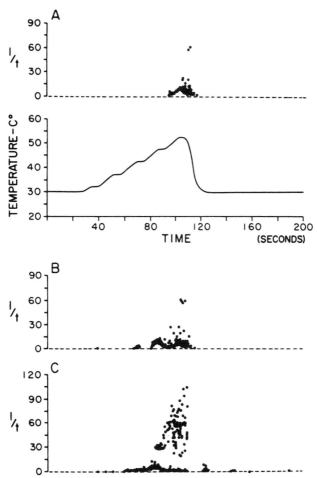

FIGURE 6   Heat sensitization of a C-fiber polymodal nociceptor from monkey hairy skin. The upper display of A and those in B and C represent "instantaneous" frequency plots: 1/interval in seconds from preceding impulses. Each dot represents a single impulse. A stepwise heating of a contact thermode centered on the receptive field of the unit is shown in the lower trace of Part A. This identical sequence of heating and sudden cooling was repeated at 200-sec intervals. (A) First heating sequence; (B) Third heating sequence; (C) Fifth heating sequence. The instantaneous frequency plots of A, B, and C were aligned by superimposing the heating cycles. Note both increased number and higher frequency of impulses produced on successive heating cycles. (Reproduced with permission from Kumazawa & Perl, 1977.)

The phenomenon of primary hyperalgesia, the increase in pain sensitivity in regions of injury or inflammation, is familiar to all. Primary hyperalgesia is in large part caused by an increase in the firing of nociceptors, the phenomenon of nociceptor sensitization (Bessou & Perl, 1969). A typical example of immediate nociceptor sensitization, in this case following mild heat injury, is shown in Figure 6. The extent to which nociceptor sensitization matches hyperalgesia has been studied in human subjects using microneurography. Under conditions of controlled skin heating, C-polymodal discharges increase in a way that closely matches the increases in subjective pain ratings (Torebjörk, LaMotte, & Robinson, 1984). Recent studies have shown that there are also similar enhanced responses in A-fiber heat-sensitive nociceptors (Goto, Kagosima, & Sakurai, in press). Thus, the peripheral basis for primary hyperalgesia is diverse.

Although direct effects of injury on the membranes of nociceptor terminals cannot be excluded in many situations, one principal mechanism of nociceptor sensitization appears to be the action of mediator chemicals released following injury of receptive tissue or other events associated with the process of inflammation. Table I indicates that there is no lack of suitable

TABLE 1   Endogenous Agents That Cause Nociceptor Sensitization or Hyperalgesia[a]

Agent	Receptor type(s)	Source	Direct or indirect action
Histamine	H1	Mast cells	Direct
Serotonin (5HT)	5HT1, 2, and 3	Mast cells, platelets	Direct
Bradykinin	B2	Plasma protein precursor	Direct and indirect
PGE1, 2	EP2, 3	Various immune system cells; sympathetic nerve terminals	Direct
PGI2	IP	Various immune system cells	—
8R, 15S–diHETE	—	Various immune system cells, sympathetic nerve terminals	Direct
Leukotrienes	—	Various immune system cells	Indirect
Hydrogen ions (acidity)	—	All cells	Direct
Unidentified platelet factor	—	Platelets	—
Interleukin 1b	—	Macrophages; other immune system cells	Indirect
ATP	P2	Platelets, various cells	Direct
Adenosine	A2	ATP breakdown	Direct
Nerve growth factor	trkA	Schwann cells, fibroblasts, keratinocytes	Direct

[a]Data from Levine & Taiwo, 1994; Rang et al., 1994; Reeh, 1994; Kumazawa et al., 1995.

candidates. The particular agents involved will vary with the nature of the injurious or inflammatory event, the time after its onset, and probably the tissue and species under study. Even in a given tissue such as the skin, polymodal nociceptors vary in their responsiveness to mediators, and not all units respond to all the agents by any means (e.g., Cohen & Perl, 1990; Lang, Novak, Reeh, & Handwerker, 1990; Reeh, 1986, 1994). Chemical mediators can act directly on the nociceptor terminal (e.g., histamine, prostaglandins), indirectly via other elements that release directly acting mediators (e.g. interleukin 1b), and through both direct and indirect means (e.g., bradykinin: see Levine & Taiwo, 1994; Rang, Bevan, & Dray, 1994). Many mediators trigger second messenger systems in the afferent terminals, and these in turn will act on proteins affecting excitability via protein kinases. Proposals of a role for cAMP, particularly in the actions of prostaglandins, has come from work on several models (Taiwo, Bjerknes, Goetzl, & Levine, 1989; Mizumura, Koda, & Kumazawa, 1993). Studies on protein kinases are not well advanced, but evidence has been obtained in different situations for involvement of protein kinase A (PKA) (Taiwo & Levine, 1991) and protein kinase C (PKC) (Mizumura, Koda, & Kumazawa, 1995).

B. Development of Sensitivity to Sympathetic Agents Following Nerve Injury

Normally nociceptors are unaffected by sympathetic efferent activity or exposure to catecholamines (e.g., Barasi & Lynn, 1986; Roberts & Elardo, 1985; Shea & Perl, 1985b). However, within a few days following nerve injury, nociceptors develop clear sensitivity to the sympathetic stimulation and to norepinephrine (Sato & Perl, 1991). This change appears to involve the de novo synthesis of α_2 adrenergic receptors (Nishiyama et al., 1993; Nishiyama, Sugiura, & Perl, 1995). These new properties of nociceptor terminals are relevant to the development of neuropathic pain following nerve injury (Perl, 1993, 1994). After partial nerve injuries that can be quite minor (for example brief stretch), the surviving nerve terminals show increased catecholamine sensitivity. In addition, many regenerating nociceptor fibers also become catecholamine sensitive (Devor, 1983), and there is increased sympathetic innervation of the ganglion cells themselves (Chung, Kim, Na, Park, & Chung, 1993; McLachlan, Jänig, Devor, & Michaelis, 1993).

Inflammation itself, at the tissue level, may also cause development of catecholamine sensitivity by nociceptor terminals (Sato, Suzuki, Iseki, Tamura, & Kumazawa, in press). In addition, the release of prostaglandins by sympathetic terminals has been suggested to play a part in nociceptor sensitization to some mediators (Levine & Taiwo, 1994).

VI. SUMMARY

It has become clear that almost all tissues are innervated by specialized afferent units that encode information about stimuli in the noxious range. The great majority of such units have small myelinated (Aδ) or unmyelinated (C) fibers. In hairy skin the C-fibers predominantly are associated with polymodal nociceptors, responding to heat, pressure, and irritant chemicals. In contrast, the fine A-fibers of hairy skin are predominantly mechanical nociceptors with negligible prompt sensitivity to stimuli other than strong pressure, although in primates mechano-heat nociceptor units are found relatively frequently. In many other tissues (e.g., muscles and joints) the present evidence for correlation of function with fiber size is less clear and in particular polymodal-type units with Aδ axons are common. In some tissues, including cornea and tooth pulp, many mechanical or polymodal units also fire to strong cooling stimuli. The presence of inflammation appears to lead to marked increases in sensitivity for all types of nociceptors. Nociceptor responses to brief skin stimuli correlate well with pain sensations in humans. With longer duration stimuli, the situation is more complex secondary to the development of changes in responsiveness of neurons of the CNS because of both locally mediated alterations in synaptic effectiveness and interaction with local and descending control networks.

References

Abrahams, V. C., Lynn, B., & Richmond, F. (1984). Organisation and sensory properties of small myelinated fibers in the dorsal cervical rami of the cat. *Journal of Physiology, 347,* 177–187.

Adriaensen, A., Gybels, J., & Handwerker, H. O. (1983). Response properties of thin myelinated (Aδ) fibers in human skin nerves. *Journal of Neurophysiology, 49,* 111–122.

Baker, D. G., Coleridge, H. M., Coleridge, J. C. G., & Nerdrum, T. (1980). Search for a cardiac nociceptor: Stimulation by bradykinin of sympathetic afferent nerve endings in the heart of the cat. *Journal of Physiology, 306,* 519–536.

Barasi, S., & Lynn, B. (1986). Effects of sympathetic stimulation on mechanoreceptive and nociceptive afferent units from the rabbit pinna. *Brain Research, 378,* 21–27.

Baumann, T. K., Simone, D. A., Shain, C. H., & LaMotte, R. H. (1991). Neurogenic hyperalgesia: The search for the primary cutaneous afferent fibers that contribute to capsaicin-induced pain and hyperalgesia. *Journal of Neurophysiology, 66,* 212–227.

Beck, P. W. Handwerker, H. O. (1974). Bradykinin and serotonin effects on various types of cutaneous nerve fibers. *Pflueger's Archives, 347,* 209–222.

Beck, P. W., Handwerker, H. O., & Zimmermann, M. (1974). Nervous outflow from the cat's foot pad during noxious radiant heat stimulation. *Brain Research, 67,* 373–386.

Beitel, R. E., & Dubner, R. (1976). The response of unmyelinated (C) polymodal nociceptors to thermal stimuli applied to the monkey's face. *Journal of Neurophysiology, 39,* 1160–1175.

Belmonte, C., & Giraldez, F. (1981). Responses of cat corneal sensory receptors to mechanical and thermal stimulation. *Journal of Physiology, 321,* 355–368.

Belmonte, C., Gallar, J., Pozo, M. A., & Rebollo, I. (1991). Excitation by irritant chemical substances of sensory afferent units in the cat's cornea. *Journal of Physiology, 437,* 709–725.

Berkley, K. J., Robbins, A., & Sato, Y. (1988). Afferent fibers supplying the uterus in the rat. *Journal of Neurophysiology, 59,* 142–163.

Berkley, K. J., Robbins, A., & Sato, Y. (1993). Functional differences between afferent fibers in the hypogastric and pelvic nerves innervating female reproductive organs in the rat. *Journal of Neurophysiology, 69,* 533–544.

Bessou, P., & LaPorte, Y. (1961). Étude des récepteurs musculaires innervés par les fibres afférentes du groupe III (fibres myelinisées fines), chez le chat. [Study of the muscular receptors innervated by the afferent fibers of group III (fine myelinated fibers) in the cat.] *Archives Italian Biology, 99,* 293–321.

Bessou, P., & Perl, E. R. (1966). A movement receptor of the small intestine. *Journal of Physiology (London), 182,* 404–426.

Bessou, P., & Perl, E. R. (1969). Response of cutaneous sensory units with unmyelinated fibers to noxious stimuli. *Journal of Neurophysiology, 32,* 1025–1043.

Bessou, P., Burgess, P., Perl, E. R., & Taylor, C. (1971). Dynamic properties of mechanoreceptors with unmyelinated (C) fibers. *Journal of Neurophysiology, 34,* 116–131.

Burgess, P. R., & Perl, E. R. (1967). Myelinated afferent fibres responding specifically to noxious stimulation of the skin. *Journal of Physiology, 190,* 541–562.

Burgess, P. R., Petit, D., & Warren, R. M. (1968). Receptor types in cat hairy skin supplied by myelinated fibres. *Journal of Neurophysiology, 31,* 833–848.

Cadden, S. W., Lisney, S. J. W., & Matthews, B. (1983). Thresholds to electrical stimulation of nerves in cat canine tooth-pulp with Aβ-, Aδ and C-fibre conduction velocities. *Brain Research, 261,* 31–41.

Campbell, J. N., & Meyer, R. A. (1986). Primary afferents and hyperalgesia. In T. L. Yaksh (Ed.), *Spinal afferent processing* (pp. 59–81). New York: Plenum.

Campbell, J. N., & LaMotte, R. H. (1983). Latency to detection of first pain. *Brain Research, 266,* 203–208.

Cash, R. M., & Linden, R. W. (1982). The distribution of mechanoreceptors in the periodontal ligament of the mandibular canine tooth of the cat. *Journal of Physiology, 330,* 439–447.

Cervero, F. (1982). Afferent activity evoked by natural stimulation of the biliary system in the ferret. *Pain, 13,* 137–151.

Cervero, F., & Jänig, W. (1992). Visceral nociceptors: a new world order? *Tins, 15,* 374–378.

Chung, K., Kim, H. J., Na, H. S., Park, M. J., & Chung, J. M. (1993). Abnormalities of sympathetic innervation in the area of an injured peripheral nerve in a rat model of neuropathic pain. *Neuroscience Letters, 162,* 85–88.

Cohen, R. H., & Perl, E. R. (1990). Contributions of arachidonic acid derivatives and substance P to the sensitization of cutaneous nociceptors. *Journal of Neurophysiology, 64,* 457–464.

Collins, W. F., Nulsen, F. E., & Randt, C. T. (1960). Relation of peripheral nerve fibre size and sensation in man. *Archives of Neurology, 3,* 381–385.

Craig, A. D. (1994). Spinal and supraspinal processing of specific pain and temperature. In J. Boivie, P. Hansson, & U. Lindblom (Ed.), *Progress in pain research and management, 3: Touch, temperature, and pain in health and disease: Mechanisms and assessments* (pp. 421–437). Seattle: IASP Press.

Devor, M. (1983). Nerve pathophysiology and mechanisms of pain in causalgia. *Journal of the Autonomic Nervous System, 7,* 371–384.

Dubner, R., Price, D. D., Beitel, R. E., & Hu, J. W. (1977). Peripheral neural correlates of behaviour in monkey and human related to sensory-discriminative aspects of pain. In D. J. Anderson & B. Matthews (Ed.), *Pain in the trigeminal area* (pp. 57–66). Amsterdam: Elsevier.

Dubner, R., Sumino, R., & Starkman, S. (1973). Responses of facial cutaneous thermosensitive and mechanosensitive afferent fibers in the monkey to noxious heat stimulation. *Advances in Neurology, 4,* 61–71.

Fitzgerald, M., & Lynn, B. (1977). The sensitization of high threshold mechanoreceptors with myelinated axons by repeated heating. *Journal of Physiology, 265,* 549–563.

Gallar, J., Pozo, M. A., Tuckett, R. P., & Belmonte, C. (1993). Response of sensory units with unmyelinated fibres to mechanical, thermal and chemical stimulation of the cat's cornea. *Journal of Physiology, 468,* 609–622.

Georgopoulos, A. P. (1976). Functional properties of primate afferent units probably related to pain mechanisms in primate glabrous skin. *Journal of Neurophysiology, 39,* 71–83.

Goto, K., Kagosima, Y., & Sakurai, K. (in press). Response properties of A-δ fibers in human skin nerves to repeated noxious heat stimuli. *Japanese Journal of Physiology, 45, Supp. 2,* S243.

Guilbaud, G., Bernard, J. F., & Besson, J. M. (1994). Brain areas involved in nociception and pain. In P. D. Wall & R. Melzack (Ed.), *Textbook of pain* (3rd ed.) (pp. 113–128). Edinburgh: Churchill Livingstone.

Habler, H. J., Jänig, W., & Koltzenburg, M. (1990). Activation of unmyelinated afferent fibres by mechanical stimuli and inflammation of the urinary bladder in the cat. *Journal of Physiology, 425,* 545–562.

Hallin, R. G., & Torebjörk, H. E. (1976). Studies on cutaneous A and C fibre afferents. Skin nerve blocks and perception. In Y. Zotterman (Ed.), *Sensory functions of the skin in primates* (pp. 137–148). Oxford: Pergamon.

Handwerker, H. O., Anton, F., & Reeh, P. W. (1987). Discharge patterns of afferent cutaneous nerve fibres from the rat's tail during prolonged noxious mechanical stimulation. *Experimental Brain Research, 65,* 493–504.

Handwerker, H. O., Forster, C., & Kirchoff, C. (1991). Discharge patterns of human C-fibers induced by itching and burning stimuli. *Journal of Neurophysiology, 66,* 307–315.

Handwerker, H. O., & Kobal, G. (1993). Psychophysiology of experimentally induced pain. *Physiological Review, 73,* 639–671.

Heppelmann, B., Messlinger, K., Neiss, W. F., & Schmidt, R. F. (1990). Ultrastructural three-dimensional reconstruction of group III and IV sensory nerve endings ("free nerve endings") in the knee joint capsule of the cat: Evidence for multiple receptive sites. *Journal of Comparative Neurology, 292,* 103–116.

Heppelmann, B., Messlinger, K., Neiss, W. F., & Schmidt, R. F. (1995). Fine sensory innervation of the knee joint capsule by group III and group IV nerve fibers in the cat. *Journal of Comparative Neurology, 351*(3), 415–428.

Hu, J. W., & Sessle, B. J. (1988). Properties of functionally identified nociceptive and non-nociceptive facial primary afferents and presynaptic excitability changes induced in their brain stem. *Exploratory Neurology, 101,* 385–399.

Jyvasjarvi, E., & Kniffki, K. D. (1989). Afferent C fibre innervation of cat tooth pulp: Confirmation by electrophysiological methods. *Journal of Physiology, 411,* 663–675.

Jyvasjarvi, E., Kniffki, K. D., & Mengel, M. K. (1988). Functional characteristics of afferent C fibres from tooth pulp and periodontal ligament. *Progress in Brain Research, 74,* 237–245.

Koltzenburg, M., & Handwerker, H. O. (1994). Differential ability of human cutaneous nociceptors to signal mechanical pain and to produce vasodilatation. *Journal of Neuroscience, 14,* 1756–1765.

Konietzny, F., Perl, E., Trevino, D., Light, A., & Hensel, H. (1981). Sensory experiences in man evoked by intraneural electrical stimulation of intact cutaneous afferent fibers. *Experimental Brain Research, 42,* 219–222.

Kress, M., Koltzenburg, M., Reeh, P. W., & Handwerker, H. O. (1992). Responsiveness and functional attributes of electrically localized terminals of cutaneous C-fibers *in vivo* and *in vitro. Journal of Neurophysiology, 68,* 581–595.

Kruger, L., Perl, E. R., & Sedivic, M. J. (1981). Fine structure of myelinated mechanical nociceptor endings in cat hairy skin. *Journal of Comparative Neurology, 198,* 137–154.

Kumazawa, T., & Mizumura, K. (1977). Thin fibre receptors responding to mechanical, chemical, and thermal stimulation in the skeletal muscle of the dog. *Journal of Physiology, 273*, 179–194.

Kumazawa, T., & Mizumura, K. (1980a). Chemical responses of polymodal receptors of the scrotal contents in dogs. *Journal of Physiology, 299*, 219–231.

Kumazawa, T., & Mizumura, K. (1980b). Mechanical and thermal responses of polymodal receptors recorded from the superior spermatic nerve of dogs. *Journal of Physiology, 299*, 233–245.

Kumazawa, T., & Perl, E. R. (1977). Primate cutaneous sensory units with unmyelinated (C) afferent fibres. *Journal of Neurophysiology, 40*, 1325–1338.

LaMotte, R. H., & Thalhammer, J. G. (1982). Response properties of high-threshold cutaneous cold receptors in the primate. *Brain Research, 244*, 279–287.

Lang, E., Novak, A., Reeh, P. W., & Handwerker, H. O. (1990). Chemosensitivity of fine afferents from rat skin. *Journal of Neurophysiology, 63*, 887–901.

Levine, J., & Taiwo, Y. (1994). Inflammatory pain. In P. D. Wall & R. Melzack (Eds.), *Textbook of pain* (3rd ed.) (pp. 45–56). Edinburgh: Churchill Livingstone.

Lewis, T. (1942). *Pain*. New York: Macmillan.

Lewis, T., & Pochin, E. (1937). The double pain response of the human skin to a single stimulus. *Clinical Science, 3*, 67–76.

Liang, Y. -F., & Terashima, S. -I. (1993). Physiological properties and morphological characteristics of cutaneous and mucosal mechanical nociceptive neurons with A-delta peripheral axons in the trigeminal ganglia of crotaline snakes. *Journal of Comparative Neurology, 328*, 88–102.

Light, A. (1992). *The initial processing of pain and its descending control: Spinal and trigeminal systems, pain and headache*. (Vol. 12). Basel, Switzerland: Karger.

Lisney, S. J. W. (1978). Some anatomical and electrophysiological properties of tooth-pulp afferents in the cat. *Journal of Physiology, 284*, 19–36.

Lloyd, D., & Chang, H. (1948). Afferent fibers in muscle nerves. *Journal of Neurophysiology, 11*, 199–208.

Lundberg, L. E., Jorum, E., Holm, E., & Torebjörk, H. E. (1992). Intra-neural electrical stimulation of cutaneous nociceptive fibres in humans: Effects of different pulse patterns on magnitude of pain. *Acta Physiologica Scandinavica, 146*, 41–48.

Lynn, B. (1984). The detection of injury and tissue damage. In P. D. Wall & R. Melzack (Eds.), *Textbook of pain* (pp. 19–33). Edinburgh: Churchill Livingstone.

Lynn, B., & Baranowski, R. (1987). A comparison of the relative numbers and properties of cutaneous nociceptive afferents in different mammalian species. In R. F. Schmidt et al. (Eds.), *Fine afferent nerve fibers and pain* (pp. 85–94). VCH Weinheim: Verlagsgesellchaft.

Lynn, B., & Carpenter, S. E. (1982). Primary afferent units from the hairy skin of the rat hind limb. *Brain Research, 238*, 29–43.

Lynn, B., Faulstroh, K., & Pierau, F. -K. (1995). The classification and properties of nociceptive afferent units from the skin of the aneasthetized pig. *European Journal of Neuroscience, 7*, 431–437.

Lynn, B., & Shakhanbeh, J. (1988). Properties of A-delta high threshold mechanoreceptors in the rat hairy and glabrous skin and their response to heat. *Neuroscience Letters, 85*, 71–76.

Lynn, B., Schütterle, S., & Pierau, F. -K. (1996). The vasodilator component of neurogenic inflammation is caused by a special sub-class of heat-sensitive nociceptors in the skin of the pig. *Journal of Physiology, 494.*

MacIver, M. B., & Tanelian, D. L. (1993). Free nerve ending terminal morphology is fiber type specific for A-delta and C fibers innervating rabbit corneal epithelium. *Journal of Neurophysiology, 69*, 1779–1783.

Mackenzie, R. A., Burke, D., Skuse, N. F., & Lethlean, A. K. (1975). Fibre function and perception during cutaneous nerve block. *Journal of Neurology, Neurosurgery, and Psychiatry, 38*, 865–873.

McLachlan, E., Jänig, W., Devor, M., & Michaelis, M. (1993). Peripheral nerve injury triggers noradrenergic sprouting within dorsal root ganglia. *Nature, 363,* 543–546.

McMahon, S. B., & Koltzenburg, M. (1990). Novel classes of nociceptors: beyond Sherrington. *Tins, 13,* 199–201.

McMahon, S. B., Lewin, G. R., & Wall, P. D. (1993). Central hyperexcitability triggered by noxious inputs. *Current Biology, 3,* 602–610.

Marchettini, P., Cline, M., & Ochoa, J. L. (1990). Innervation territories for touch and pain afferents of single fascicles of the human ulnar nerve. Mapping through intraneural microrecording and microstimulation. *Brain, 113* (Oct Pt 5), 1491–1500.

Mengel, M. K., Jyvasjarvi, E., & Kniffki, K. D. (1992). Identification and characterization of afferent periodontal C fibres in the cat. *Pain, 48,* 413–420.

Mengel, M. K., Jyvasjarvi, E., & Kniffki, K. D. (1993a). Identification and characterization of afferent periodontal A delta fibres in the cat. *Journal of Physiology, 464,* 393–405.

Mengel, M. K., Stiefenhofer, A. E., Jyvasjarvi, E., & Kniffki, K. D. (1993b). Pain sensation during cold stimulation of the teeth: Differential reflection of A delta and C fibre activity? *Pain, 55,* 159–169.

Mense, S. (1993). Nociception from skeletal muscle in relation to clinical pain. *Pain, 54,* 241–289.

Mense, S., Light, A. R., & Perl, E. R. (1981). Spinal terminations of subcutaneous high-threshold mechanoreceptors. In A. G. Brown & M. Réthelyi (Eds.), *Spinal cord sensation. Sensory processing in the dorsal horn* (pp. 79–86). Edinburgh: Scottish Academic Press.

Mense, S., & Meyer, H. (1985). Different types of slowly-conducting afferent units in cat skeletal muscle and tendon. *Journal of Physiology, 363,* 403–417.

Mense, S., & Stahnke, M. (1983). Responses of muscle afferent fibres of slow conduction velocity to contractions and ischaemia in the cat. *Journal of Physiology, 342,* 383–397.

Meyer, R. A., Cohen, R. H., Davis, K. D., Treede, R. -D., & Campbell, J. N. (1991). Evidence of cutaneous afferents that are insensitive to mechanical stimuli. In M. R. Bond, J. C. Charlton, & C. J. Woolf (Eds.), *Proceedings of the VIth world congress on pain* (pp. 71–75). Amsterdam: Elsevier.

Michaelis, M., Goder, R., Habler, H. J., & Jänig, W. (1994). Properties of afferent nerve fibres supplying the saphenous vein in the cat. *Journal of Physiology, 474,* 233–243.

Mizumura, K., Koda, H., & Kumazawa, T. (1993). Augmenting effects of cyclic AMP on the heat response of canine testicular polymodal receptors. *Neuroscience Letters, 162,* 75–77.

Mizumura, K., Koda, H., & Kumazawa, T. (1995). Different effects of protein kinase-C activation on the responses to bradykinin and to heat of visceral polymodal receptors *in vitro. Society for Neuroscience Abstracts, 21,* 643.

Nahri, M., Yamamoto, H., Ngassapa, D., & Hirvonen, T. (1994). The neurophysiological basis and the role of inflammatory reactions in dentine hypersensitivity. *Archives of Oral Biology, 39*(Suppl.) 23S–30S.

Nishiyama, K., Brighton, B. W., Bossut, D. F., & Perl, E. R. (1993). Peripheral nerve injury enhances alpha-2-adrenergic receptor expression by some DRG neurons. *Society for Neuroscience Abstracts, 19,* 499.

Nishiyama, K., Sugiura, Y., & Perl, E. R. (1995). α2-adrenergic receptor expression in DRG neurons after peripheral nerve injury and sympathectomy. *Fourth IBRO World Congress of Neuroscience Abstracts,* 115.

Ochoa, J., & Torebjörk, E. (1989). Sensations evoked by intraneural microstimulation of C nociceptor fibres in human skin nerves. *Journal of Physiology, 415,* 583–599.

Paintal, A. (1960). Functional analysis of group III afferent fibres of mammalian muscles. *Journal of Physiology, 152,* 250–270.

Perl, E. R. (1968). Myelinated afferent fibres innervating the primate skin and their response to noxious stimuli. *Journal of Physiology, 197,* 593–615.

Perl, E. R. (1984). Pain and nociception. In I. Darian-Smith (Ed.), *Handbook of physiology—The nervous system III* (pp. 915–975). Bethesda, MD: American Physiological Society.

Perl, E. R. (1993). Causalgia: Sympathetically-aggravated chronic pain from damaged nerves. *Pain Clinical Updates, 1,* 1–4.

Perl, E. R. (1994). Causalgia and reflex sympathetic dystrophy revisited. In J. Boivie, P. Hansson, & U. Lindblom (Ed.), *Touch, temperature, and pain in health and disease: Mechanisms and assessments. Progress in pain research and management* (Vol. 3, pp. 231–248). Seattle: ISAP Press.

Pozo, M. A., Gallego, R., Gallar, J., & Belmonte, C. (1992). Blockade by calcium antagonists of chemical excitation and sensitization of polymodal nociceptors in the cat's cornea. *Journal of Physiology, 450,* 179–189.

Price, D. D., Hu, J. W., Dubner, R., & Gracely, R. H. (1977). Peripheral supression of first pain and central summation of second pain evoked by noxious heat pulses. *Pain, 3,* 57–68.

Rang, H. P., Bevan, S., & Dray, A. (1994). Nociceptive peripheral neurons: Cellular properties. In P. D. Wall & R. Melzack (Eds.), *Textbook of pain* (3rd ed.) (pp. 57–78). Edinburgh: Churchill Livingstone.

Raymond, S. A., & Gissen, A. J. (1987). Mechanisms of differential nerve block. In G. R. Strichartz (Ed.), *Handbook of experimental pharmacology. Local anesthetics.* (Vol. 81, pp. 95–164). Berlin: Springer-Verlag.

Reeh, P. W. (1986). Sensory receptors in mammalian skin in an in vitro preparation. *Neuroscience Letters, 66,* 141–146.

Reeh, P. W. (1994). Chemical excitation and sensitization of nociceptors. In L. Urban (Ed.), *Cellular mechanisms of sensory processing* (NATO ASI Series, H 79.) (pp. 119–131). Berlin: Springer-Verlag.

Roberts, W. J., & Elardo, S. M. (1985). Sympathetic activation of A-delta nociceptors. *Somatosensory Research, 3,* 33–44.

Sato, J., & Perl, E. R. (1991). Adrenergic excitation of cutaneous pain receptors induced by peripheral nerve injury. *Science, 251,* 1608–1610.

Sato, J., Suzuki, S., Iseki, T., Tamura, T., & Kumazawa, T. (in press). Increasing impulse activity in the nociceptive afferent fibers by sympathetic activity during chronic inflammatory states. *Japanese Journal of Physiology, 45,* Supp. 2, S238.

Schaible, H. G., & Grubb, B. D. (1993). Afferent and spinal mechanisms of joint pain. *Pain, 55,* 5–54.

Schaible, H. G., & Schmidt, R. F. (1983). Responses of fine medial articular nerve afferents to passive movements of knee joint. *Journal of Neurophysiology, 49,* 1118–1126.

Schaible, H. G., & Schmidt, R. F. (1985). Effects of an experimental arthritis on the sensory properties of fine articular afferent units. *Journal of Neurophysiology, 54,* 1109–1122.

Schmidt, R. F., Schaible, H. G., Messlinger, K., Heppelmann, B., Hanesch, U., & Pawlak, M. (1994). Silent and active nociceptors: Structure, function and clinical implications. In G. F. Gebhart, D. L. Hammond, & T. S. Jensen (Eds.), *Progress in pain research and management, 2: Proceedings 7th world congress on pain.* Seattle: IASP Press.

Schmidt, R., Schmelz, M., Forster, C., Ringkamp, M., Torebjörk, E., & Handwerker, H. (1995). Novel classes of responsive and unresponsive C nociceptors in human skin. *Journal of Neuroscience, 15,* 333–341.

Sengupta, J., & Gebhart, G. (1994). Mechanosensitive properties of pelvic nerve afferent fibers innervating the urinary bladder of the rat. *Journal of Neurophysiology, 72,* 2420–2430.

Sengupta, J. N., Saha, J. K., & Goyal, R. K. (1990). Stimulus–response function studies of esophageal mechanosensitive nociceptors in sympathetic afferents of opossum. *Journal of Neurophysiology, 64,* 796–812.

Sengupta, J. N., Saha, J. K., & Goyal, R. K. (1992). Differential sensitivity to bradykinin of esophageal distension-sensitive mechanoreceptors in vagal and sympathetic afferents of the opossum. *Journal of Neurophysiology, 68,* 1053–1067.

Shea, V. K., & Perl, E. R. (1985a). Sensory receptors with unmyelinated (C) fibers innervating the skin of the rabbit's ear. *Journal of Neurophysiology, 54,* 491–501.

Shea, V. K., & Perl, E. R. (1985b). Failure of sympathetic stimulation to affect responsiveness of rabbit polymodal nociceptors. *Journal of Neurophysiology, 54,* 513–519.

Simone, D. A., Caputi, G., Marchettini, P., & Ochoa, J. (1991). Muscle nociceptors identified in human intraneural recordings, microstimulation and pain. *Society for Neuroscience Abstracts, 17,* 546.5.

Sugiura, Y., Lee, C. L., & Perl, E. R. (1986). Central projection of identified, unmyelinated (C) afferent fibers innervating mammalian skin. *Science, 234,* 358–361.

Sugiura, Y., Hosoya, Y., Ito, R., & Kohno, K. (1988). Ultrastructural features of functionally identified primary afferent neurons with C (unmyelinated) fibers of the guinea pig: Classification of dorsal root ganglion cell type with reference to sensory modality. *Journal of Comparative Neurology, 276,* 265–278.

Sugiura, Y., Terui, N., Hosoya, Y., Tonosaki, Y., Nishiyama, K., & Honda, T. (1993). Quantitative analysis of central terminal projections of visceral and somatic unmyelinated (C) primary afferent fibers in the guinea pig. *Journal of Comparative Neurology, 332,* 315–325.

Szolcsányi, J. (1987). Selective responsiveness of polymodal nociceptors of the rabbit ear to capsaicin, bradykinin and ultra-violet radiation. *Journal of Physiology, 388,* 9–23.

Szolcsányi, J., Anton, F., Reeh, P. W., & Handwerker, H. O. (1988). Selective excitation by capsaicin of mechano-heat sensitive nociceptors in rat skin. *Brain Research, 446,* 262–268.

Taiwo, Y. O., & Levine, J. D. (1991). Further confirmation of the role of adenyl cyclase and of cAMP-dependent protein kinase in primary afferent hyperalgesia. *Neuroscience, 44,* 131–135.

Taiwo, Y. O., Bjerknes, L. K., Goetzl, E. J., & Levine, J. D. (1989). Mediation of primary afferent hyperalgesia by the cAMP second messenger system. *Neuroscience, 32,* 577–580.

Torebjörk, H. E. (1974). Afferent C units responding to mechanical, thermal and chemical stimuli in human non-glabrous skin. *Acta. Physiologica Scandinavia, 92,* 374–390.

Torebjörk, H. E., & Hallin, R. G. (1973). Perceptual changes accompanying controlled preferential blocking of A and C fibres responses in intact human skin nerves. *Experimental Brain Research, 16,* 321–332.

Torebjörk, H. E., LaMotte, R. H., & Robinson, C. J. (1984). Peripheral neural correlates of magnitude of cutaneous pain and hyperalgesia: Simultaneous recordings in humans of sensory judgments of pain and evoked responses in nociceptors with C-fibers. *Journal of Neurophysiology, 51,* 325–339.

Torebjörk, H. E., Lundberg, L. E. R., & LaMotte, R. H. (1992). Central changes in processing of mechanoreceptive input in capsaicin-induced secondary hyperalgesia in humans. *Journal of Physiology, 448,* 765–780.

Torebjörk, H. E., & Ochoa, J. L. (1980). Specific sensations evoked by activity in single identified sensory units in man. *Acta Physiologica Scandinavia, 110,* 445–447.

Torebjörk, H. E., Schady, W., & Ochoa, J. (1984). Sensory correlates of somatic afferent fibre activation. *Human Neurobiology, 3,* 15–20.

Torebjörk, H. E., Vallbo, A. B., & Ochoa, J. L. (1987). Intraneural microstimulation in man, its relation to specificity of tactile sensations. *Brain, 110,* 1509–1529.

Tuckett, R. P., & Wei, J. Y. (1987). Response to an itch-producing substance in the cat. II. Cutaneous receptor populations with unmyelinated axons. *Brain Research, 413,* 95–103.

Uchida, Y., & Murao, S. (1974a). Excitation of afferent cardiac sympathetic nerve fibers during coronary occlusion. *American Journal of Physiology, 226,* 1094–1099.

Uchida, Y., & Murao, S. (1974b). Potassium-induced excitation of afferent cardiac sympathetic nerve fibers. *American Journal of Physiology, 226,* 603–607.

Uchida, Y., Kamisaka, K., Murao, S., & Ueda, H. (1974). Mechanosensitivity of afferent cardiac sympathetic nerve fibres. *American Journal of Physiology, 226,* 1088–1093.

van Hees, J., & Gybels, J. M. (1972). Pain related to single afferent C fibres from human skin. *Brain Research, 48,* 397–400.

van Hees, J., & Gybels, J. (1981). C nociceptor activity in human nerve during painful and non painful skin stimulation. *Journal of Neurology and Neurosurgery, 44,* 600–607.

Wall, P. D., & McMahon, S. B. (1985). Microneurography and its relation to perceived sensation. A critical review. *Pain, 21,* 209–229.

Wall, P. D., & McMahon, S. B. (1986). The relationship of perceived pain to afferent nerve impulses. *Tins 9,* 254–255.

Woolf, C. J. (1994). The dorsal horn: State-dependent sensory processing and the generation of pain. In P. D. Wall & R. Melzack (Eds.), *Textbook of pain* (3rd ed.) (pp. 101–112). Edinburgh: Churchill Livingstone.

CHAPTER 6

Measurement of Pain Sensation

Richard H. Gracely
Bruce D. Naliboff

The experience of pain is critical for survival. Pain is the primary warning system of tissue damage. Almost any stimulus that injures results in pain sensation, including heat, cold, pressure, pulling movements, electrical current, and chemical irritants. Unlike other sensory systems, the sensory apparatus for pain is widespread; a painful sensation may be initiated in almost any part of the body or in the central nervous system (CNS) itself. These varied sites are matched by the variety of pain sensations. The perception of pain is clearly a rich, multidimensional experience, varying in sensory quality, sensory intensity, and in affective–motivational characteristics. However, most studies treat pain as a single dimension varying only in magnitude.

This chapter will first focus on the evaluation of pain as a single univariate variable. Like the overall brightness of a complex visual stimulus, the concept of overall or integrated pain can be meaningful. For example, consider a situation in which a subject is presented with two stimuli, simultaneously varying in quality (e.g., electrical and thermal) and duration, and must choose which stimulus is preferred for a third stimulation. Such choices would be made on the overall integrated impression related to the overall aversiveness of the two stimulus-evoked experiences. In addition, the issues and caveats associated with univariate methods apply also to the

Pain and Touch

separate assessment of important pain dimensions. The following sections will first focus on methods that separately assess two primary dimensions of the intensity and unpleasantness of pain sensations and conclude with a brief critique of methods that evaluate all of the multiple dimensions of pain experience.

I. PAIN AS A UNIVARIATE VARIABLE

A. The Pain Threshold

Figure 1 contrasts conventional sensory-detection thresholds and the special case of determining a pain threshold. In each case the threshold is not a discrete value of stimulus magnitude, but rather a band of stimulus magnitudes over which the probability of a positive response (detection, pain) increases from 0 to 1. The major difference is that for sensory detection, the decision is based on the presence or absence of a sensation, whereas the pain threshold usually involves the determination of the quality of a sensation that is always present. In addition, the Method of Limits is often modified to present only ascending series, thereby avoiding excessive suprathreshold stimulation. Such suprathreshold stimulation could possibly be beyond the tolerance limit of the subject, or likely result in altered thresholds due to mechanisms of receptor sensitization, receptor suppression, or central summation (Meyer & Campbell, 1981; Price, Hu, Dubner, & Gracely, 1977). The use of only ascending series obviously lacks the balanced controls for psychophysical errors that are incorporated into the Method of Limits. The resultant thresholds theoretically should be vulnerable to both errors of anticipation, of indicating the presence of pain before the sensation becomes

-->

FIGURE 1   Comparison of detection threshold and pain threshold. (A) Increasing stimulus intensity results in a transition from no sensation (white area) to a nonpainful sensation (light hatching) to a pain sensation (darker hatching). Thus a detection threshold is a judgment of "stimulus present," whereas a pain threshold is a judgment of the attributes of a sensation that is always present. The text inside the graph show examples of the Method of Limits. The bottom left shows the evaluation of detection threshold. Stimulus intensity is increased in successive discrete presentations on trial one until a positive response (yes) is made. Intensity is decreased on trial 2 until a negative response (no) is made. Several trials are run with varied initial starting intensities. The threshold is defined as the mean of the response transitions for each trial. A common modification of the Method of Limits to measure pain threshold is shown at the upper right of the top panel. Only ascending series are used to avoid excessively painful stimulation. (B) An example of the Method of Constant Stimuli, in which a range of stimulus intensities about the threshold are presented in random sequence. The graph shows the probability of a positive response over stimulus intensity. The result typically is an ogive and the threshold is defined as the stimulus intensity corresponding to a specific response probability (in this case 0.5). This method emphasizes that the transition between no sensation, nonpainful sensation and pain sensation shown in (A) are not distinct and vary over trials.

A

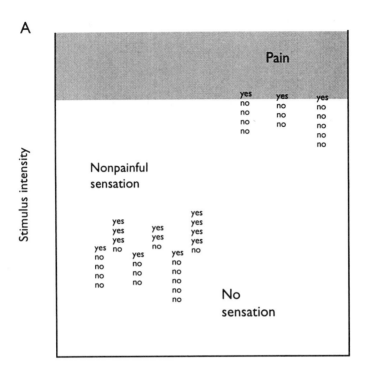

Trials

B

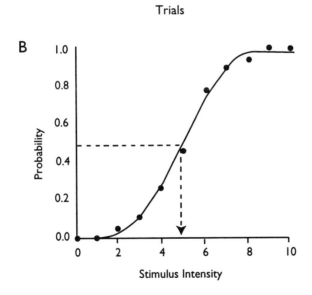

painful, and errors of habituation, of continuing to use the nonpainful responses after the stimulus has become painful (Engen, 1971a). These and additional errors are discussed in the next section.

Figure 2 illustrates the use of the Method of Limits to obtain both detection and pain threshold to sensations evoked by electrical stimulation of the tooth pulp (McGrath, Gracely, Dubner, & Heft, 1983). In this method, the subject holds a cathode imbedded in a nylon cylinder against the incisal edge of an upper central or lateral incisor, with contact made with ordinary toothpaste. An ear clip serves as an anode. In this study, 1-sec trains of 1-msec constant current electrical stimuli were delivered at seven randomly presented frequencies, as shown in the figure. In each session, first detection thresholds then pain thresholds were determined by the Method of Limits. The results shown in Figure 2 clearly illustrate a band of prepain sensation between the detection and pain threshold for electrical tooth pulp stimulation. In addition, Figure 2 shows that detection threshold was not influenced by stimulation frequency, whereas the pain thresholds showed evidence of

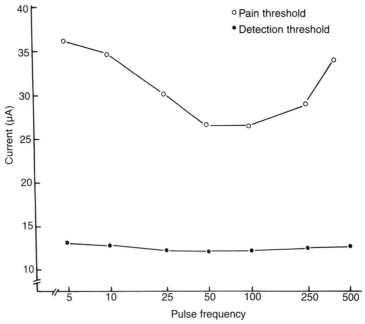

FIGURE 2   Effect of stimulus frequencies on detection and pain thresholds to electrical tooth pulp stimulation. Mean detection thresholds (closed circles) and pain thresholds (open circles) are plotted in microamperes as a function of pulse frequency. Monopolar stimulation consisted of 1-sec trains of 1-msec cathodal pulses delivered to the upper central incisor. Each point is the mean of six determinations from each of 15 subjects. (From McGrath et al., 1983.)

temporal summation at frequencies below 100 Hz and temporal suppression at frequencies above 100 Hz.

This is a significant result because the tooth pulp had been regarded as a source of pure pain sensations without the typical band of nonpainful sensation between the detection and pain threshold. This study identified such a band, and because only the pain thresholds showed temporal effects, further suggested that the nonpainful and painful sensations were mediated by separate neuronal systems. The presence of large-diameter Aβ primary afferents in the tooth pulp, which usually convey touch and not pain sensation, has been verified by anatomical techniques (Byers, 1984).

Figure 3 shows how the problems associated with ascending series may be avoided by the use of adaptive "titration," methods such as the method of double random staircases developed by Cornsweet (1962). These methods present stimuli both above and below the threshold, and are very efficient because most of the responses contribute to the analysis. The double random staircase is a variant of a simple staircase strategy in which a response of "painful" would lower the intensity of the next presented stimulus, whereas a response of "not painful" would increase the intensity of the next presented stimulus. Because the next stimulus is always response-contingent, subjects can become aware of the staircase rule. The double random staircase reduces this contingency by randomly alternating between two separate, independent staircases. The response to a stimulus determines the intensity of the next presentation from that particular staircase the next time that staircase is randomly chosen.

1. Issues in the Determination of the Pain Threshold

There are a number of sources of error in the assessment of pain thresholds. When considering these, it is important to keep in mind that subjects or patients may have goals or hidden agendas quite different from those of the experimenter. They may wish to minimize pain and discomfort, or appear to be "tough" to an opposite sex experimenter (Levine & DeSimone, 1991). They may lack confidence in their ability and strive to give the "correct" answer. In the assessment of an analgesic, they may use perceived side effects as a clue that an active agent has been delivered and change their responses accordingly. Emotional factors such as anxiety also may have an effect.

a. Confound with number of trials or time

Extraneous factors that assist subjects in reproducing their responses may allow them to consciously or unconsciously bias their responses to achieve these goals. For example, many applications of the ascending Method of Limits confound the threshold with the number of presentations of a dis-

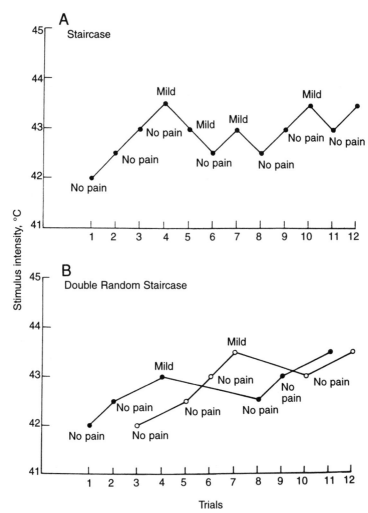

FIGURE 3   Staircase threshold assessment. (A) A simple staircase assessment of pain threshold. Stimulus intensity is plotted against sequential stimulus trials. A stimulus is presented on each trial and the choice of two possible responses determines the intensity of the stimulus presented on the next trial. A response of no pain increases the intensity of the next stimulus, a response of mild pain decreases the intensity of the next stimulus. This method tracks pain threshold efficiently, as most of the responses contribute to threshold determination. However, the relation between the response and the change in the stimulus intensity is readily apparent to the subject. (B) A double random staircase threshold assessment. On each trial, one of two staircases is randomly selected and the stimulus intensity indicated by that staircase presented. The response determines the intensity of the next stimulus presented by that staircase the next time it is again randomly selected. This method reduces the contingency between the response and the intensity of the next stimulus, and the method is not readily apparent to the subject.

crete stimulus or the elapsed time for a continuously increasing stimulus. If subjects desire to respond consistently over repeated trials, they can emit a response at the same time or after the same number of presentations. Subjects who wish to respond appropriately after administration of an analgesic can refrain from responding for a longer time or greater number of trials.

b. Judgment based on irrelevant quality or quantity

The band of prepain sensation shown in Figure 1 may also assist subjects in producing artifactual responses. As stimulus intensity increases, the evoked sensation may change in quality, location, or duration. For example, warm sensations may change from a diffuse feeling to a more confined locus as stimulus temperature increases, and sensations evoked by electrical stimulation may progress from a light tap to a sharp radiation. There is nothing to prevent subjects from using some characteristic change in any aspect of the stimulus-evoked sensations as their personal criterion for pain, and produce orderly responses by using this point as their pain threshold. In contrast, detection thresholds are immune to this problem, because there is no prepain sensation to use as an artifactual criterion.

c. Psychophysical errors

The pain threshold is also subject to the same measurement errors that influence other psychophysical measurements. In the case of the modified ascending Method of Limits, the primary error can be either the error of anticipation or habituation, a tendency to respond either too early or late that is not counterbalanced by descending trails (Engen, 1971a). In the case of the Method of Constant Stimuli, which assesses the probability of a positive response to a range of fixed stimuli (see Figure 1B), the choice of fixed stimuli is known to influence the threshold (Woodworth & Schlosberg, 1954).

d. Shift in the response criterion: The use of SDT

More generally, all psychophysical judgments involve a "response criterion," which in the case of the pain threshold can be defined as the specific attributes of a stimulus-evoked sensation that result in the label "painful." This criterion can vary from liberal to conservative, both across different subjects, and most importantly, within the same subject over time. An individual's response criterion can be assumed to reflect several factors, including both trait characteristics (culture, stoicism, etc.) and state variables, such as demand characteristics, response contingencies, and anxiety. Thus, a pain threshold includes both a sensory magnitude and the labels used to describe this magnitude. Two fundamental problems in pain evaluation are the static case (e.g., "Do differences in pain reports from victims of an auto accident represent different sensory magnitudes or just different labels applied to similar sensory experiences?") and the dynamic case (e.g., "Does a reduction in pain after a therapeutic treatment represent a reduction in pain sensation, or a reduction only in the labels used to describe an essentially unchanged pain sensation?"). These problems immediately spawn a group of important empirical questions: What is the relative influence of psychophysical bias (response criterion) in comparisons between groups or evaluation of a pain-reducing intervention? Can this influence be evaluated or controlled? If not evaluated directly, can variation or predilection be predicted by other measured variables?

The role of the response criterion and the separate assessment of the criterion and sensory performance is explicitly addressed by the use of signal detection or sensory decision theory (SDT). Developed during World War II to assess performance of radar operators, SDT methods result in two measures of performance (Swets, 1964). The first, often represented by the parametric statistic d', is a measure of sensory discrimination, the ability to distinguish between two different sensations. One of these stimuli can be a "blank" stimulus, a condition of no stimulation; thus d' also described the ability to detect the presence of a stimulus-evoked sensation. The second parameter, often represented by the parametric statistic β, is the response criterion, which, as described above, is the magnitude of the label attached to a sensation, or the point at which a sensation is classified as present. Inherent in SDT is the concept of sensory noise, such that a detection-threshold judgment is actually a discrimination between a noise condition (no stimulus delivered) and a signal-plus-noise condition (stimulus delivered). Furthermore, because noise is present, the distribution of sensations produced by these two conditions overlap (i.e., in some instances the sensation during "noise" might be greater than "noise-plus-signal"). Basically, SDT analytical procedures consider a threshold judgment in terms of a 2 × 2 table, as shown in Figure 4. The four cells of the table describe the possible combinations of whether the stimulus is present and whether it is reported as present. The two correct responses are a hit (a stimulus is reported present when it is present) and a correct rejection (stimulus reported absent when it is absent). The two errors are called a miss (stimulus reported absent when present) and a false alarm (stimulus reported present when absent). In statistical terms, SDT treats a threshold determination like a t-test. The statistic d' is directly analogous to the t statistic; it is the measure of difference in the means of two sensory distributions divided by their variability. The response criterion parameter is directly related to the alpha level set by the experimenter. It is the probability of a type I error, which is the same (and more meaningfully described) as a false alarm or false positive. Similar to the t-test, the setting of this criterion to minimize this error increases the other type of error, referred to as type II or beta error (or as a miss or false negative).

Stimulus Reported

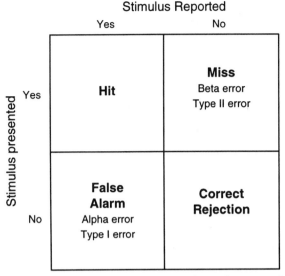

FIGURE 4   Sensory Decision Theory (SDT). The four cells present the four possible outcomes of a stimulus-detection task in which a stimulus may be presented and a subject must determine if it has been presented. If the stimulus is reported present when it is present the response is scored as a hit, and if the stimulus is reported absent when it is absent the response is termed a correct rejection. The other two cells describe the two possible errors. Reporting the stimulus absent when present is a miss, and reporting the stimulus present when it is absent is termed a false alarm. Other common terms for these errors in statistical testing are shown. Two parameters are computed from these cells. Discrimination performance is a function of the relative proportion of hits and false alarms. The response criterion is related to the relative proportion of false alarms to misses.

In terms of the original application to radar operators, d' describes the overall sensitivity of the system, including the visual acuity of the operator, and the characteristic of the radar receiver. The response criterion parameter is set by the operator and in this case determined by the demand characteristics. During wartime it is obviously beneficial to adapt a liberal criterion, to call anything remotely resembling an enemy plane an enemy plane. This vigilant strategy minimizes the misses at the expense of increased false alarms.

SDT has been applied to several studies examining the influence of placebo, nitrous oxide, the minor tranquilizer diazepam, the narcotic analgesic morphine, acupuncture, and transcutaneous electrical nerve stimulation (TENS) (Chapman & Feather, 1973; Chapman, Gehrig, & Wilson, 1975; Chapman, Murphy, & Butler, 1973; Chapman, Wilson, & Gehrig, 1976; Clark & Yang, 1974; Feather, Chapman, & Fisher, 1972; Yang, Clark, Ngai, Berkowitz, & Spector, 1979). Pain in these studies was evoked by either electrical stimulation of the tooth pulp or by heat applied to the skin.

The design of these studies was modified from the usual SDT assessment of detection in which the two stimulus conditions are stimulus present and not present (blank). Applications of SDT to pain assessment, elaborated in the following section on suprathreshold scaling, have usually presented several stimulus intensities and calculated parameters at both threshold and supra-threshold levels. The results of these studies were consistent with an interpretation of d' as pain sensitivity and the response criterion as response bias. Administration of placebo resulted in only a shift in the response criterion without a change in d', whereas the active interventions reduced d' with variable effects on the response criterion.

e. Interpretation of SDT variables

These findings created considerable interest in this technique, and focused renewed attention on the evaluation of analgesia and response bias (Chapman, 1977; Clark & Yang, 1983; Coppola & Gracely, 1983; Rollman, 1977). This focus identified a number of issues that limit the interpretation of results using SDT analytical procedures. For example:

1. Discriminability among sensations is not equivalent to sensory magnitude (Coppola & Gracely, 1983; Rollman, 1977). It includes both sensory magnitude, variability (noise) in the sensory system, and variability in choosing labels to describe sensation.

2. Discriminability between sensations could be based on sensory qualities unrelated to pain magnitude (Gracely, 1994) such as radiation or temporal characteristics.

3. A change in the response criterion could represent analgesia. The International Association for the Study of Pain (Merskey & Bodduk, 1994) defines analgesia as an elimination of pain[1] without anesthesia (i.e., without a change in sensitivity to nonpainful sensation). Because pain is defined as a feeling of hurt (with both sensory and affective components), an intervention that reduces the affective, unpleasant aspect of a pain sensation could conceivably result in a change in the response criterion without a change in the discriminability of the stimulus–evoked sensations.

The results of SDT studies provide considerable evidence that analgesic interventions can alter both sensations and response behavior, but also that the SDT parameters of discrimination and response criterion cannot be interpreted as direct measures of these constructs. Rather, the influence of response bias must be controlled as in other psychophysical experiments; through careful experimental designs that consider the many factors that can influence response behavior.

---

[1] *Hypalgesia* formally describes the reduction but not the elimination of pain. The term *analgesia* will be used loosely to describe both complete analgesia and hypalgesia.

f. Relation to Suprathreshold Sensation: Static and Dynamic

Clinically, threshold-level pain sensations are of little interest. The major medical problem is the management of intense acute pain following injury or medical procedures, and most importantly, the control of relentless pain associated with chronic conditions. Thus, the utility of the pain threshold depends on the relation of this measure to suprathreshold pain sensitivity. In the best case, this relation would be described by Figure 5A. The relation of

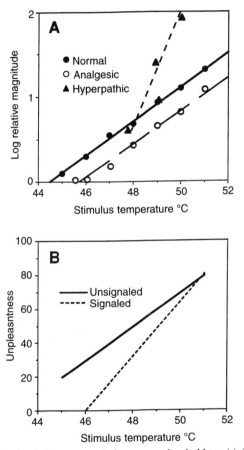

FIGURE 5   Pain thresholds may not indicate suprathreshold sensitivity. Response magnitude is plotted against the intensity of 3-sec thermal stimulus delivered by a 0.8-cm diameter contact thermode with a baseline temperature of 37°C. (A) A typical normal function before (filled circles) and after (open circles) administration of an analgesic such as the opioid fentanyl. The triangles show the type of function that could be obtained in the clinical condition of hyperpathia in which the threshold is raised and suprathreshold sensitivity is increased. (B) Another result in which a signal indicating a restricted range of the most intense stimuli results in lowered ratings of unpleasantness as shown by the dashed line. (Adapted from Gracely, 1991a, and Price et al., 1980.)

pain magnitude to stimulus intensity would be described by a uniform psychophysical function of constant slope in which the threshold, the point at which the function intercepts the abscissa, would completely predict suprathreshold response. This prediction also would be accurate for the dynamic response to an analgesic intervention, resulting in a parallel shift in the function as shown in the figure.

There is a body of evidence supporting both uniform psychophysical pain functions and essentially parallel shifts following analgesic interventions (Gracely, 1991b; Gracely, Dubner, & McGrath, 1979; Price, 1988). However, other effects are certainly possible. For example, Figure 5A also shows the type of function observed with hyperpathia, in which the threshold has been elevated and suprathreshold sensitivity is increased. Figure 5B shows that the dynamic response to an intervention does not have to result in a parallel shift in the pain psychophysical function. In each of these cases, there is a rotation of the function such that suprathreshold pain magnitude is not predicted by the pain threshold.

In addition, in cases in which the threshold does predict suprathreshold responses, as shown in Figure 5A, the usefulness of the pain threshold will depend on other factors. In instances in which pain sensitivity has changed dramatically, the threshold may provide an efficient clinical indicator. In other situations, it may be less sensitive than suprathreshold measures that collect many more responses, reducing random scaling error. Threshold-level measures also may be influenced more by psychophysical biases than by procedures that present a full range of painful stimulation.

B. Suprathreshold Scaling

Psychophysical scaling techniques are described in many secondary sources (Anderson, 1974; Atkinson, Herrnstein, Lindzey, & Luce, 1988; Bock & Jones, 1968; Bolanowski & Gescheider, 1991; Engen, 1971b; Falmagne, 1986; Gescheider, 1985; Luce & Krumhansl, 1988; Marks, 1974; Stevens, 1975; Torgerson, 1958), including volume 2 of this handbook (Carterette & Friedman, 1974). Briefly, most suprathreshold methods can be divided into three classes: (a) discrete numerical or verbal category scales, (b) bounded or confined continuous-measure equivalents of bounded scales, such as the 10-cm visual analog scale (VAS), and (c) unbounded scales, such as the well-known method of Magnitude Estimation.

1. Discrete Categorical Scales

Discrete numerical (e.g., numerals 0 to 10) or verbal (none, mild, moderate, intense) category scales can be analyzed in at least three ways:

1. The categories can be considered to be equal steps and assigned in-

tegers—1, 2, 3, etc.,—a procedure formally titled the Method of Equal Appearing Intervals (Engen, 1971b).

2. The category values can be determined empirically from the response behavior of the subject simultaneously with the actual scaling of stimulus attributes. In this analysis, formalized by Thurstone (1959), the spacing of two categories is inversely related to the amount they overlap, the relative proportion that each response is used to describe the same stimulus.

3. Category values may be determined independently from their use in describing evoked sensation.

a. Analysis of discrete category scales

i. Method of Equal Appearing Intervals   A large number of pain studies have used simple category scales, such as the 0–10 numerical scale (Brennum, Arenndt-Nielsen, Secher, Jensen, & Bjerring, 1992). Four-point category scales of pain (none, mild, moderate, severe) or pain relief (none, some, lots, complete) have been the standard in clinical trials of analgesic agents (Max & Laska, 1991). Even the widely used McGill Questionnaire (MPQ) (see section III), with empirically defined category values, is usually analyzed by assigning integer ranks to each category subscale. In each case, the analysis assigns integers to these categories, thus implicitly defining an interval scale in which the psychological distance between successive categories is the same for each adjacent category pair.

The major criticism of this analysis is that it may greatly distort the scale, arbitrarily assigning equal distances between, for example, severe pain and moderate pain, and mild pain and no pain. Such distortions could have a number of consequences. In the assessment of an analgesic, a drug that alleviates moderate pain may appear more efficacious than a drug that is capable of reducing severe pain to moderate levels. In the assessment of varying chronic pain, this analysis would indicate greater overall pain in individuals with constant moderate pain than in those experiencing debilitating attacks of severe pain half of the time with little pain at other times. Nevertheless, the Method of Equal-Appearing Intervals has been used extensively with satisfactory results, and until studies specifically identify limitations, its simplicity practically assures continued use by clinical investigators.

ii. Empirical determination of category values during scaling   Thurstone (1959) developed a scaling analysis based on the distribution of category responses to a stimulus set. In the case of the Method of Successive Categories (Engen, 1971b; Thurstone, 1959), category values are assigned based on the discriminal dispersion, or the variability of category responses to each stimulus. This method is related to other procedures that use discriminability of stimuli as the unit of psychological measurement, such as the concept of just noticeable differences (JNDs), and the analysis can be applied to category scaling data and Thurstone's method of Paired Comparisons. The use of the sensitivity measure of SDT also is directly related to the Method of Successive Categories.

This analysis has been applied to pain measurement (LaMotte & Campbell, 1978) and could be applied to many more studies currently using the Method of Equal-Appearing Intervals to analyze categorical data. Indeed, a post hoc parallel analysis comparing the applied utility of each method is readily available to those who have used, or plan to use, categorical pain scales to rate multiple presentations of a stimulus set. A related method termed item response theory (IRT) has recently been used to analyze specific items in categorical scales of clinical pain. Items with broad discriminal dispersions (i.e., that are used to describe a broad range of pain magnitudes) can be identified by IRT, which can be used for item weighting or for item selection during further scale development (McArthur, Cohen, & Schandler, 1989).

iii. Independent determination of category values   In contrast to the Thurstone methods of category value determination simultaneous with scaling, several methods have been used to quantify category values in a procedure separate from the use of the scale for pain assessment. The extensive work on independent techniques for category scaling is due to both the importance of having a technology for sensitive pain assessment and the fact that clinical pain studies cannot provide the stimulus control needed for Thurstone analysis. The widely used MPQ is composed of 20 subscales. The two to six items within each subscale were quantified by the use of both five and seven-point numerical scales and a Thurstone analysis, and by the use of five to seven quantified descriptors based on this analysis (Melzack & Torgerson, 1971).

Tursky (1976) requantified some of the MPQ items using a ratio–scaling paradigm advocated by Stevens (1975).[2] A set of descriptors were presented as visual stimuli and subjects used a ratio–scaling method to rate the relative magnitude of the descriptor property implied by each descriptor. Subjects also used the same ratio–scaling method to rate the lengths of randomly presented lines. The psychophysical function describing the particular ratio-scaling response method to line length was then used as a calibration function to convert the mean response made to each descriptor from the unit of the response used to a common unit of line length, referred to as *relative magnitude*.

[2] Ratio-methods are described briefly in a later section and more extensively in the references cited at the beginning of this section.

This method was noteworthy for several reasons. It allowed the use of more than one response method, because the results of each were converted to a common unit. The use of multiple methods permitted a check of internal consistency, because each method should result theoretically in the same scale. Verification that such scales were similar provided validation for both the methods and for the general concept that verbal descriptors, with no physical metric of their own, could be rated as reliably as physical stimuli, and the accuracy of ratings verified by statistical techniques. Once scales generated from different response methods are demonstrated to be similar, these scales can be combined into an overall scale of magnitude. Figure 6 shows an example in which descriptors of sensory intensity were quantified by cross-modality matching to handgrip force and to time duration (Gracely, McGrath, & Dubner, 1978a).

The calibration procedure has further benefits. In addition to converting different response methods to a single unit, it also normalizes eccentric use of a single-response method across individuals. For example, the power function exponent describing time-duration judgments to line length should approximate unity, yet may vary considerably between individuals. If an individual is consistent in describing both verbal stimuli and line length, any eccentric use should cancel out, resulting in a common scale

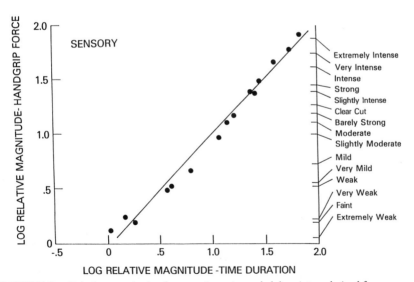

FIGURE 6   Relative magnitude of sensory intensity verbal descriptors derived from cross-modality matching to handgrip force (ordinate) and to time duration (abscissa). Each point is the geometric mean of 80 responses of 40 subjects to a single descriptor. The mean of the two magnitudes for each descriptors is shown on the right vertical axis. (From Gracely et al., 1978a.)

across individuals. Furthermore, Stevens (1975) proposed this method as one means of reducing the regression bias inherent in psychophysical scaling. This bias refers to the findings that the range of a response modality is usually contracted in comparison to the stimulus modality. For example, the exponent describing the power function derived from drawing line lengths to numbers should be the reciprocal of the exponent derived from using numbers (Magnitude Estimation) to describe line lengths. In practice, both exponents are reduced from the theoretical expectation. Assuming that this response regression would apply equally to ratings of descriptors and calibration stimuli (in this case, line length), this regression effect should cancel out with this paradigm.

This method is described briefly here (for further information see Duncan, Bushnell, Lavigne & Duquette, 1988; Gracely, McGrath, & Dubner, 1978a,b; Tursky, 1976). The main point of these procedures is that they provide a method to assess the consistency and reliability of verbal category values. For Gracely et al.'s (1978a) quantification of pain sensory-intensity descriptors, the properties were acceptable, mean individual reliabilities were 0.96 over 1 week, and the mean of correlations between each individual's scale and mean scale from the remaining subjects in a group was also 0.96. The equivalence of these correlations indicated objectivity; an individual's quantification of the verbal descriptors was predicted equally well by that individual or by a group norm. This objectivity indicates general agreement about the ordering and spacing of English words describing magnitudes of sensory intensity, validating the use of a group normative scale. The results of these pain-related investigations are in general agreement with previous psychological studies of semantic meaning. For example, the adverbs "very" and "slightly," when used to modify adjectives of pain magnitude (Gracely & Dubner, 1987) or pain unpleasantness (Walther & Gracely, 1986) appear to serve as multiplicative modifiers independent of the adjective modified (e.g., pain, distressing, unpleasantness). This property was described by Cliff (1959) and the constants found in that study using a Thurstone analysis of paired comparisons (1.30 for very, 0.55 for slightly) are in general agreement with those found with ratio-scaling procedures (1.44 for very, 0.42 for slightly; Gracely & Dubner, 1987).

These methods and psychometric properties, however, do not address the variability in the absolute magnitude of the words, for example how much sensation is represented by the word *mild*. This is only accomplished in an indirect way by equating the full range of words across individuals and relying on the face validity of the use of a common language. The issue reduces to what is the best subjective yardstick over a population; the use of language, the number "4" on a 0–10 numerical scale, or a mark placed 4 cm along a 10-cm VAS, described below. Another approach involves the use of evoked pain sensations as the standard, for example, instructing patients to match their clinical pain to the pain evoked by a 47°C heat stimulus applied to their arm, or to match the intensity of their pain to that of a nonpainful stimulus, such as the brightness of a light. (Duncan, Feine, Bushnell, & Boyer, 1988). Another approach anchors a subjective scale by using the same scale to rate both clinical and experimental pain sensations (Gracely, 1979; Heft, Gracely, Dubner, & McGrath, 1980). The validity of this approach is based on the assumption that pain sensitivity is equivalent across the population. Certain stimuli, such as heat, approach this ideal, whereas others show a wide variability across individuals. For example, electrical tooth pulp stimuli reveal a broad range of sensitivities that probably mainly reflect differences in tooth geometry and not differences in pain sensitivity (Gracely, 1985). However, in the best case, there is considerable variation in experimental pain sensitivity across a population, and always the possibility of a disease or other process that alters sensory sensitivity in a particular individual. Thus, the ultimate validity of an "absolute" pain scale is based heavily on the face validity of a common language (or number, VAS use, etc.), possibly buttressed by additional information gathered from responses to controlled standard stimulation.

b. Category scale formats and paradigms

The quantification of category values, however, does not constitute a pain scale. Establishing the values for the "items" does not address the issue of how these categories should be used to measure pain. Categorical scales can be presented in ordered sequences as a simple checklist, with the patient required to choose the most appropriate category to describe pain magnitude. Quantification of the category values permits additional formats. Categories may be placed at the appropriate points along a continuum to form what has been called a "graphic rating scale" (Gracely, 1991b; Heft & Parker, 1984).

i. Clinical scales using multiple items Quantification of category words also allows presentation of individual category items as comparison stimuli. This is advantageous because the common practice in clinical pain assessment is to obtain only a single rating. This single assessment contrasts sharply with psychophysical methods in which responses can be gathered to a hundred stimuli, reducing error variance and allowing for evaluation of measurement consistency. If psychophysical methods can be considered to be the collection of multiple stimulus–evoked responses in relation to a single subjective standard, quantified categories can be used to collect responses comparing a single clinical stimulus to multiple subjective standards. At least one clinical scale has used this approach by presenting 12 quantified descriptors and instructing subjects to rate if their clinical pain is equal to that implied descriptors, or how much less or greater on a +10/

−10 scale. This Descriptor Differential Scale (DDS) has been administered to patients in acute or chronic pain in several studies, revealing adequate reliability, internal consistency, validity, and sensitivity (Doctor, Slater & Atkinson, 1995; Gracely & Kwilosz, 1988; Hargreaves, Dionne, Mueller, Goldstein, & Dubner, 1986; Kwilosz, Schmidt, & Gracely, 1986).

The use of multiple, comparatively rated, descriptor items permits additional analyses and utilities not possible with single-item scales. For example, the pattern of item responses provides a second, independent quantification of the descriptor values that can be compared to either group or individual norms to provide a measure of scaling consistency. Preliminary studies have shown generally high consistency that improves from the first to the second administration (Gracely & Kwilosz, 1988). These consistency measures can be used to identify subjects who obviously are not attending to the scaling task. Eliminating these subjects may improve psychometric properties and increase the sensitivity of an analgesic assay.

The use of multiple items also allows the construction of alternative forms, each with different descriptor items, that are theoretically equivalent. Kwilosz, Gracely, and Torgerson (1984) have validated such forms and used them in a study demonstrating the accuracy of memory of acute postsurgical pain over a 1-week period. The use of alternative forms is critical in such studies because otherwise the demonstrated memory for pain may actually be the memory of a previous pain rating.

Finally, the use of multiple items decreases the influence of random scaling errors. The reliability and homogeneity of the overall score of the DDS has been shown to be greater than the mean score obtained from each individual item (Gracely & Kwilosz, 1988). This improvement is consistent with recent findings that averages of multiple responses to a single-item pain diary are superior to single responses or to the mean of only a few responses (Jensen & McFarland, 1993).

ii. Randomized category scales   Quantified categorical items also allow the use of a novel scaling paradigm in which the categories are presented in random order and subjects quantify stimulus–evoked sensation by choosing the appropriate category from the randomized list (Gracely et al., 1978a). The use of randomized category items forces category choices based on the semantic content of the item rather than on its spatial location in a list. Although possibly more difficult, there are at least two good reasons to use randomized scales. First, basing choices on the meaning of a word or phrase may facilitate the distinction between different dimensions of pain experience, discussed below in the section on dual dimensions of pain. Second, choices from a randomized list may reduce or completely avoid the family of rating biases associated with choosing responses from a bounded scale.

There is considerable evidence indicating that subjects tend to spread repeated responses evenly over a category scale. This response behavior results in scales that are sensitive to all aspects of the distribution of delivered stimulus intensities, including range, spacing, and frequency of presentation (Parducci, 1974). Varying any of these parameters has been shown to distort the category scale. Most psychophysical treatments of this effect have striven for a "true" scale void of these distortions. Approaches include iterative scaling techniques in which distortions are systematically eliminated (Pollack, 1965; Stevens, 1975) and the use of open-ended scales described below. Although an undistorted scale is also desirable for pain measurement, applied studies of pain may be more concerned with the influences of these response behaviors in the study of analgesic interventions. In contrast to the static goal of a true category scale, such studies are concerned with an accurate evaluation of a change in pain produced by the intervention. In this dynamic situation, the tendency to produce a uniform response distribution would diminish the response to a real effect, degrading the sensitivity of the assay. This normalizing effect is recognized statistically as regression to the mean, physiologically as The Law of Initial Values, and informally as floor—ceiling effects.

Randomized category scales can be expected to eliminate or reduce this response behavior because the task is changed from (a) matching a stimulus space to a response space to (b) matching perceived intensity to the intensity implied semantically by a specific verbal description. The degree of influences of spatial biases and the possible improvement of randomized category paradigms have not been explored in either pain or general psychophysical studies. A preliminary study has found that the response distributions of a 0–6 category scale showed a reduced response range immediately after a pain-reducing intervention, which quickly spread to again incorporate the higher categories by the end of the postintervention scaling trials. In contrast, a randomized verbal descriptor scale showed a reduction that remained stable throughout the postintervention scaling trials (Gracely, Taylor, Schilling, & Wolskee, 1984).

2. Bounded Continuous Scales: The Visual Analog Scale

The VAS is presently one of the most widely used pain measurement tools. The VAS is usually a 10-cm line labeled at the ends with descriptors such as "no pain" and "worst pain imaginable." Patients or experimental subjects indicate pain magnitude by simply marking the line, and a ruler is used to quantify the measurement on a 0–100-mm scale. Variations include vertical or horizontal alignments, placing descriptors along the scale, and scales of different lengths (Jensen, Karoly, & Barver, 1986; Price, 1988).

Although the ease of administration and scaling have been sufficient to ensure the clinical popularity of the VAS, there is growing evidence to support the reliability and validity of this method. It is used increasingly in clinical pharmacology, a field long dominated by the consistent use of four-point categorical scales of pain intensity and pain relief. VAS scales have been shown to be equal or better than these scale, and there is some consensus that the VAS scale of pain relief (e.g., labeled "none" or "complete") may be the most sensitive measure of analgesia in the acute situation (Wallenstein, 1991).

The validity of the VAS is also supported by psychophysical studies described below in the sections on ratio scaling and measurement of dual pain dimensions. The continuous nature of the scale allows a greater response resolution, limited by the observers discriminative capacity rather than by the scale. It also allows for increased independence of multiple responses in designs requiring repeated measures, because the use of a new VAS for each measure reduces the influence of previous responses and increases the probability that each response will be based on the present subjective experience. In contrast, subjects using a category scale can appear consistent by repeatedly choosing the same category without attending to subjective sensation, resulting in artifactual reliability.

Other bounded continuous scales have been applied to pain assessment. Squeezing a handgrip dynamometer is a natural response to a brief pain stimulus, and the method has produced reliable results to verbal stimuli (Gracely, McGrath, & Dubner, 1978a; Tursky, 1976), controlled electrical (Gracely et al., 1978b) and thermal (Price, Hu, Dubner, & Gracely, 1977; Wolskee, Gracely, Dorros, Schilling, & Taylor, 1983) skin stimulation. Other measures have included a finger span device, a mechanical slide rule, and a slider along the back of a meter stick (Cooper, Vierck, & Yeomans, 1986; McGrath et al., 1983; Price, 1988; Price, Bush, Long, & Harkins, 1994).

Because the VAS and similar scales are bounded (i.e., present a finite response space), they also should be vulnerable to the spatial rating biases demonstrated with category scales. They should be sensitive to stimulus range, spacing, and frequency of presentation. Subjects may spread out their responses after an analgesic intervention, reducing sensitivity. A few studies have addressed these biases with scales such as the VAS (Fernandez, Nygren, & Thorn, 1991), but generally these effects have not been investigated in any detail.

3. Unbounded Scales: Ratio-Scaling Methods

Stevens (1975) distinguished between four main classes of scales: (a) nominal, identifying names with no metric information, such as common names or football jersey numbers; (b) ordinal, ranks with no information about distances between values, as finishing first, second, and third in a race; (c) interval, a scale with equal units with no meaningful zero point, such as the Fahrenheit temperature scale; and (d) ratio scales, which have both a true zero-point and a defined interval, corresponding to physical scales such as length and mass.

Stevens's advocacy of ratio-scaling methods has had a definite impact on the field of psychophysics. This influence, and the details of these methods form an extensive literature (for review see Bolanowski & Gescheider, 1991; Carterette & Friedman, 1974; Engen, 1971b; Gescheider, 1985; Falmagne, 1986; Marks, 1974; Stevens, 1975). These methods share four characteristics:

a. Characteristics of ratio scales

i. Instructions The instructions to subjects emphasize ratio or proportional judgments. The following is an example by S. S. Stevens (1975, p. 30.):

> You will be presented with a series of stimuli in irregular order. Your task is to tell how intense they seem by assigning numbers to them. Call the first stimulus any number that seems appropriate to you. Then assign successive numbers in such a way that they reflect your subjective impression. There is no limit to the range of numbers that you may use. You may use whole numbers, decimals or fractions. Try to make each number match the intensity as you perceive it.

The instructions can include a modulus, "rate the sensation produced by the first stimulus a '10' and then make proportional judgments thereafter" or be modulus-free, allowing any number for the first response. The first stimulus delivered can be a "standard" or be randomly chosen.

ii. Open-ended response A touted feature of many ratio-scaling methods is that they avoid the biases and problems of a bounded scale by offering an unlimited response continuum. This is literally true for measures such as magnitude estimation (number choice), time duration (given patience), and line production (given enough paper!), and virtually true for brightness (13-log unit range). Other ratio responses, which can be any adjustable continuum, may be relatively unbounded in practice but ultimately bounded due to physical (handgrip force) or safety (loudness) limitations.

iii. Power functions The relation between response and stimulus intensity is described by a power function of the form $R = S^n$ in which R is the response, S the stimulus, and n the power function exponent (ignoring additive constants). These power relations are usually expressed in logarithmic plots ($\text{Log } R = n + \text{Log } S$) because these are linear, and the exponent is equal to the slope of the function. Ratio scaling assumes theoretical relations between the stimulus and response modalities and the resulting exponent. Essentially, each modality is associated with its own specific exponent, and the exponent derived from a psychophysical function is the ratio of the response-specific exponent divided by the stimulus-specific exponent. The most common response method is Magnitude Estimation, in which numbers are assigned to stimulus-evoked sensations. Number use has been arbitrarily assigned an exponent of 1; thus the specific exponent for any stimulus modality assessed by magnitude estimation would be equal to the reciprocal of the exponent of the psychophysical function, because the function exponent is equal to the response exponent (1) divided by the stimulus exponent.

Magnitude Estimation is a specific instance of cross-modality matching, which describes the general case in which any adjustable response is used to match responses to any controllable stimulus modality. The empirically derived exponent from this match will be related to the theoretical value computed from the modality-specific exponents described above. It will likely not be exactly the same because the empirical exponent is influenced by a number of other factors. These include the influence of regression bias described above in the quantification of category scales. This effect has been explained as a tendency to contract the response continuum, resulting in an empirical exponent always less than the theoretical exponent computed from the ratio of the response and stimulus-specific exponents. Other factors, such as stimulus range and individual differences among subjects, can also influence the empirical exponent.

iv. Stimulus set   The stimuli administered during a ratio-scaling task and other response methods usually follow certain norms. A fixed set of stimulus intensities are administered more than once in a randomized sequence. For example, 7 stimuli are repeated three times each and presented randomly in a block of 21 stimuli that may be repeated to allow analysis over blocks. The spacing of the stimuli are chosen to produce subjectively equal distances. Given Weber's and Fechner's law, this usually means equal logarithmic steps between stimuli. Stimulus spacing also should be small enough to cause confusion about the number and identity of the stimuli. Large spacing can result in stimulus identification, and thus artifactual reliability resulting from attempts to give the same response to the same, identified stimulus.

Although the open-ended nature of ratio-scaling methods such as Magnitude Estimation have been shown to reduce biases due to stimulus range, frequency, and spacing, there is considerable evidence that power function exponents can be influenced by factors such as stimulus range (Poulton, 1979). Indeed, the variation of modality-specific exponents reflects inherent ranges of various modalities, (Teghtsoonian, 1971).

Ratio-scaling methods have been applied to experimental pain assessment, with resulting power function exponents of 1.0 for radiant heat, 2.1–2.2 for contact heat, and 1.8 for electrical skin stimulation (Gracely, 1977; Hardy, Wolf, & Goodell, 1952; Price, 1988; Tursky, 1974). These applications raise two critical issues. First, what are the factors that influence these exponents, and second, what is the purpose of establishing exponents in pain research?

b. Factors that can influence exponents derived from pain stimuli

Interpretation of exponents must always consider various (and arbitrary) physical measures of stimulus intensity, because different measures of the same modality may be nonlinearly related. For example, the difference in exponents for radiant and contact heat may reflect the differences in unit; the exponent of 1.0 found for radiant heat is based on a unit of power (watts), whereas the exponent of about 2.2 found for contact heat is based on a unit of temperature.

Similarly, the exponents of close to 2.0 found for electrical stimulation of the skin are based on the quantification of stimulus magnitude in terms of current (mA). If electrical power arbitrarily was used instead, the obtained exponents would be approximately 1.0 because electrical power is equal to the current squared times the electrical resistance.

To avoid scale distortions near threshold, Stevens (1975) incorporated a threshold correction into the power law, which subtracted the threshold from the stimulus intensity. In this case the power law, $R = Sn$ becomes $R = (S - T)n$, in which R is the response, S is stimulus intensity, T is the threshold correction, and n is the exponent. The use of this correction with stimulus modalities with a large stimulus range below detection threshold (and most pain modalities with a large stimulus range below pain threshold) ensures that the psychophysical function begins near the origin, and that increases in stimulus intensities near threshold result in a corresponding increase in sensory magnitude.

The use of this correction can significantly influence the value of the obtained exponent. Thus, a critical comparison of exponents must ascertain whether such a correction is used. In addition, the choice of the correction factor may also influence the reported exponent. The threshold correction can be based on an empirically determined threshold, and has been determined by iterative techniques that maximize the goodness of fit to a power function (maximize Pearson correlation in a log × log plot). The influence of this correction factor is an important issue in the studies of Price, McGrath, Rafii, and Buckingham (1983), who used a correction of 34°C, which is distinctly different than the observed pain thresholds of about 45°C. They used this correction to compare their psychophysical results with power functions generated from neural responses, in which the value of 34°C is a good estimate of the threshold at which primary afferents begin firing to a thermal stimulus. Although necessary for these types of comparisons, this correction would certainly influence the power function exponent in comparison to other psychophysical studies of pain that used the pain threshold of approximately 45°C as the logical correction factor.

In addition to choice of threshold correction, obtained exponents can be influenced by the stimulus and response range (Poulton, 1979). The use of a small stimulus range, not uncommon when delivering painful stimuli, will increase the value of the exponent. Similarly, constriction of the response scale, such as using a bounded VAS scale, will usually lower the exponent to about one-half of that obtained with an unbounded scale. These lowered exponents have been termed *virtual exponents*, because they reflect both the exponent of the underlying unrestrained continuum, and the compressive effect of limiting the top of the scale (Stevens, 1975). It is interesting to note that the exponents obtained by Price et al. (Price, 1988; Price et al., 1983) for VAS measures of thermal pain intensity (2.1–2.2) are similar to those found with line production, because the compressive effect should have resulted in a VAS exponent approximately one-half that observed for line production. Price et al. (1983) explain this effect as resulting from their specific instruction to subjects, the use of the longer than standard (10 cm) 15-cm scale, and pre-experience with the stimulus range. They posit that this training converts the bounded VAS to essentially a line production response, because subjects make appropriately small responses from the beginning, responses not influenced by the scale limit of 15 cm. This is considered to be desirable because the bounded scale has been converted essentially to an unbounded scale that is minimally influenced by factors such as stimulus spacing known to distort category scales. The exponents obtained with the Price et al. (1983) VAS method can also be considered to be an excellent example of how the power function exponent can be manipulated by subject instructions; in this case it is increased by instructions and demonstrations designed to avoid the confining effect of the top of the scale. This becomes an important issue when exponents are compared across studies that may use different instructions and contain subjects with varying comprehension of these instructions. The successful avoidance of scale end effects may also be compromised in the evaluation of interventions that increase the range of stimulus-evoked sensations.

c. The utility of power function exponents in pain measurement

In traditional Steven's psychophysics, the size of this exponent is determined by properties of both stimulus and the response, and can be influenced further by factors such as the stimulus range (Poulton, 1979). What must be stressed is that this and other scaling methods have been concerned primarily with the relative effects of different stimulus intensities. For example, the Steven's ratio-scaling methods focus on the slope, which describes the rate that perceived intensity increases with a given increase in stimulus intensity. Much of the literature is concerned with the variation of slope with different stimulus modalities or response methods. The results of ratio-scaling methods also are compared to the results obtained with other procedures, such as category scales, in an effort to find the "true" function relating stimulus energy to the magnitude of evoked sensation.

However, there is a fundamental difference between the applied needs of a pain measurement tool and the goals of most psychophysical studies. Pain measurement is concerned primarily with the absolute level of sensation, with issues such as, Does this group experience greater pain than this group? or, Did this intervention significantly reduce pain magnitude in comparison to placebo?

Unfortunately, most psychophysical studies have focused on the shape, and not the overall height, of the psychophysical function. Indeed, in many paradigms this height is purposefully meaningless. For example, in the free modulus situation, in which subjects use their own "unit" of measurement, the height of this function reflects arbitrary assignments of the size of the scale unit. In situations in which the initial stimulus (which can be a fixed standard) is described by a fixed modulus, such as 10, this procedure stabilizes the units between individuals but adds no meaning to overall magnitude.

The emphasis on the shape of the function pervades much of the psychophysical literature, including comparisons of the different functions generated by bounded category scales and unbounded ratio-scaling procedures. Because of this emphasis on slope rather than height, much of this literature may have little applied utility to pain evaluation.

However, there have been a few applications of power function exponents to investigations of pain. For example, Mountcastle (1967) compared psychophysical functions of subjective estimates of stimulus-evoked sensations to functions constructed from the neural responses to the same stimuli. He interpreted similar exponents as evidence that subjective sensations resulted from peripheral sensory transduction and invariant processing from primary afferent receptor to consciousness. This conclusion was controversial, because it diminished the role of known spinal and thalamic processing of sensory input (Kruger & Kenton, 1973; Warren & Warren, 1963). Recently, several studies of painful stimulation have renewed the interest in the relationship between neural data and conscious reports. These studies have examined not only primary afferent responses to brief ($\approx$3 sec) noxious contact heat stimuli, but also neural responses in the spinal dorsal horn, spinal projection neurons, superior colliculus, and cerebral cortex. In each case, the power function exponent was between 2.1 and 2.5, which is also consistent with the exponents of 2.1–2.2 observed in human judgments of the same stimuli (see Price, 1988, for review). Thus, for at least the limited case of phasic thermal stimuli, the power function obtained from psychophysical judgments describes a neural response relation that appears to be maintained throughout the entire afferent nervous system.

Power function exponents have also been used to support the validity of pain scales. Following the method of Dawson and Brinker (1971), Gracely (1977) plotted handgrip force responses to verbal descriptor stimuli against time duration responses and found a cross-modality power function exponent of 0.66, which agreed with the exponent (0.63) found comparing these responses to physically measurable line length stimuli. Because the descriptor stimuli have no physical metric of their own, this consistent cross-modality relation provided one line of evidence that each descriptor of pain intensity implied a reliable magnitude of pain intensity.

Whether these ratio-scaling methods truly provide ratio-level measurement is an issue that has been debated in both the pain (Gracely & Dubner, 1981; Hall, 1981) and general (Anderson, 1974; Ekman, 1964; Warren & Warren, 1963; Zinnes, 1969) psychophysical literature. For applications to pain assessment, ratio level judgments permit meaningful statements about changes in pain that are independent of the unit of psychological measurement. At the very least, it is reasonable to pursue scales that contain as much information as possible (Gracely & Dubner, 1981). In addition, there are converging lines of evidence consistent with ratio-level measurement (Gracely & Dubner, 1981), including a study by Price et al. (1983), who asked subjects to determine stimulus intensities that evoked sensations that were a specified ratio to a standard sensation produced by standard stimulus. They found that these stimuli were closely predicted by VAS psychophysical functions; for example, 46.5°C produced a sensation judged to be twice as intense as that produced by a 43°C stimulus, whereas an independently determined psychophysical function predicted a temperature of 46.4°C (response value double the response to 43°C).

d. Inferring pain magnitude from ratio scales

The traditional emphasis on the exponent (slope) and not the height of the psychophysical function limits the utility of commonly used ratio procedures. To be useful for pain assessment, scales such as Magnitude Estimation and line production need to be anchored to a subjective standard similar to the concept of response criterion in SDT analyses (Clark & Yang, 1983; Swets, 1964). Just as the response criterion is known to be variable and somewhat arbitrary, the concept of subjective standard is an ideal that can be approached but never obtained with complete certainty. It is an important, even essential, property of pain measurement that can never be completely isolated.

As with bounded scales, subjective anchoring of ratio scales can be approached in a number of ways. In the bounded VAS scale, the finite line is assumed to represent the possible range of pain magnitude, with the consequence that a mark of 4 cm on a 10-cm scale represents a moderate amount of pain and a 6-cm mark indicates a pain magnitude approaching more intense levels. Numerical category scales also represent the range of possible pain magnitude in a response space. Verbal categories similarly represent a response range, with the addition of linguistic labels that may further specify a subjective standard. One issue for verbal category scales is the relative contribution of these two sources of subjective standards. Subjects could use the meaning of the words irrespective of the number or spacing of the categories, or alternatively treat the scale as a numerical scale without attending to the meaning of the words, or some combination of both. One feature of randomized scales discussed above is that this ambiguity is reduced because information provided by spatial location in a list is purposefully omitted.

Subjective standards have been applied to ratio-scaling methods in two fundamental ways. In the first, a subjective standard is associated with a modulus in much the same way a stimulus standard is used. For example, Hilgard, in a study of hypnotic modification of cold pressor pain, instructed subjects to use the modulus "10" to describe a "critical level of pain," one that "they would very much wish to terminate." The other method allows subjects to make free modulus judgments of both pain stimuli and intermingled verbal stimuli that imply pain magnitudes (e.g., moderate, intense), pain experiences (childbirth, a headache), or a behavioral consequence, ("pain severe enough to take a minor or major medication"). This second paradigm would simultaneously scale the stimuli, the subjective standard, and interlock the responses to these two stimulus sets, effectively anchoring the pain responses. This method has been used to scale pain produced by electrical tooth pulp stimulation and assess the action, and interaction, of analgesic agents (Gracely, Wolskee, Deeter, & Dubner, 1981; Gracely & Wolskee, 1983).

This second class is related to the use of randomized verbal descriptors quantified by ratio (or other techniques). In this method the subjective anchors are also presented as stimuli and quantified, although in a session separate from scaling pain sensations. Once quantified, the descriptors are formed into a response scale and used to directly measure pain sensations effectively. Further consideration of these two measures reveals that they are functionally equivalent, each with specific advantages. Pain scaling with the descriptor method is efficient, requiring less stimuli because each response anchors the stimulus-evoked sensations. A separate descriptor quantification session may not be needed, because consistency has been observed across individuals and can be easily checked by simple procedures, such as ranking (see Gracely et al., 1978a). On the other hand, descriptor choices force each response into a discrete category, and subjects can exhibit poor response behaviors, such as perseverance or use of a limited number of descriptors. Although longer, the use of ratio-scaled responses to both pain and semantic stimuli may provide greater resolution and eliminate problems associated with choosing descriptors.

4. Other Suprathreshold Scaling Techniques: Discrimination and Single-Stimulus, Single-Response Designs

a. Discrimination

SDT analysis of pain responses was discussed in the section on threshold measure because SDT is traditionally used and explained in terms of detection. The applications to pain measurement, however, have been in the form of suprathreshold paradigms in which subjects receive a random series of stimuli and rate them using a category scale. Several measures of d' are computed, corresponding to pairs of adjacent stimuli (see Chapman et al., 1975; Clark & Yang, 1983). Briefly, the analysis of detection examines the proportion of positive and negative responses to presentations of a stimulus and blank stimulus (i.e., between a stimulus pair and a response pair). Analyses in the suprathreshold paradigms consider each possible stimulus pair and response pair, generating measures of discriminability and response criterion in the cases allowed by the data (i.e., sufficient hits, misses, false alarms, and correct rejections for the particular combinations of stimulus and response pairs).

Whether a direct scaling or SDT paradigm is used may not be obvious to the subject. In each case a response scale is being used to rate randomly presented stimulus intensities. This similarity is shown in Figure 7, which contrasts common scaling and SDT designs. In each method, multiple blocks of stimuli are presented to subjects; the difference is only the relative number of stimuli, stimulus repetitions, and number of possible response categories. The two methods are actually formally equivalent, differing not in the method but in the analyses. These procedures form a class of methods that could be described as single-stimulus, single-response (SSSR) designs. All SSSR designs basically provide the same information, the mean response to each stimulus, and a measure of discrimination between stimulus-evoked sensations. Any method shown in Figure 7 can be used to compute a mean response to a stimulus. Similarly, discrimination analyses are not limited to the traditional SDT paradigms shown in the figure. The bottom panel shows an alternative, nonparametric pain-rating paradigm developed by Richard Coppola (Buchsbaum, Davis, Coppola, & Naber, 1981) that turns the SDT paradigm around; a few responses (4) are used to describe a large number of stimulus intensities (31) rather than a large number of responses (12–14) used to describe a few (4) stimulus intensities.

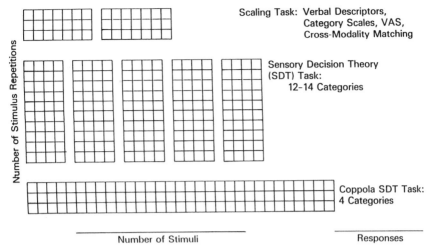

FIGURE 7   Summary of three psychophysical pain paradigms. A direct scaling task is shown on the top (VAS, visual scaling analog), a sensory decision theory (SDT) task is shown in the middle, and a variant of an SDT task is shown on the bottom. The horizontal dimension of each block represents the number of stimulus intensities used; the vertical dimension represents the number of times each intensity is administered. The order of stimulus presentations within each block are randomized. More than one block may be delivered. The sensations evoked by these stimuli usually are quantified by the methods shown on the right. The scaling tasks use a wide variety of response measures; the SDT tasks use several or a few discrete categories. (From Gracely, 1984.)

Figure 8 shows that discrimination can also be computed from the standard direct scaling paradigm. For each subject, discrimination between any two stimulus pairs can be computed by parametric or nonparametric tests, such as the t-test or Mann Whitney U. This measure is analogous to d' because both represent the difference in means of distributions divided by the variance of these distributions. In addition, an overall measure of discrimination can be computed simply by a one-way analysis of variance (ANOVA) using the stimulus intensities as the levels of the factors and the repetitions as pseudo-subjects. The resultant F statistic is related to discrimination; high slopes and/or small standard errors around each stimulus will produce large F ratios, whereas low slopes and/or large standard errors will result in small F ratios. This method was applied to a study of the effect of naloxone (Gracely, Deeter, Wolskee, & Dubner, 1980) on pain evoked by electrical tooth pulp stimulation. Naloxone is a narcotic-antagonist that is used as an indirect measure of endogenous opioid activity. An increase in pain magnitude following administration of naloxone is interpreted as an antagonism of ongoing endogenous analgesia produced by activation of endogenous opioid systems. In this study, 10-mg naloxone, in comparison to placebo, did not alter any measure of pain magnitude determined by

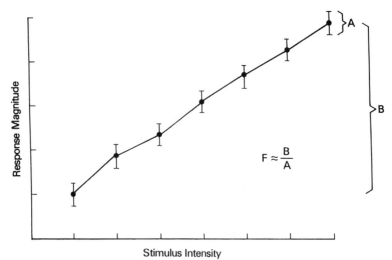

Stimulus Intensity

FIGURE 8   Derivation of discriminability from a direct scaling method. This idealized figure for a single subject shows response magnitude from any method plotted against seven stimulus intensities. Each intensity has been presented randomly N times. Good performance (high discrimination) is characterized by small variability of responses to repetitions of the same stimulus (A), and high slopes (B). Thus performance is related to the slope and inversely related to the variability at each point. Several measures can be used to express this relation, including the F ratio derived from a simple one-way analysis of variance. This overall measure can be derived from a relatively few presentations of many stimuli, in contrast to the multiple presentations of few stimuli in sensory decision theory assessment of discrimination. In addition, a measure directly analogous to d' can be derived by comparing the responses at any two stimulus intensities by either parametric (e.g., t test) or nonparametric (e.g., Mann Whitney U) methods.

choosing quantified verbal descriptors. Naloxone, however, significantly increased the mean F-ratio measure (inverse of discriminability) computed for each subject, indicating that it interfered with scaling ability. This result suggests that previous demonstrations of naloxone hyperalgesia, based in part on discrimination measures, may have reflected a disruption in scaling performance and not antagonism of putative endogenous analgesia.

Modifications of this method of computing discrimination include normalization by using the square root of F (analogous to the t statistic), and use of the interclass correlation coefficient (ICC) (Weiffenbach, Cowart, & Baum, 1986).

Thus, SSR designs share common features and provide (a) information about the mean response applied to a stimulus and (b) the ability to discriminate stimulus-evoked sensations. The limitations of these parameters have been discussed above in the section on SDT. The first parameter is a measure of perceived pain magnitude, influenced by a subject's response criterion. The second parameter is a measure of the intensity of the pain sensation, noise in the neural afferent system, and most importantly, general cognitive ability in consistently choosing the same labels to describe the same sensory magnitude. Neither is a pure measure of pain magnitude.

b. Stimulus integration: Double stimulus, single response designs

Two additional measurement methods applied to pain assessment can be grouped into a class termed double-stimulus, single-response (DSSR) designs. These methods, termed *functional measurement* and *conjoint measurement* are very similar. In each method a pair of stimuli is administered on each trial and the subject chooses a single response to describe some integrated impression of the two stimuli. The result is an independent scale for each stimulus set, and both individual and group measures of scaling performance. Rather than passively collect responses to stimuli, these procedures treat psychological measurement as an experiment with a positive or negative outcome. If the outcome is positive, the scales are accepted. A negative outcome indicates some sort of failure in scaling, integrating, or reporting the stimulus attributes, or some combination of such errors. The methods basically differ in the analysis. Functional Measurement treats the responses as interval-level data and performs a parametric analysis. Conjoint measurement assumes only ordinal properties in the responses, and performs a nonparametric analysis.

This description is of necessity oversimplified, as is the use of the term DSSR. In most applications to pain measurement, two distinct stimuli are presented and the subject is required to provide a single response using a specified integrating operation such as averaging. Other possibilities include summation, subtraction, or ratios. In some paradigms not involving pain stimuli, two attributes of a single stimulus, rather than two separate stimuli, are delivered and the integration may be implicit rather than explicitly specified. For example, subjects may be presented weights varying independently in size and mass, and the overall judgment of heaviness will integrate the influence of size without specific instructions. Such methods could be applied to painful stimuli by varying both intensity and another attribute, such as duration or stimulation area. In addition, the design need not be limited to two stimulus sets. Subjects may integrate any number of delivered stimuli or stimulus attributes. Thus the term DSSR used here actually refers to multiple-stimulus or stimulus-attributes, single response.

The few applications of these methods to pain assessment have demonstrated the variety and potential utility of these DSSR techniques. Jones (1980) presented 14 subjects with all possible pairs of two identical sets of four electrocutaneous stimuli varying from 1 to 4 mA. Responses to the average painfulness of these stimuli were analyzed by methods used for functional measurement. A two-way ANOVA with repeated measures showed significant main effects for each stimulus set, but the interaction between the sets was insignificant. Graphically, this result can be plotted as the overall response against one of the two stimulus sets with the other stimulus set as the parameter (i.e., four separate lines). The lack of interaction indicates that these lines are parallel.

In a study related to pain assessment. Algom, Raphael, and Cohen–Raz (1986) administered two sets of stimuli from qualitatively different stimulus modalities. Subjects used magnitude estimations to judge the overall aversiveness of all possible pairs of six electrocutaneous stimuli and six loud auditory stimuli. The lack of an interaction provided evidence that subjects can integrate unpleasantness across sensory modalities.

In addition to stimuli from different modalities, these methods have been used also to integrate the magnitude of pain sensations with the magnitude implied by a verbal descriptor. Gracely and Wolskee (1983) instructed subjects to match handgrip force to the average intensity of pain sensations produced by all possible pairs of five electrical tooth pulp stimuli and five verbal descriptors (weak, mild, moderate, strong, intense). Figure 9 shows mean log handgrip force plotted against the five verbal descriptors for each of the tooth pulp stimuli. An ANOVA indicated significant main effects for the tooth pulp and the verbal stimuli and a nonsignificant interaction. The parallelism shown in this figure and documented by the nonsignificant interaction indicates that subjects could (a) scale each stimulus set, (b) successfully integrate (average) somatic and semantic intensity, and (c) use handgrip force to rate this integrated impression. A lack of parallelism demonstrated by a significant interaction would have indicated a failure at one or more of these three stages. This integration of somatic and verbal stimuli was replicated by Heft and Parker (1984) using electrical skin stimulation and a conjoint measurement analysis.

The use of stimulus integration techniques for pain assessment is potentially advantageous because these methods provide an independent validity criterion that tests whether the measurement results can be accepted. This criterion can also be applied to individual subjects, identifying those who perform adequately in the task. The methods may also distinguish between analgesic effects and effects on general cognitive performance. For example, Gracely et al. (1981) designed a study in which subjects used a 20-point category scale to rate the average intensity of five electrical tooth pulp stimuli (four intensities and one blank) and five verbal descriptors (no sensation, weak pain, moderate pain, strong pain, very intense pain) both before and after two experimental intravenous infusions. The first infusion was either the minor tranquilizer diazepam or placebo, whereas the second infusion was either the short-acting narcotic fentanyl or placebo. Thus the design was a 2 × 2 factorial in which four separate groups received either one of the drugs, both drugs, or neither drug. The results of this complex

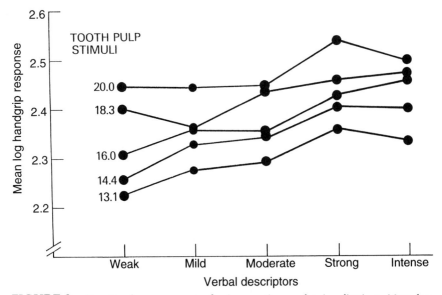

FIGURE 9 Functional measurement of pain sensations and pain adjectives. Mean log handgrip force is plotted against five verbal descriptors of sensory intensity for five tooth pulp stimuli that ranges in equal log steps from pain threshold to pain tolerance for each of 10 subjects. Each point is the mean of 20 responses for the average of the intensity produced by the tooth pulp stimulus and implied by the descriptors. The parallelism shown in this figure indicates that these subjects could perform this scaling and averaging task, validating the scales of verbal and tooth pulp magnitude derived from this method. (From Gracely & Wolskee, 1983).

design were simple. The narcotic reduced the magnitude of the pain sensations, whereas the minor tranquilizer interfered with the ability to do the task without altering the magnitude of the pain sensations. Neither drug altered the verbal scale, evidence that there was no effect on the response criterion. This result suggests that DSSR methods can distinguish between analgesic action and disruption of scaling performance. This finding is particularly interesting in comparison to results found with SDT analysis of analgesics. Both narcotics and tranquilizers have been shown to reduce discrimination of the experimental pain sensations, an effect interpreted as analgesia (Yang et al., 1979). However, as discussed above, discrimination is a measure of both analgesia and a measure of general scaling performance. The results with functional measurement suggest that the narcotic effect represents true analgesia, whereas the reduction of discrimination by the tranquilizer represents a general cognitive effect and not analgesia. This result is consistent with clinical anecdotal evidence that minor tranquilizers are ineffective analgesics.

To briefly summarize, SSSR methods are capable of providing specific information regardless of how the responses are analyzed. These methods constitute the bulk of the literature addressing subjective measurement of suprathreshold pain sensations. Measures of discrimination and response magnitude do not provide pure measures of analgesia, but are influenced by other factors, such as cognitive ability and variable response criterion. These factors cannot be isolated by an SSSR method, but rather must be controlled by careful experimental design. Preliminary results with other procedures, such as DSSR methods, however, may be capable of providing more information than SSSR methods, although the capabilities and limitations of these methods have not been examined in detail. Another new class of procedures termed adaptive or titration methods reverse the role of the stimulus and the response in the SSSR design.

c. Stimulus-dependent methods: Adaptive or titration methods

Psychophysical methods are usually classified according to threshold or suprathreshold assessment, or by the method (e.g., direct or indirect) used to quantify the psychological unit of magnitude. Another simple scheme classified procedures by the dependent variable used in the analysis. The direct scaling methods presented above are examples of response-dependent methods; subjects make variable responses to a fixed set of stimuli and these responses are used for analysis. In contrast, threshold measures such as the Method of Limits can be considered to be stimulus-dependent, because a variable stimulus intensity corresponding to a fixed subjective standard (e.g., pain threshold) is the dependent variable used for the analysis.

Although traditionally confined to threshold assessment, stimulus-dependent methods have been developed and applied to suprathreshold pain assessment. The most common are measures of pain tolerance to continuous stimulation increasing in severity, such as the submaximum effort tourniquet technique and cold pressor stimulation. Other variants of this technique also request subjects to indicate when the increasing pain magnitude reaches predetermined subjective levels (El Sobky, Dostrovsky, & Wall, 1976; Goldberger & Tursky, 1976). This variant shares the lack of psychophysical controls and vulnerabilities to bias noted for tolerance measures above.

A stimulus-dependent method using presentations of discrete stimuli has been applied to the evaluation of pain sensations analgesics. As shown previously in Figure 3, this method is based on a staircase rule in which a positive response (e.g., stimulus is detected) lowers the intensity of the next stimulus presented, and a negative response (e.g., stimulus not detected) increases the intensity of the next stimulus presented. The ideal result is a series of stimulus intensities titrated about the threshold value. Unfortunately, human subjects often become aware of the contingency between each response and the value of the next stimulus intensity. As mentioned in the section on thresholds, Cornsweet (1962) reduced this contingency by developing the Double Random Staircase method, which uses two separate, independent staircases. As shown in the figure, for each trial one of the two staircases is randomly chosen and the predetermined stimulus intensity is delivered. The response determines the next predetermined stimulus intensity to be delivered by that staircase the next time it is randomly chosen.

Figure 10 shows how this staircase method has been applied to suprathreshold scaling of experimentally evoked thermal pain sensation. If a threshold is defined as a subjective interval in between two responses, the suprathreshold paradigms simply use a category response scale with additional staircases titrated about intervals defined by specific pairs of responses. As shown in the figure, three double random staircases are titrated about (a) the threshold interval—no pain and mild pain—and two supra-

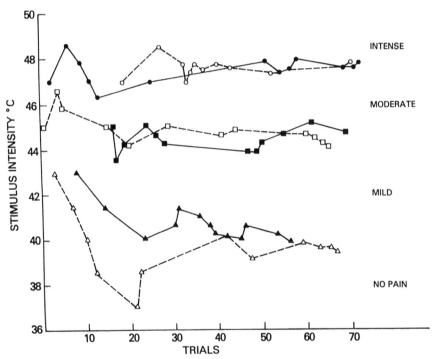

FIGURE 10   Three double random staircases. This figure shows stimulus intensity plotted against 72 stimulus trials for a single subject. On each trial, one of the six staircases is selected randomly and the stimulus temperature indicated by that staircase is presented. The subject selects one of four response categories to rate the intensity of the sensation evoked by the stimulus. Each double staircase is associated with a boundary located between two response categories. Responses above the boundary reduce the intensity of the next stimulus presented by that staircase. Responses below the boundary increase the next stimulus. For example, stimuli in the lower pair of staircases are increased after a response of "no pain" and decreased after the other three responses. (From Gracely, 1988).

threshold intervals, between (b) mild pain and moderate pain, and between (c) moderate pain and intense pain. On each trial one of these six staircases is randomly chosen and the current stimulus intensity associated with that interval is presented. A response above the interval lowers the intensity of the next stimulus presented by that particular staircase, and a response below the interval increases it. For example, if a staircase for the interval between mild pain and moderate pain was chosen on a trial, a response of either moderate pain, intense pain, or extremely intense pain would lower the intensity of the next stimulus presented by that staircase, whereas a response of either nothing, warm, not painful, hot, not painful, or mild pain would increase it.

The use of three double random staircases is just one of many possible combinations of staircases and the response intervals. For example, Figure 11 also shows an example in which six individual staircases were titrated about response intervals ranging from no sensation and warm for the lowest staircase and slightly intense pain and intense pain for the highest staircase (Gaughan, Gracely, & Friedman, 1990).

Once the number of staircases and their associated response intervals are chosen, the experimenter must chose the starting intensities, the decision rule, and the response increment. In the applications discussed here, the decision rule was simple alternation. Any response above the interval lowered the next stimulus intensity delivered by that staircase, and any response

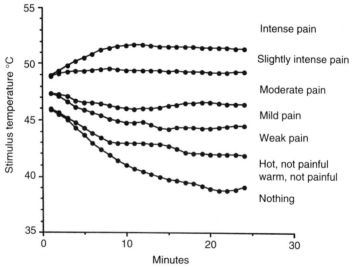

FIGURE 11   Six random staircases. Stimulus temperature is plotted against time. Stimuli were delivered at 20-sec intervals for a total of 72 trials. Staircases were initiated at 46, 47.5, and 49°C and titrated between the response categories shown. Each point is the mean of observations from 28 subjects. (Data from Gaughan et al., 1990).

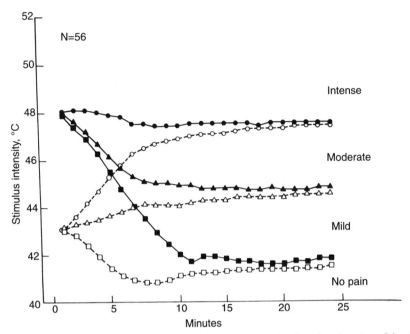

FIGURE 12   Staircase convergence. Stimulus temperature is plotted against time. Stimuli were delivered at 20-sec intervals for a total of 72 trials. One staircase from each pair was initiated at 43°C and one was initiated at 48°C. Staircases quickly converged, superimposing within 0.3°C by the termination of the session. Each point is the mean of observations from 56 subjects. (From Gracely et al., 1988.)

below the interval increased the intensity of the next stimulus delivered by that staircase. The response increment (the amount the stimulus intensity is changed) is a critical factor in these studies. If this increment is too large, the responses will simply alternate on every trial, providing no measurement. If too small, an excessive number of trials would be required to reach a reversal point. The ideal increment is a compromise between resolution obtained by small increments and speed of adjustment obtained by larger increments. One solution provides a dynamic increment that varies in size depending on the behavior of the subject. This method specifics an ideal number of steps before a response reversal, and the increment size is automatically adjusted throughout the measurement. If a response reversal occurs in the ideal number of steps, the increment is unchanged. If it occurs in less than the number of ideal steps, the increment is assumed to be too large and is reduced by half. If the reversal does not occur until after the number of ideal steps have been exceeded, the increment is assumed to be too small and is doubled in size.

Once these parameters are chosen, the method runs itself. Figure 12 shows the ability of the parameters described above to quickly seek a stable response level (Gracely, Lota, Walther, & Dubner, 1988). In this example, the two staircases in each double random staircase were started at high- and low-stimulus temperatures, with an initial increment of 1.6°C. These staircases converged to the appropriate levels over 36 trials and the increment was reduced to 0.2 to 0.4°C after stable response levels were achieved.

Figure 13 demonstrates one use of staircase techniques, the determination of stable values that can be compared among different groups. For example, this method has been used to compare the cutaneous pain sensitivity of patients suffering from angina pain, suggesting that patients with coronary artery disease are generally more sensitive than patients with pain but no evidence of coronary pathology (Cannon et al., 1990). The staircase rules are not the only adaptive method that can be used for this application. Rather than base response change on the immediate response history, other methods use probabilities computed from past responses to continuously update the current best estimate of the stimulus intensity corresponding to a specific response interval (Duncan, Miron, & Parker, 1992). These probability methods may provide a good estimate of static response levels, because all the data contribute to the estimate. However, these methods may be relatively insensitive when used to assess altered pain sensitivity, such as studies of analgesic agents described immediately below. Thus each method may be best suited for specific purpose; staircase methods may be more useful for tracking changing pain sensitivity, whereas probability methods may provide a better assessment of static or unchanging sensitivity. The interested reader will appreciated that these attributes of each method represent extremes. The dynamic response of probability methods can be increased by shortening the "response window" used to compute probabilities, at the expense of including less data in the resulting estimate. Similarly, the use of a dynamic increment in staircase designs is another means of including the response history in the estimate of pain sensitivity. Both methods may be very similar over a range of parameters chosen for an intermediate dynamic response.

The staircase assessment has been used to assess the action of analgesic effects. Figure 14 shows that the fast adjustment has allowed evaluation of the kinetics of a fast-acting analgesic in cross-over designs in a single session. Oxygen, and nitrous-oxide and oxygen in combination, were admin-

-->

FIGURE 13   Multiple random staircase evaluation of thermal cutaneous pain sensitivity. Threshold and suprathreshold values determined over a 34-min period are plotted for patients with chest pain and normal coronary arteries (triangles), hypertrophic cardiomyopathy (open circles), or coronary artery disease (closed circles). Stimulus temperatures required to produce these three subjective levels of pain sensation are shown in the three panels. These temperature differences indicate decreased cutaneous pain sensitivity in the chest pain, normal coronary artery group, suggesting that their chest pain does not result from a generalized pain sensitivity. (Adapted from Cannon et al., 1990.)

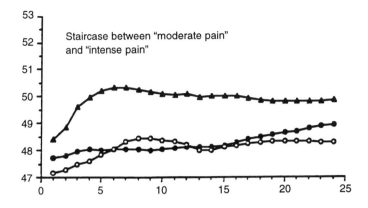

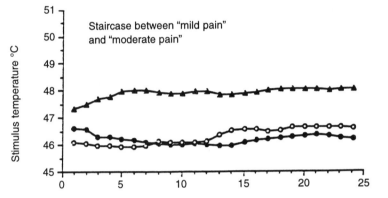

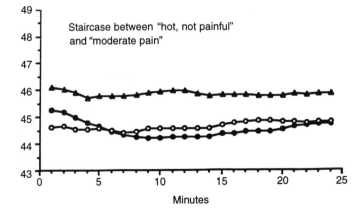

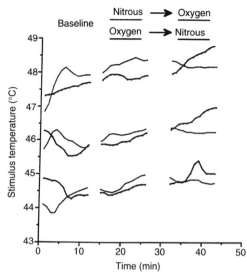

FIGURE 14   Results of multiple random staircase assessment of 50% nitrous-oxide/oxygen and 100% oxygen. Stimulus temperature in °C is plotted against minutes for three staircases titrated between no pain and mild pain (bottom), "mild pain" and "moderate pain" (middle), and "moderate pain" and "intense pain" (top). The breaks in the connecting lines separate the 12-min baseline assessment (left) from the two 12-min experimental periods. Ten subjects receiving nitrous oxide-oxygen in the first period and oxygen in the second period are shown by the dashed lines, and the ten subjects receiving oxygen followed by nitrous oxide-oxygen are shown by the solid lines. The left break in the lines indicates a 2-min induction period, the right break indicates a 3-min washout with 100% oxygen followed by a 2-min induction period. (From Kaufman et al., 1992).

istered to dental patients in a counterbalanced crossover design (Kaufman, Chastain, Gaughan, & Gracely, 1992). The time course of the significant changes in thermal pain sensations required to maintain the same level of response closely matched the known time course of avaelor concentrations of nitrous oxide.

Figure 15A shows dose–response curves for the opioid fentanyl for the staircases titrated between the responses "mild pain" and "moderate pain" in a study administering three doses of the narcotic fentanyl or placebo in a double-blind design (Gracely, 1988). Figure 15B shows the results of a study (Gracely & Graughan, 1991) in which an initial intravenous infusion of fentanyl was followed 12 min later by a second infusion of either fentanyl or placebo. The method was sensitive enough to detect the incremental analgesia produced by the second dose of fentanyl.

These recent applications of the stimulus-dependent methods have revealed several potential advantages over the more common response-dependent methods. First, as the name implies, the response is quantified in units

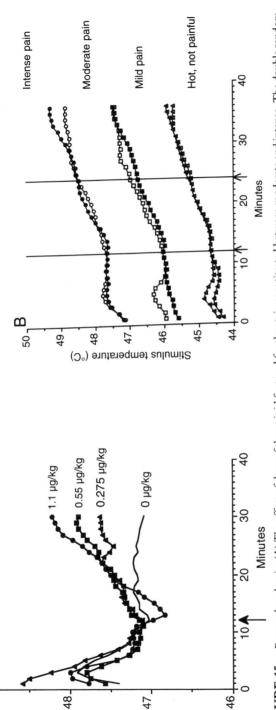

FIGURE 15 Fentanyl analgesia. (A) The effect of dose of the opioid fentanyl for the staircases titrated between moderate and intense. The double random staircases titrated at this interval were combined into one mean staircase to show the results of four groups (n = 16 each). Stimulus intensity is shown on the ordinate, and 5 trials before and 30 trials after a 2-min double-blind intravenous infusion administered at the arrow are shown on the abscissa. Subjects received either placebo, shown by solid line, or 0.275, 0.55, or 1.1 µg/kg fentanyl shown respectively by the triangles, squares, and circles. Fentanyl significantly increases staircase temperatures, indicating analgesia. Both rate of onset and final temperature were dose-related. (B) Multiple Random Staircase (MRS) assessment of thermal analgesia after infusion of the short-acting opiate fentanyl or placebo. Stimulus temperature in °C is plotted against minutes for the group (n = 27) receiving 0.55 µg/kg fentanyl during infusions at 12 and 24 min, shown by the filled symbols, and for the group (n = 30) receiving 0.55 µg/kg fentanyl during the first infusion and saline placebo during the second infusion, shown by the open symbols. The MRS method maintained one-third of the stimuli at pain threshold, shown by the triangles, one-third of the stimuli between subjective levels of mild and moderate pain, shown by the squares, and one-third of the stimuli between subjective levels of moderate and intense pain, shown by the circles. The first infusion resulted in a significant increase in stimulus temperatures, consistent with analgesia previously shown with this dose. The second infusion resulted in a significant increase in stimulus temperatures (t(55) = 2.24, p < .05) in only the top staircase.

of stimulus intensity, permitting direct comparisons between individuals and even separate experiments.

Second, experimental pain studies using the more common fixed stimulus sets must always be concerned with individual differences in pain range. The stimuli must be sufficiently intense to evoke pain responses, yet not exceed the tolerance levels of subjects or cause tissue damage. For some pain modalities, such as heat-induced pain, these parameters are consistent enough across a population to allow the same fixed stimulus set for all individuals. However, in this case the variability of individual pain ranges can result in scaling biases that may significantly influence the results. In other studies of stimuli such as electrical tooth pulp stimulation (Gracely et al., 1979) the differences in individual pain ranges may be so great that individual stimuli must be chosen for each subject, complicating both the procedure and the analysis. In either case, the administration of analgesic interventions can drastically reduce the perceived range of the stimuli with resultant biases due to range effects and the knowledge that the reduced range indicates that an analgesic has been delivered. The use of adaptive stimulus-dependent methods obviates these problems because the interactive paradigm quickly determines the appropriate stimulus intensities for each subject. Within the limits of variable response criterion(s), each subject is presented with a constant subjective range of sensations rather than a constant physical range of stimulus intensities. In addition, this range is maintained following administration of an analgesic intervention, after a transient adjustment of the stimulus intensities to the new values required to evoke the old responses.

Third, the use of multiple staircases, especially several pairs of double random staircases, provides an internal check of scaling consistency or performance. A pair of double random staircases should superimpose if the subject is attending to the rating task and making consistent judgments of stimulus-evoked sensations. Poor performance can be identified by a "random walk" taken by these staircases.

The separate measurement of scaling performance is an important concept that has received only minimal attention in pain measurement. Scaling performance can be treated like a skill, with individual differences in both skill level, and in compliance with instructions. It makes sense to *a priori* choose skillful subjects for an expensive clinical trial, as does the *post hoc* elimination of subjects (before double-blind codes are broken) that could not, or did not, perform the task.

Fourth, stimulus-dependent procedures such as the staircase method automatically track the time course of pain sensitivity, providing information about the dynamic response of subjects and the kinetics of analgesic interventions. These methods also may provide dose–response information for single subjects in single sessions.

Fifth, when used to assess analgesics, stimulus-dependent, adaptive methods expose the subjects to less aversive pain sensations. To illustrate, suppose an investigator wishes to assess the efficacy of a powerful analgesic on intense pain sensations, such as those produced by 51°C thermal stimuli. In a conventional response-dependent paradigm the subject is administered 51°C stimuli before the analgesic and a response is noted, which is compared to the lower response evoked by the same 51°C stimulus after the analgesic, perhaps corresponding to the response made to 49°C before the analgesic was delivered. In a staircase method, the subject determines a stimulus intensity between intense and moderate pain, which turns out to be 49°C. After the analgesic, it requires 51°C to produce this same response. Both methods assess the analgesia to 51°C stimulation. However, subjects in the response-dependent method must experience the severe pain produced by 51°C, which is then reduced by the analgesic. In contrast, subjects in the staircase experiment only experience the sensation produced by 49°C. They never experience the sensations produced by the higher stimulus intensity in the absence of analgesia.

It is interesting to note that these advantages are made possible by "interactive" paradigms that are greatly facilitated by computerized delivery systems. They represent a second generation of computerized methods (the first being simply the automation of classical methods, such as direct scaling) that can be administered easily without a computer.

II. PAIN AS DUAL VARIABLES OF INTENSITY AND UNPLEASANTNESS

Although most of the literature still refers to a single dimension of pain, there is a growing recognition that pain is not a simple, single variable such as the brightness of a light. This recognition has a long history; Aristotle and Plato referred to the "passion of the soul," while more recent influential authors distinguished between original pain and the "reaction component," or between sensory intensity and a feeling state that has been termed "unpleasantness," "discomfort," "distress," or the "affective" or "evaluative" component.

A. Precedents from Other Sensory Modalities

This feeling state certainly is not unique to pain. The concepts of drive, motivation, and emotion are psychological staples in studies of both animals and humans. In the psychophysical literature, it has been recognized that sensory modalities with homeostatic significance (such as temperature and the chemical senses of taste and olfaction) are accompanied by a feeling state or "hedonic" that can be either positive or negative. Investigations of these modalities reveal several commonalities.

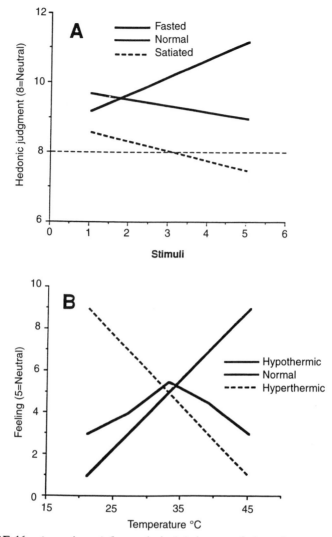

FIGURE 16 Internal state influences hedonic judgments of odor and temperature. Panel A shows the results, in the same subset of subjects, of category ratings of the pleasantness or unpleasantness of various concentrations of orange odorant under conditions of 22-hr fast, normally fed, and after satiation with a glucose drink. (Adapted form Mower, Mair, & Engen, 1977.) As concentration increased, hunger in the fasted condition increased, and satiety decreased the pleasantness of the odors in comparison to the normal condition. (B) The results in the same subjects of category ratings of pleasantness or unpleasantness of the sensations produced by immersion of a hand in a water bath with temperatures varying from 21–45°C under conditions of hypothermia, normal body temperature, and hyperthermia. (Adapted from Mower, 1976.) In the normal condition the extreme temperatures were rated as unpleasant, whereas intermediate temperatures were rated as less unpleasant to pleasant. Manipulating body temperature by either prolonged immersion in cool water or exercise while heavily clothed produced dramatic shifts in the function. All cool temperatures became pleasant under conditions of hyperthermia and all warm temperatures became pleasant under conditions of hypothermia. In each of these studies, manipulation of internal temperature did not alter ratings of sensory intensity.

First, increasing stimulus intensity results in a monotonic increase in sensory intensity, but usually a nonmonotonic change in what can be referred to as hedonic tone, as the overall pleasantness or unpleasantness of the sensation. For example, increasing the sweetness of a sugar solution results in increasing judgments of "sweetness" and increasing judgments of pleasantness that first peak and then decline as the solution becomes syrupy, ultimately becoming unpleasant at high concentrations (Mower, Mair, & Engen, 1977). Temperature is a special case in which stimulus intensity can be increased in two different directions, with ratings of pleasantness near the baseline values and ratings of decreasing pleasantness or unpleasantness at both extremes (Mower, 1976).

Second, hedonic judgments, but not judgments of sensory magnitude, are influenced by the internal state of the observer. Figure 16A shows how hunger can shift hedonic judgments such that previously neutral food odors now become pleasant, whereas satiety can result in the opposite shift, in which previously pleasant odors become less pleasant or unpleasant. This result is hardly surprising, give the large amount of anecdotal evidence that starving individuals will find shoe leather appetizing, and the common unappealing nature of many foods after one has overeaten. Figure 16B shows a similar hedonic shift for judgments of thermal stimuli after manipulations of body core temperature, a result again consistent with common experiences, such as the preference for an excessively chilled room in hot summer weather that would be considered too cold in winter. In all of these cases, manipulation of the internal, hemostatic environment altered the hedonic consequences of stimulation with little, if any, effect on judgments of sensory intensity.

These examples of hedonic psychophysics provide an excellent reference point to address the obvious hedonic associations with pain judgments. Although not homeostatic in the W. B. Cannon (1939) definition of maintaining biological balance, pain shares a common but even more important role in the maintenance of the individual. It is the front line of defense, signaling potentially life-threatening injury and promoting postinjury behaviors that maximize healing.

B. Univalent versus Bivalent Hedonic

However, before addressing the commonalities, it is important to note one important difference between the experience of pain and the experience of smell, taste, and thermal sensation. In the psychophysical examples shown in Figure 16, the hedonic is bivalent, the experience can be either pleasant or unpleasant, and the same stimulus may be perceived as either pleasant or unpleasant after a manipulation of internal state. In contrast, the hedonic associated with pain sensation is univalent, it is (with rare exceptions) only unpleasant and disagreeable, motivating behaviors of escape and avoidance.

One consequence of this univalent hedonic is that the separate dimensions of intensity and unpleasantness may be more difficult to discriminate. In the sense of temperature, taste, and olfaction, the bivalent, shifting nature of the hedonic distinguishes it from the concept of sensory magnitude, and separate ratings of these dimensions are made without difficulty. In the case of pain, however, both the intensity and the negative hedonic always grow monotonically with stimulus intensity, and it may be much more difficult to distinguish between these two dimensions.

Several authors have promoted the use of language to facilitate the evaluation of these and other dimensions of pain experience. Melzack and Torgerson (1971) categorized and quantified a list of words compiled by Dallenbach (1939). These words were sorted into discrete categories of meaning and the values of the words within each category were determined by ranking techniques and a Thurstone scaling analysis (1959). This study formed the basis of the widely used MPQ, described in the next section (Melzack, 1975). Tursky (1976) used a subset of these words as stimuli in a cross-modality matching evaluation described in the previous section on separate quantification of category descriptors. Tursky's approach is significant for two reasons. First, it reduced the large number of categories of the Melzack and Torgerson study and the subsequent MPQ to a few dimensions common to all of pain experience. Second, it applied the previously described cross-modality methods to generate separate ratio-level scales with different response measures, the comparison of which provides a method to validate scaling consistency to verbal descriptors or other subjective material with no underlying physical measure. (See previous section on quantification of categorical responses.)

Gracely et al. (1978a, 1979) refined Tursky's methods by reducing the scales to dimensions of sensory intensity and unpleasantness and changing some of the descriptors such that the sensory intensity class could be generalized for use with any sensation. After reliability and scaling consistency was demonstrated for descriptors of both sensory intensity and unpleasantness, these scales were used in a series of studies evaluating pain sensations evoked by experimental stimulation and the effect of analgesic agents.

C. Validity of Hedonic Scales

The validity of hedonic scales of olfaction and temperature has been based on two findings: (a) that scales of sensory intensity and pleasantness or unpleasantness result in different psychophysical functions, and (b) that these scales are differentially altered by manipulations of internal state. Gracely et al. (1978b) extended these findings to hedonic assessment of pain sensations. Figure 17 shows a summary of initial studies in which subjects

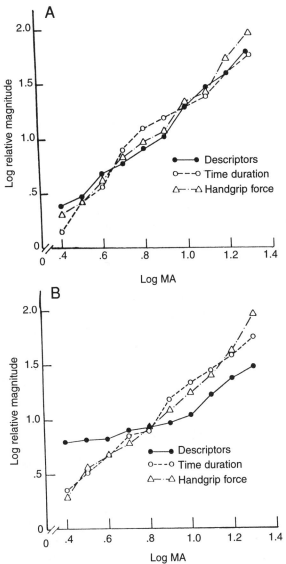

FIGURE 17   Verbal descriptor and cross-modality matching functions for sensory intensity and affective unpleasantness responses. Log relative magnitude is plotted against stimulus intensity in log milliamperes. (A) Psychophysical functions derived from choosing sensory descriptors from a random list and from cross-modality matching time duration and handgrip force to the intensity of the pain sensations evoked by the stimuli. (B) Psychophysical functions derived form choosing unpleasantness descriptors from random lists and from cross-modality matching of time duration and handgrip force to the unpleasantness of the stimulus-evoked sensations. Each point is the mean of four observations from each of 16 subjects. All of the functions are similar except for the unpleasantness descriptor responses. (From Gracely et al., 1978b).

chose verbal descriptors from randomized lists and made cross-modality matches to handgrip force and to time duration to assess the sensory intensity and unpleasantness of pain sensations produced by electrical stimulation of the skin. Only the descriptor scales showed different psychophysical functions for sensory intensity and unpleasantness. In addition, both the cross-modality scales of sensory intensity and unpleasantness were similar to the function derived from the intensity verbal descriptors, suggesting that the subjects used the cross-modality methods to rate the salient dimension of sensory intensity regardless of instructions to rate either intensity or unpleasantness. This result provided evidence that the use of language, reinforced by forcing choices based on meaning of words in a randomized list, may facilitate the discrimination of pain unpleasantness from pain intensity.

Although the finding of different psychophysical functions is one line of evidence that the two scales measure different dimensions, two caveats must be kept in mind. First, the lack of different functions is not strong negative evidence, because separate scales of these different dimensions could conceivably result in the same function (Gracely, 1992). Second, the finding of different functions with dissimilar types of scaling methods for sensory intensity and unpleasantness could reflect a difference only in method variance and not the separate evaluation of separate dimensions (Gracely et al., 1978b). Given these concerns, the finding of different functions with the verbal descriptor scales, structurally similar except for the specific words used, supports both the concept of two pain dimensions and the discriminate validity of the descriptor method.

D. Validity of Scales of Pain Unpleasantness

The finding of different functions can be considered to be "static" evidence, obtained from a single measurement. Stronger evidence can be obtained by evaluating the dynamic response of these scales to interventions that can be assumed to alter preferentially one dimension. Figures 18 and 19 summarize the results of a study in which different groups of experimental subjects rated either the intensity or unpleasantness of sensations produced by electrocutaneous stimuli before and after the intravenous administration of 5-mg diazepam (™ Valium). Unpleasantness descriptor responses were reduced following diazepam, whereas verbal descriptor responses of sensory intensity and cross-modality matches to both sensory intensity and unpleasantness were not altered by the drug.

Figure 20 shows the effects of the potent short-acting narcotic fentanyl on verbal descriptor responses of the intensity and unpleasantness of sensations evoked by electrical stimulation of the tooth pulp. Sensory intensity responses were significantly reduced after fentanyl in comparison to placebo;

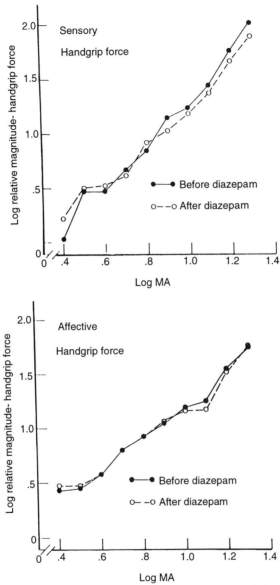

FIGURE 18   Influence of diazepam on cross-modality matching to the intensity and affective unpleasantness of sensations evoked by electrocutaneous stimulation. Administration of 5 mg of intravenous diazepam had no effect on cross-modality matches made to either dimension. Each point is the mean of two responses from each of 16 subjects. (From Gracely et al., 1978b.)

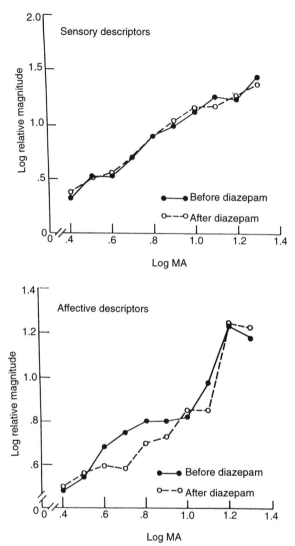

FIGURE 19   Influence of diazepam on verbal descriptor responses to the intensity and affective unpleasantness of sensations evoked by electrocutaneous stimulation. Administration of 5 mg of intravenous diazepam had no effect on responses made by choosing sensory intensity descriptors from a random list, but did significantly reduce responses made by choosing affective unpleasantness responses from a random list. Each point is the mean of 2 responses from each of 16 subjects (From Gracely et al., 1978b).

the magnitude of this effect was a reduction of about 51% such that a stimulus-evoked sensation originally rated as barely strong was rated as mild after fentanyl administration, whereas a stimulus-evoked sensation originally rated as mild was now called weak. This in itself was a significant result because the prevailing opinion at the time of this study was that narcotics produced analgesia primarily by reducing unpleasantness and not by reducing sensory intensity (Jaffe, 1975). A growing body of evidence now indicates that narcotics powerfully attenuate the sensory-discriminative component of pain sensation. Figure 20 also shows that, unlike sensory intensity, descriptor responses of unpleasantness tended to increase after administration of fentanyl. This result likely does not generalize to other situations because the subjects were required to walk to another room after the administration of the narcotic, which resulted in dysphoria and nausea in most subjects. The result, however, further validated the separate scales of intensity and unpleasantness and demonstrated the need to measure these two dimensions. Two different scale dimensions, used interchangeably to measure pain in other studies, yielded opposite results.

The finding of differential effects on the separate dimensions of intensity and unpleasantness has been observed in additional studies. These scales have been disassociated after administration of a combination of fentanyl and diazepam (Gracely, Dubner, & McGrath, 1982), after administration of naloxone (Hargreaves et al., 1986), and after nonpharmacological pain control interventions such as meditation (Gaughan, Gracely, & Friedman, 1990) and hypnosis (Malone, Kurts, & Strube, 1989).

These dimensions also have been disassociated using variations of the VAS. Price et al. (1977) asked subjects to draw lines (line production) on 28-cm strips of paper in a study in which carefully trained subjects could receive either a full range of painful thermal stimuli or only the upper portion of the range on any given trial. A visual signal indicated if the next stimulus was from the restricted upper range. The signal had no effect on ratings of sensory intensity but significantly reduced ratings of unpleasantness (see Figure 5). This effect can be interpreted by examining the response to the lowest stimulus of 45°C in the restricted upper range. During the unsignaled trials, this stimulus was more painful than many lower stimuli, and received an appropriate rating of unpleasantness. During the signaled trials, however, when the subject expected only intense stimuli, this stimulus produced the lowest possible pain magnitude, and thus was rated as relatively less unpleasant. In addition to emphasizing the role of cognitive expectancies and sets in pain judgments, this result suggests that the complex methods of verbal descriptor scaling may be unnecessary for the evaluation of sensory intensity and unpleasantness. Simple VAS or line production scales and appropriate instructions may be sufficient. However, this result contrasts with the findings in Figures 17 and 18, which show that instructions and cross-modality scales did not distinguish between these two dimensions. A study by Duncan, Bushnell, and Lavigne (1989) suggests that this discrepancy may represent difference in degree, not in kind. These investigators directly compared both verbal descriptor and VAS scales of both dimensions and concluded that both methods distinguish between these dimensions, although the verbal descriptor scales appeared more sensitive. Thus it is possible that VAS and sufficient training or instructions can provide adequate measurement of these dimensions. The relative discriminative power of these methods is an obvious topic of future research. This issue bears directly on the interpretation of results in which both scales show the same effects. Were sensory intensity and unpleasantness responses influenced similarly, or were the scales incapable of discriminating between these dimensions?

E. Interpretations of Changes in Pain Sensory Intensity and Unpleasantness

Figure 20 illustrates another important point in the interpretation of the results of dual scales of sensory intensity and unpleasantness. In the experimental situation, a reduction in the magnitude of both sensory intensity and unpleasantness psychophysical functions can be correctly described as a reduction in both intensity and unpleasantness response. However, assum-

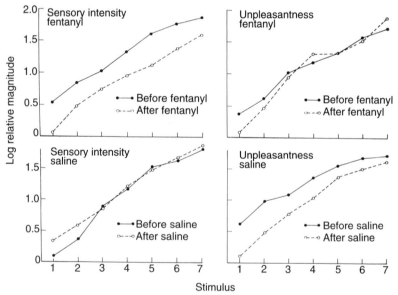

FIGURE 20   Sensory intensity and affective unpleasantness verbal descriptor responses to the sensations produced by electrical tooth pulp stimuli presented in seven equal log step currents between individually determined pain threshold and pain tolerance. Each pain is the mean of six observations from each of ten subjects before and after the double-blind intravenous infusion of either 1.1 μg/kg fentanyl, a fast-acting opioid, shown at the top, or saline placebo, shown on the bottom. The left panels show that the sensory intensity responses were significantly reduced after fentanyl in comparison to placebo. The right panels show a trend in which the unpleasantness responses were reduced after placebo and essentially unaltered following administration of fentanyl. (Adapted from Gracely et al., 1979.)

ing adequate measurement, it is incorrect to describe an equal effect on sensory intensity and unpleasantness, for the mechanisms responsible for unpleasantness may not have been altered. This issue is clarified by considering the use of dual scales to measure clinical pain and some means of modulating the clinical "stimulus." Because stimulus intensity cannot be measured, the relation between these scales can be described by plotting unpleasantness responses against sensory intensity responses, yielding a measure of the amount of unpleasantness associated with a specific sensory intensity. This relation is consistent with a model in which sensory input is fed into an "affective amplifier," and the "affective gain" of this amplifier determines the function shown in this figure, the amount of unpleasantness associated with a specific sensory intensity (Gracely, 1977, 1992; Mower et al., 1977). Using this model, the result shown in Figure 21A, in which both sensory intensity and unpleasantness responses are reduced equally after a manipulation, is interpreted as only a reduction in sensory intensity without any change in the affective gain. Unpleasantness responses have been reduced because of decreased sensory input, but the central hedonic processing has not been altered. In contrast, Figure 21B shows a situation in which sensory intensity responses are reduced slightly and unpleasantness responses greatly. The right panel shows that a given sensory intensity is now less unpleasant, indicating a reduction in affective gain.

A comparison of the findings with pain and those of other sensory modalities such as taste or temperature reveal parallel mechanisms. In each case interventions have altered the hedonic response, or "affective gain" to stimuli without altering the perceived magnitude of the stimulus-evoked sensations. These mechanisms may be the only pain treatment available for syndromes such as poststroke pain in which the pain sensations cannot be attenuated. Treatments such as meditation, relaxation, hypnosis, and cognitive therapy may offer relief by reducing the affective response to an unchanged pain sensation. These treatments can target several types of affective responses. Pain unpleasantness is a primary type that refers to the immediate negative feelings associated with pain sensations that motivate escape and avoidance behaviors. It can be considered to be a somatic distress like hunger and thirst, and to be present in animals as well as in humans. Another class can be characterized by emotional responses to thoughts about the pain experience, such as concern for the future, that would be attributed primarily to human pain problems. A third, human type is also possible, the negative thoughts and emotions associated with being a patient with a medical problem (Gracely, 1992).

There is an additional property that pain shares with the chemical senses, but not with the sense of temperature. Pain is not a simple entity varying only in intensity and discomfort. Like odor, pain comes in a wide variety of qualities. It can be sharp or dull, deep or superficial, localized or diffuse, stationary or moving or spreading. It can be burning, squeezing, throbbing,

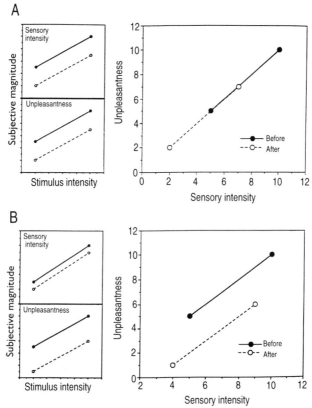

FIGURE 21   Distinction between effects on ratings of sensory intensity and unpleasantness and effects on the unpleasantness of a sensation. The small panels in A show the results of an intervention that reduces both ratings of sensory intensity and unpleasantness by the same amount. The large panel plots the unpleasantness responses against the sensory intensity responses. This figure shows that the effect only reduced the intensity of the pain sensations, because a specific sensory intensity was equally unpleasant before and after the intervention. The small panels in B show the results of another hypothetical intervention in which unpleasantness responses are reduced to a greater degree than sensory intensity responses. The large panel on the right shows that in addition to the reduction in the sensory intensity shown in the small panel, this effect represents an independent reduction of the unpleasantness of a specific sensory intensity.

or aching. The evaluation of these qualities goes beyond the measurement of just sensory magnitude and unpleasantness to consider all of the relevant dimensions of the pain experience.

III. MULTIDIMENSIONAL PAIN ASSESSMENT

As with other sensory events, pain can be described by four attributes: location, duration, intensity, and quality. Pain location and duration are recognized as important aspects of the pain experience, and although they are not always easily quantifiable, they are at least conceptually straightforward. Unidimensional pain intensity measurement as described earlier in this chapter also has received considerable attention, and a variety of reasonable measurement approaches have been developed for clinical and experimental pain situations. There is no similar agreement about the definition or approach to measurement of pain quality. In most research and clinical settings pain is scaled along a single or at most two dimensions (sensory and affective magnitude). The complexity of the pain experience as well as the importance of accurate assessment of clinical pain has promoted exploration of a variety of more elaborate multidimensional pain scaling schemes. For this section we will use multidimensional to apply to scaling along three or more pain quality dimensions. Underlying this approach is the assumption that pain is inherently multidimensional; that is, pain is perceived as simultaneously varying in intensity along several qualitatively different dimensions. The scaling system for measurement of multiple dimensions of a single perception are obviously complex. In addition, once multiple dimensions are introduced, the allure is to open the flood gates separating the measurement of pain from the measurement of the response to pain (including emotions, cognitions, behaviors, and physiological responses). This is particularly true for strategies in which the true nature of pain is "discovered" through multivariate analysis of multiple responses to a single stimulus. At the heart of the matter is a question of whether one is interested in scaling the experience associated with a painful stimulus or the painfulness of a particular sensory experience. The latter is the intended domain for pain psychophysics, but, as the definition of painfulness is expanded to include multiple qualitative dimensions, it becomes harder to separate judgments of the stimulus from judgments of general feeling states.

A. Criterion for Utility of Multidimensional Scaling of Pain

The utility of a multidimensional pain scaling method rests on its incremental value over a unidimensional or two-dimensional (sensory/affect) model. In addition, a multidimensional approach also needs to provide incremental utility over assignment of a simple pain quality judgment to a stimulus (e.g., "Is it a burning pain?"). Thus, although there may be multiple measurable qualities in a particular perception, and in fact there always are, the existence of these dimensions does not automatically imply usefulness.

A multidimensional system may increase utility if it does the following:

1. Leads to an increase in accuracy of pain reports. If, for example, a rating of intensity and affect misses or blurs critical aspects of a pain sensation, then a patient or subject's pain may change due to treatment or experimental manipulation and this change could be missed. This is essentially an issue of reliability.

2. Increases greater diagnostic sensitivity. If, for example, the amount of prickliness of a pain is a clear marker of certain types of tissue pathology, then assessment of only sensory and affective intensity (painfulness) may yield poorer diagnostic discrimination. Similarly, pain ratings with very unusual patterns of multidimensional ratings might indicate malingering or confusion.

3. Increases communication about pain, and therefore empathy with patients suffering, and

4. Improves the correspondence between neurophysiological and psychological data. With the dramatic increase in sensitivity in brain imaging we might expect to see more specificity in terms of which brain areas correspond to which pain dimensions.

Clark, Janal, and Carroll (1989) have made the distinction between restrictive and nonrestrictive measurement systems. The two-dimensional magnitude estimation methods discussed above can be classified as "restricted" techniques because the number and quality of the dimensions are predetermined before the measurement is made. Unrestricted multidimensional methods allow the subjects to determine the number and type of dimensions and are therefore purported to have less bias and better represent the reality of the pain experience. In actuality, this differentiation in approach is not so clear because the objective of even "unrestrictive" methods is to discover the real pain dimensions and therefore lead to restrictive measurement. Conversely, the system of sensory and affective dimensions was at least in part "discovered" through empirical categorizing of reports.

B. Methodology in Multidimensional Pain Psychophysics

The methods used in defining, validating, and assessing multiple pain dimensions can be characterized by the raw data collected, the analytical procedure used, and the validation criteria employed. The raw data for multidimensional scaling (MDS) is theoretically no different than that used in single- or two-dimensional pain models. This includes verbal descriptors, VAS, other cross-modality matching techniques, and potentially, physiological responses and observed behavior. However, verbal descriptors contain far more nuances of stimulus judgment and therefore lend themselves better to multidimensional approaches than numerical or cross-modality matching responses. By far the most widely used and studied instrument for multidimensional analysis of pain responses has been the MPQ. The MPQ is a descriptor checklist consisting of 78 pain-related adjectives grouped in 20 categories (see Figure 22). In the initial development work of Melzack and Torgerson (1971), a list of 102 pain descriptors gathered from the clinical literature were classified by physician and nonphysician groups based on similarity of meaning in terms of pain quality.

Sensory

1. Flickering Quivering Pulsing Throbbing Beating Pounding	2. Jumping Flashing Shooting	3. Pricking Boring Drilling Stabbing Lancinating	4. Sharp Cutting Lacerating	5. Pinching Pressing Gnawing Cramping Crushing
6. Tugging Pulling Wrenching	7. Hot Burning Scalding Searing	8. Tingling Itchy Smarting Stinging	9. Dull Sore Hurting Aching Heavy	10. Tender Taut Rasping Splitting

Affective

11. Tiring Exhausting	12. Sickening Suffocating	13. Fearful Frightful Terrifying	14. Punishing Grueling Cruel Vicious Killing	15. Wretched Blinding

Evaluative   Miscellaneous

16. Annoying Troublesome Miserable Intense Unbearable	17. Spreading Radiating Penetrating Piercing	18. Tight Numb Drawing Squeezing Tearing	19. Cool Cold Freezing	20. Nagging Nauseating Agonizing Dreadful Torturing

FIGURE 22   Pain descriptors from the McGill Pain Questionnaire. The 20-word sets are grouped into four major classes; 1–10 are sensory descriptors, 11–15 are affective descriptors, 16 is evaluative; and 17–20 are miscellaneous. Descriptors are ordered within each set based on increasing intensity of the particular pain quality. (Adapted from McGill Pain Questionnaire, copyright © Ronald Melzack, 1975.)

The adjectives were grouped into 16 subclasses under three general classes. The general classes were Sensory words describing pain in terms of temporal, spacial, pressure, and thermal features: Affective words, describing tension, fear, and autonomic qualities; and Evaluative words, which describe the general intensity of the pain experience. The two to six words grouped in each subclass were those rated as qualitatively very similar but differing in intensity or perhaps subtly in terms of quality (e.g., tugging and pulling). The words of each subclass where then scaled for intensity. After the initial analysis four new or "miscellaneous" categories were added to include descriptors that did not fit in any of the three general categories. Any specific pain experience can therefore be scaled for intensity on each of the 20 pain quality categories using the MPQ. In practice however, the MPQ is typically scored for only the three general dimensions (Sensory, Affective, Evaluative) by summing the scores of the respective subclasses.

Much of the current interest in MDS procedures is focused on the analytical techniques. Three approaches, factor analysis, ideal-type analysis, and INDSCAL (Individual Differences Scaling) have been at least preliminarily applied to pain responses and merit some discussion. Factor-analytic methods are commonly used in psychometric scale development to classify items into homogeneous scales. This approach aims to maximize intrascale correlations and minimize interscale correlations or, in other words, to define a set of independent dimensions with high internal consistency. Factor analysis methods have been used both to "discover" the number of independent dimensions involved in pain ratings and to validate the a priori three-dimensional model of the MPQ (Holroyd et al., 1992; Melzack, & Katz, 1992). In these studies the item responses of groups of patients rating their clinical pain are subjected to exploratory and sometimes confirmatory factor analysis. In general, factor-analytic research, including a recent large multicenter study, has supported the multidimensional structure of the MPQ with a large first factor of sensory items and one to three other factors including affective and evaluative qualities (Holroyd et al., 1992).

Some have questioned, however, the relevance of these factor-analytic studies for our understanding of the dimensions inherent in the pain experience as assessed by the MPQ. As pointed out by Torgerson, BenDebba, and Mason (1988) and Gracely (1992), the grouping of words in the development of the MPQ emphasized "semantic" similarity, whereas factor-analytic studies evaluate overlap in occurrence, or "associative" similarity. For example, butter and margarine go together in terms of semantic similarity because they have similar ingredients, are part of the same food group, and so on. On the other hand, bread and butter are related by frequency of association. A factor analysis of meals (analogous to a factor analysis of clinical pain ratings) would therefore yield scales of bread/butter, hamburger/french fries, and so on, and not scales consisting of items from the same food groups. There are other difficulties in defining dimensions of pain using factor-analytic procedures. Factor approaches strive to identify independent dimensions and therefore can miss important interactive dimensions. For example, height and weight are highly correlated dimensions of body size. Their high correlation might lead to their being treated as redundant in a factor-analytic approach, leading to a combined "dimension" of body size or even elimination of one measure in favor of using the item most highly correlated with the dimension (Gracely, 1992). Obviously, this would result in loss of important information because there are clear and significant medical distinctions between persons with the same body size due to being tall and thin versus short and heavy. In addition, it is the interaction of height and weight that often yields the most important information (i.e., the relationship between weight and a third variable when controlling for height). A third difficulty with most of the factor-analytic research on pain dimensions stems from the static nature of the data analyzed. Because data are gathered at one time point from multiple subjects, the validity of the dimensions and instruments are based on discrimination among pain subgroups or other static demographic variables. Given the importance of the variability in pain (due to treatments, daily patterns, physiological interactions, etc.) it would be very important to study the utility of any multidimensional schema as a measure of the dynamic response of pain perception to pain control (or pain-increasing) interventions. Overall then, the factor-analytic approach to the development and validation of multidimensional pain assessment is problematic. These techniques do help to identify how individuals and groups use a particular scale and therefore may help in the evaluation of the diagnostic utility of a pain instrument.

A very different approach to MDS of the MPQ has been proposed by Torgerson and colleagues (Torgerson et al., 1988). Pain descriptors can be organized in terms of both general intensity and a variety of qualities. In the original MPQ design each word was assumed to represent only one quality (in addition to general intensity). In the Torgerson model a word may be associated to varying degrees with multiple qualities. To investigate this model Torgerson et al. utilize an ideal-type schema in which each descriptor is scaled by its similarity to a set of ideal or pure types, as well as a quantitative dimension of intensity. As an analogy, the ideal types can be thought of as similar to primary colors and each descriptor (or pain described by that descriptor) as a particular color mix. A color mix can be located on a color wheel by its distances from the primary colors, determined by the percentage of each in the mix (e.g., red/orange is mostly pure red with a touch of yellow). In the initial work with this approach, Torgerson et al. used 17 stimuli (16 from the original Melzack–Torgerson set) and identified four ideal types: bright, slow–rhythmic, thermal, and, vibratory–arrhythmic. As with most multidimensional techniques the meaning of a particular ideal type is derived from the descriptors that are closest in the mathematical space to the ideal type. The determination of meaning for some qualities can therefore be difficult if no descriptor falls unambiguously close to a particular ideal type. Figure 23 illustrates from two studies (one examining sensory descriptors and one affective descriptors) the projection of stimulus words on surfaces defined by three ideal types (see Torgerson et al., 1988). Torgerson et al. (1988) has summarized multiple studies using various overlapping subsets of the original Melzack and Torgerson descriptors. In all these studies groups of judges rate the similarity of pairs of descriptors in order to provide data for derivation of the underlying structure of how these descriptors are used. Nineteen ideal-type pain qualities were obtained as well as an

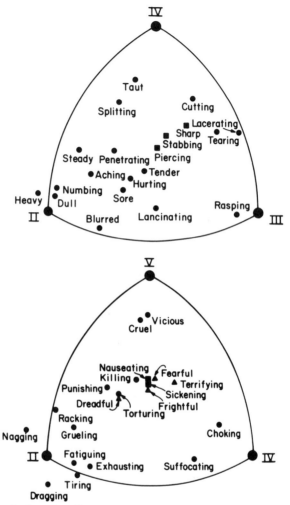

FIGURE 23   Surface plots illustrating the ideal-type structures obtained from a set of sensory (upper plot) and affective (lower plot) words. The numbers at the vertices of the hyperspherical triangle correspond to three of the ideal types identified in each analysis. A total of four ideal types were identified for the sensory word and five for the affective words. (From Torgerson et al., 1988.)

overall quantitative pain-intensity scale. Torgerson et al. argued that these studies confirm the general intensity dimension of the initial MPQ organization but that the ideal-type schema better reflects the organization of pain descriptors than the initial grouping into discrete subclasses. Fourteen of the qualities appear to be sensory and five affective. Examples of descriptors best indicating the sensory ideal types, include Sharp, Tugging, and Pounding. Examples of the affective types include, Frightful, Sickening, and Cruel. Figure 24 shows a list of the two pain descriptors closest to each of the nineteen ideal types. Initial application of this approach to measurement of clinical pain in back pain patients indicates a correlation between scores for the sensory qualities and diagnosis, and between the affective qualities and psychological distress (BenDebba, Torgerson, & Long, 1993).

Clark, Janal, and collaborators have proposed a third MDS approach to pain, which also uses proximity judgments as the raw data (Clark et al., 1989). They use INDSCAL procedures combined with PREMAP (Preference Mapping) to both define a multidimensional "global pain space" and place a particular judgment within it. In this system the number and type of dimensions inherent in ratings of stimuli are abstracted from the proximity judgments made. In describing this system Clark et al., use a geographical analogy involving similarity judgments of all the possible pairwise combinations of six cities. If judges were only basing their similarity judgments on population, a single dimension would emerge. However, if distance between cities, elevation, crime rate, and population were factors used in similarity judgments, then four dimensions would emerge from the INDSCAL analysis. As with other techniques, the meaning (or labeling) of the dimensions is primarily based on examination of the configuration of the stimuli at the poles of the dimensions. The PREMAP model aids in this process by having judges rate each stimuli on bipolar property scales (e.g., sea level—high elevation). The dimensions derived from INDSCAL can then be related to each property scale providing clues as to the meaning of the dimension. The INDSCAL/PREMAP method also allows for examination of how important each dimension is to the judgments of individual subjects.

I. Sensory Domain

1. Sharp Cutting	2. Boring Drilling	3. Stabbing Pricking	4. Dull Numbing	5. Rasping Tearing
6. Taut Splitting	7. Tugging Pulling	8. Squeezing Crushing	9. Hot Burning	10. Flickering Quivering
11. Pounding Throbbing	12. Itching Tickling	13. Radiating Spreading	14. Flashing Shocking	

II. Affective domain

15. Frightful Fearful	16. Dragging Fatiguing	17. Sickening Nauseating	18. Choking Suffocating	19. Cruel Vicious

FIGURE 24 Two pain descriptors closest to each of the nineteen ideal types identified by Torgerson et al. (Adapted from Torgerson et al., 1988.)

The INDSCAL/PREMAP method has been applied to analysis of pain descriptors, clinical pain ratings, and experimental pain stimuli. In a study of similarity judgments of the 36 possible pairs of nine pain descriptors (mild pain, annoying, cramping, sickening, miserable, burning, unbearable pain, intense pain, shooting) three dimensions emerged: evaluative, aversive, and somatosensory (Clark et al., 1989). Interpretation of the meaning of these dimensions can be difficult (Syrjala, 1989). For example, the evaluative dimension seems to have both a pain attribute (from mild pain to intense pain) and an affective attribute (from annoying to sickening). The aversive dimension ranges from mild pain and sickening at one pole to intense pain and shooting at the other. This is interpreted by the authors as indicating an affective dimension going from apathy and malaise to active and threatening. The somatosensory dimension also has two attributes reflecting deep versus surface pain and intermittent versus continuous pain. In a clinical application of this technique, Clark et al. (1989) compared the judgments of pain descriptors by cancer pain suffers and nonpain controls. The results indicated the cancer pain group placed less importance on the affective dimension compared to the controls. Because this method allows

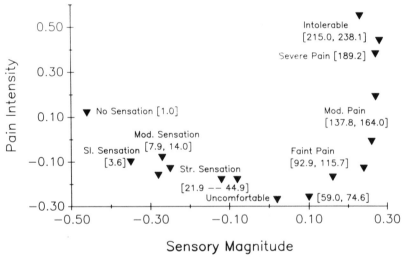

Sensory Magnitude

FIGURE 25   Sensory magnitude versus pain intensity dimensions from Individual Difference Scaling (INDSCAL) analysis of electrocutaneous stimuli. Category labels for each stimulus are determined from the mean response from a method of limits procedure. Numbers in brackets refer to the mean power (mW) of the stimulus. Note that the stimuli labeled "no sensation" to "faint pain" order well on the abscissa (sensory magnitude dimension). Higher intensities are clustered together on the sensory magnitude dimension but separate out on the pain intensity dimension (the ordinate) from "faint pain" to "intolerable." (From Janal et al., 1991).

for determining individual weights, the interpretation of each dimension can also be assessed for each individual. In this study the authors suggested that patients that assigned more importance to the affective dimension may respond better to antianxiety medications and patients that focused on somatosensory dimensions would respond better to analgesics. This analysis and application is therefore reminiscent of the sensory–affective two-dimensional model presented earlier.

Janal, Clark, and Carroll (1991, 1993) have also applied this approach to the evaluation of electrocutaneous and thermal stimuli. In these studies subjects rated the similarity of pairs of stimuli and not stimulus intensity. In all cases the analyses reveal a predominant dimension of intensity along with secondary dimensions that vary depending on the stimuli used. For electrocutaneous stimuli varying in frequency this was a frequency dimension (fast–slow) and for thermal stimuli a warm–hot dimension. Electrocutaneous stimuli of a single frequency yielded an general intensity dimension from lowest intensity to faint pain and a second dimension of painfulness from faint pain to intolerable (see Figure 25). These results indicate that the scaling methods produce dimensions corresponding to qualities specific to the stimuli presented (as well as a general intensity dimension).

C. Summary and Recommendations Regarding Multidimensional Scaling

This section identified four areas of potential utility for MDS approaches to pain. The first was an increase in reliability or accuracy of pain measurement. Similar to increasing assay sensitivity in biochemistry, better measurement should allow a clearer and less distorted assessment of the complex perceptual event we label pain. Although it might seem obvious that assessment of multiple pain qualities or dimensions would increase our accuracy in pain assessment, there is little empirical data to support this conclusion. Although an increase in the amount of raw data collected (e.g., from a simple four-category scale to a large set of verbal descriptors) probably increases reliability of measurement this may not hold true for increases in the number of dimensions abstracted from that data. Simply because pain can be described on multiple dimensions (or in terms of similarity to a variety of ideal quality types) does not mean a more global measurement is less accurate. Other than perhaps the major two dimensions of sensory and affective intensity, the number and organization of pain qualities may be highly situation- and method-specific. That is, the qualities derived from a proximity analysis of verbal descriptors in patients rating their clinical pain may not correspond to those from a proximity analysis of electrocutaneous stimuli in healthy subjects. Validation of these approaches in specific situations in which pain quality (and not just sensory or affective intensity) varies may be the key to the further evaluation and evolution of MDS. In the clinical situation, a refined analysis of pain quality may not increase accuracy. However, in some specific circumstances, such as evaluation of sensory nerve blocks or changes in patterns of neuropathic pain, it may be quite important to be able to accurately characterize subtle changes in pain such as rotational quality (i.e., from stabbing to drilling).

A second potential advantage of multidimensional assessment is increasing diagnostic sensitivity. During routine exams, the differentiation of etiologically and neurophysiologically distinct pain types is often based on patients' reports of pain quality. For example, burning and electrical feeling pains are often associated with neurological pathology rather than with myofascial pathology. MDS potentially may improve clinical judgment by better defining and scaling these discriminating qualities. Torgerson's ideal-type approach seems especially well suited for diagnostic discrimination because the ideal types or combinations of these types might parallel various pain pathologies. However, at present there is little empirical data linking specific diagnostic entities and these pain qualities or the dimensions derived from INDSCAL or factor analysis. A variety of studies, mostly using the original scoring of the MPQ, have shown some differentiation of pain reports between diagnostic groups using very heterogeneous groups (e.g., back pain vs. headache) and statistical tests showing mean group differences. The critical issue, however, is demonstrating incremental diagnostic validity over clinical exams and single- or two-dimensional approaches. In addition, in order to be of diagnostic utility, future studies need to address some standard psychometric issues relevant to any assessment procedure. These include cross-validation in a variety of settings, assessment of sensitivity and specificity of the measure, and consideration of prediction in reference to overall base rates of each category (for a general discussion of psychometric issues see Anastasi, 1968). As indicated for general sensitivity, the utility of multidimensional psychophysical assessment in diagnosis will probably be extremely situation-specific. In cases in which subtle changes in pain have great meaning it should be possible to validate instruments that are sensitive to changes in pain quality and therefore provide unique diagnostic information. One example is a study by Grushka and Sessle (1984) that found that the MPQ differentiated between the pain from a tooth with a reversibly inflamed pulp and the pain from a tooth in which the pulp inflammation was irreversible.

Multidimensional analysis of pain ratings have also been proposed as a way to assess broader issues of psychological functioning in pain patients. Clark et al. have suggested that questions such as, Is this patient really in pain or merely depressed? can be addressed by an analysis of what are the patient's coordinates in the multidimensional global pain space and which dimensions are most relevant for him. The extension of this view is that treatment would follow from an analysis of an individual's multidimensional diagnosis. A patient rating clinical pain high on a somatosensory burning dimension would receive a different medication than one with a similar pathophysiology but a greater score on a somatosensory tingling dimension or an affective dimension. To date there is no empirical support for the validity of using pain reports in this manner or for the assumption that assessment of subjective pain quality offers a more accurate and useful assessment of psychological functioning than more traditional mood or personality instruments.

Improved quantification of a patients pain report might also increase a physician's and other providers' ability to communicate, express empathy, and enlist cooperation from a suffering patient. Patients often complain that their pain report is misunderstood or dismissed by providers and perhaps a better common language could help this process. This may be especially relevant for changes in pain over the course of treatment and adjustment of pharmacological interventions. A potential difficulty with the current approach to MDS is that the dimensions are often difficult for even the researcher to interpret. Application to the clinical setting requires translation into language with face validity for both clinicians and patients. This application of multidimensional assessment reinforces the need to examine pain as a dynamic process and to validate the ability of any multidimensional framework to monitor pain as it changes with time and treatment. In addition, the multidimensional analytical techniques applied to pain have to date emphasized the "discovery" and validation of painfulness dimensions, and not analytical tools for assessing pain in a clinical situation over time.

Perhaps the most promising area for application of multidimensional pain assessment is in basic research on neurophysiology and neuroanatomy of pain. There has been and will continue to be rapid growth in techniques for study of ongoing CNS processes such as functional brain imaging and CNS stimulation and recording techniques. Better quantification of a subject's subjective experience along multiple dimensions may allow for specificity in the psychological domain corresponding to specific neurophysiological processes. For example, Lenz et al. (1993, 1994) have found that electrical stimulation of the thalamus in patients undergoing stereotactic neurosurgery evokes sensations that can be referred to distinct body locations. When these locations coincide with the locus of a chronic pain syndrome, the quality of the evoked pain is similar to the quality of the pain experienced chronically by the patient. Evoked sensations at other locations either were not painful or did not resemble the patient's clinical pain. These studies rely heavily on valid multidimensional pain assessment. In this case, the investigators used a tree-branching questionnaire that efficiently evaluated pain quality in an intraoperative environment.

IV. CONCLUSION

The measurement of pain sensation in large part draws upon psychophysical concepts and techniques. However, both the multidimensional nature of pain perception and the fact that critical medical decisions can depend on assessment of changes in suprathreshold pain magnitude has led to novel psychophysical problems. In addition, the International Association for the Study of Pain defines pain as (Merskey & Bodduk, 1994) "an unpleasant sensory and emotional experience associated with actual or potential tissue damage, or described in terms of such damage" (p. 210). This emphasizes the affective–motivational qualities of this sensation as well as the uncertain link between peripheral physiology and pain perception. As reflected in the work described in this chapter, the continuing evolution of pain measurement is an exciting process of adapting the basic framework and empiricism of traditional psychophysics to the unique and important problem of measuring this complex perceptual experience.

References

Algom, D., Raphaeli, N., & Cohen-Raz, L. (1986). Integration of noxious stimulation across separate somatosensory communications systems: A functional theory of pain. *Journal of Experimental Psychology: Human Perception and Performance, 12,* 92–102.

Anastasi, A. (1968). Psychological testing (3rd ed.). New York: Macmillan.

Anderson, N. H. (1974). Algebraic models in perception. In E. C. Carterette & M. P. Friedman (Eds.), *Handbook of perception* (Vol. 2, pp. 215–298). New York: Academic Press.

Atkinson, R. C., Herrnstein, R. J., Lindzey, G., & Luce, R. D. (1988). *Stevens handbook of experimental psychology* (2nd ed.) (Vol. 1) *Perception and Motivation.* New York: John Wiley and Sons.

BenDebba, M., Torgerson, W. S., & Long, D. M. (1993, August). *Measurement of affective and sensory qualities of back pain.* Paper presented at the 7th World Congress on Pain, August, Paris, France.

Bock, R. D., & Jones, L. V. (1968). *The measurement and prediction of judgment and choice.* San Francisco: Holden-Day.

Bolanowski, S. J., & Gescheider, G. A. (1991). *Ratio scaling of psychological magnitude.* Hillsdale, NJ: Lawrence Erlbaum Associates, Inc.

Brennum, J., Arendt-Nielsen, L., Secher, N. H., Jensen, T. S., & Bjerring, P. (1992). Quantitative sensory examination during epidural anesthesia in man: Effects of lidocaine. *Pain, 51,* 27–34.

Buchsbaum, M. S., Davis, G. C., Copopola, R., & Naber, D. (1981). Opiate pharmacology and individual differences. I. Psychophysical pain measurements. *Pain, 10,* 357–366.

Byers, M. R. (1984). Dental sensory receptors. *International Review of Neurobiology, 25,* 39–94.

Cannon, R. C., Quyyumi, A. A., Schenke, W. H., Fananapazir, L., Tucker, E. E., Gaughan, A. M., Gracely, R. H., Cattan, E. L., Epstein, S. (1990). Abnormal cardiac sensitivity in patients with chest pain and normal coronary arteries. *Journal of the American College of Cardiology, 16,* 1359–1366.

Cannon, W. B. (1939). *The wisdom of the body.* New York: W. W. Norton & Company, Inc.

Carterette, & Friedman, M. P. (Eds.) (1974). *Handbook of perception* (Vol. 2). New York: Academic Press.

Chapman, C. R. (1977). Sensory decision theory methods in pain research: A reply to Rollman. *Pain, 3,* 295–305.

Chapman, C. R., & Feather, B. W. (1973). Effects of diazepam on human pain tolerance and sensitivity. *Psychosomatic Medicine, 35,* 330–340.

Chapman, C. R., Gehrig, J. D., & Wilson, M. E. (1975). Acupuncture compared with 33 percent nitrous oxide for dental analgesia: A sensory decision theory evaluation. *Anesthesiology, 42,* 532–537.

Chapman, C. R., Murphy, T. M., & Butler, S. H. (1973). Analgesic strength of 33 percent nitrous oxide: A signal detection evaluation. *Science, 179,* 1246–1248.

Chapman, C. R., Wilson, M. E., Gehrig, J. D. (1976). Comparative effects of acupuncture and transcutaneous stimulation on the perception of painful dental stimuli. *Pain, 2,* 265–283.

Clark, W. C., Janal, M. N., & Carroll, J. D. (1989). Multidimensional pain requires multidimensional scaling, In C. R. Chapman & J. D. Loeser (Eds.), *Issues in pain measurement* (pp. 285–325). New York: Raven Press.

Clark, W. C., & Yang, J. C. (1974). Acupunctural analgesia? Evaluation by signal detection theory. *Science, 184,* 1096–1098.

Clark, W. C., & Yang, J. C. (1983). Applications of sensory decision theory to problems in laboratory and clinical pain. In R. Melzack (Ed.), *Pain measurement and assessment.* New York: Raven Press.

Cliff, N. (1959). Adverbs as multipliers. *Psychological Review, 66,* 27–44.

Cooper, B. Y., Vierck, C. J., Jr., & Yeomans, D. C. (1986). Selective reduction of second pain sensation by systemic morphine in humans. *Pain, 24,* 93–116.

Coppola, R., & Gracely, R. H. (1983). Where is the noise in SDT pain assessment? *Pain, 17,* 257–266.

Cornsweet, T. N. (1962). The staircase-method in psychophysics. *American Journal of Psychology, 75,* 485–491.

Dallenbach, K. M. (1939). Pain: history and present status. *American Journal of Psychology, 52,* 331–347.

Dawson, W. E., & Brinker, R. P. (1971). Validation of ratio scales of opinion by multimodality matching. *Perception and Psychophysics, 5,* 413–417.

Doctor, J. N., Slater, M. A., & Atkinson, J. H. (1995). The descriptor differential scale of pain intensity: An evaluation of item & scale properties. *Pain, 61,* 251–260.

Duncan, G. H., Bushnell, M. C., & Lavigne, G. J. (1989). Comparison of verbal and visual analogue scales for measuring the intensity and unpleasantness of experimental pain. *Pain, 37,* 295–303.

Duncan, G. H., Bushnell, M. C., Lavigne, G. J., and Duquette, P. (1988). Le développement d'une échelle descriptive verbale française pour mésurer l'intensité et l'aspect désagréable de la douleur. *Douleur Analg., 1,* 121–126.

Duncan, G. H., Feine, J. S., Bushnell, M. C., & Boyer, M. (1988). Use of magnitude matching for measuring group differences in pain perception. In R. Dubner, G. R. Gebhart, & M. R. Bond (Eds.), *Proceedings of the Vth World Congress on Pain* (pp. 383–390). Amsterdam: Elsevier.

Duncan, G. H., Miron, D., & Parker, S. R. (1992, July). *Yet another adaptive scheme for tracking threshold.* Paper presented at the Meeting of the International Society for Psychophysics. Stockholm, Sweden.

Ekman, G. (1964). Is the power law a special case of Fechner's law? *Perceptual and Motor Skills, 19,* 730.

El Sobky, A., Dostrovsky, J. O., & Wall, P. D. (1976). Lack of effect of naloxone on pain perception in humans. *Nature, 263,* 783–784.

Engen, T. (1971a). Psychophysics I. Discrimination and Detection. In J. W. Kling & L. A. Riggs (Eds.), *Experimental psychology* (3rd ed.) (pp. 11–46). New York: Holt.

Engen, T. (1971b). Psychophysics II. Scaling Methods. In J. W. Kling & L. A. Riggs (Eds.), *Experimental Psychology* (3rd ed.) (pp. 47–86). New York: Holt.

Falmagne, J. C. (1986). Psychophysical measurement and theory. In K. R. Boff, L. Kaufman, & J. P. Thomas *Handbook of perception and human performance* (Vol. 1) *Sensory Processes and Perception* (pp. 1–66). New York: John Wiley & Sons.

Feather, B. W., Chapman, C. R., & Fisher, S. B. (1972). The effect of placebo on the perception of painful radiant heat stimuli. *Psychosoma. Medicine, 34,* 290–294.

Fernandez, E., Nygren, T. E., & Thorn, B. E. (1991). An "open-transformed scale" for correcting ceiling effects and enhancing retest reliability: The example of pain. *Perception and Psychophysics, 49,* 572–578.

Gaughan, A. M., Gracely, R. H., & Friedman, R. (1990). Pain perception following regular practice of meditation, progressive muscle relaxation and sitting. *Pain* (suppl. 5).

Gescheider, G. A. (1985). *Psychophysics: Method, theory, and application.* Hillsdale, NJ: Lawrence Erlbaum Associates.

Goldberger, S. M., & Tursky, B. (1976). Modulation of shock-elicited pain by acupuncture and suggestion. *Pain, 2,* 417–429.

Gracely, R. H. (1977). *Pain psychophysics.* Unpublished doctoral dissertation, Department of Psychology, Brown University, Providence, Rhode Island.

Gracely, R. H. (1979). Psychophysical assessment of human pain. In J. J. Bonica, J. C. Liebeskind, & D. G. Able-Fessard (Eds.), *Advances in pain research and therapy* (Vol. 3, pp. 805–824). New York: Raven Press.

Gracely, R. H. (1985). Pain psychophysics. In S. Manuk (Eds.), *Advances in behavioral medicine* (Vol. 1, pp. 191–231). New York: JAI Press.

Gracely, R. H. (1988). Multiple random staircase assessment of thermal pain perception. In R. Dubner, M. Bond, & G. Gebhart (Eds.), *Proceedings of the Fifth World Congress on Pain* (pp. 391–394). Amsterdam: Elsevier.

Gracely, R. H. (1991a). Theoretical and practical issues in pain assessment in central pain syndromes. In K. L. Casey (Ed.), *Pain and central nervous system disease* (pp. 85–101). New York: Raven Press.

Gracely, R. H. (1991b). Experimental pain models. In M. Max, R. Portenoy, & E. Laska (Eds.), *Advances in pain research and therapy* (Vol. 18, pp. 33–47). The Design of Analgesic Clinical Trials. New York: Raven Press.

Gracely, R. H. (1992). Affective dimensions of pain: How many and how measured? *APS Journal, 71,* 243–247.

Gracely, R. H. (1992). Evaluation of multi-dimensional pain scales. *Pain, 48,* 297–300.

Gracely, R. H. (1994). Methods of testing pain mechanisms in normal man. In P. D. Wall, & R. Melzack (Eds.), *Textbook of pain* (3rd ed.). (pp. 315–336). London: Churchill Livingstone.

Gracely, R. H., Deeter, W. R., Wolskee, P. J., & Dubner, R. (1980). Does naloxone alter experimental pain perception? *Society for Neuroscience Abstracts, 6,* 246.

Gracely, R. H., & Dubner, R. (1981). Pain assessment in humans: A reply to Hall. *Pain, 11,* 109–120.

Gracely, R. H., & Dubner, R. (1987). Reliability and validity of verbal descriptor scales of painfulness. *Pain, 29,* 175–185.

Gracely, R. H., Dubner, R., & McGrath, P. A. (1979). Narcotic analgesia: Fentanyl reduces the intensity but not the unpleasantness of painful tooth pulp stimulation. *Science, 203,* 1261–1263.

Gracely, R. H., Dubner, R., McGrath, P. A. (1982). Fentanyl reduces the intensity of painful tooth pulp sensations: Controlling for detection of active drugs. *Anesthesia and Analgesia, 61,* 751–755.

Gracely, R. H., & Gaughan, A. M. (1991). Staircase assessment of simulated opiate potentiation. In M. R. Bond, J. E. Charlton, & C. J. Woolf, (Eds.), *Proceedings of the Sixth World Congress on Pain* (pp. 547–551). Amsterdam: Elsevier.

Gracely, R. H., & Kwilosz, D. M. (1988). The descriptor differential scale: applying psychophysical principles to clinical pain assessment. *Pain, 35*, 279–288.

Gracely, R. H., Lota, L., Walther, D. J., & Dubner, R. A. (1988). A multiple random staircase method of psychophysical pain assessment. *Pain, 32*, 55–63.

Gracely, R. H., McGrath, P. A., & Dubner, R. (1978a). Ratio scales of sensory and affective verbal pain descriptors. *Pain, 5*, 5–18.

Gracely, R. H., McGrath, P., & Dubner, R. (1978b). Validity and sensitivity of ratio scales of sensory and affective verbal pain descriptors: Manipulation of affect by diazepam. *Pain, 5*, 19–29.

Gracely, R. H., Taylor, F., Schilling, R. M., & Wolskee, P. J. (1984). The effect of a simulated analgesic on verbal descriptor and category responses to thermal pain. *Pain* (suppl. 2), 173.

Gracely, R. H., & Wolskee, P. J. (1983). Semantic functional measurement of pain: Integrating perception and language. *Pain, 15*, 389–398.

Gracely, R. H., Wolskee, P. J., Deeter, W. R., & Dubner, R. (1981). Functional measurement of dental pain and analgesia. *Journal of Dental Research, 60* (Special Issue A), *384*, 1981.

Grushka, M., & Sessle, B. J. (1984). Applicability of the McGill Pain Questionnaire to the differentiation of 'toothache' pain. *Pain, 19*, 49–57.

Hall, W. (1981). On "ratio scales of sensory and affective pain descriptors." *Pain, 11*, 101–107.

Hardy, J. D., Wolff, H. C., & Goodell, H. S. (1952). Pain sensations and reactions. New York: Hafner.

Hargreaves, K. M., Dionne, R. A., Meuller, G. P., Goldstein, D. S., & Dubner, R. (1986). Naloxone, fentanyl and diazepam modify plasma beta-endorphin levels during surgery. *Clinical Pharmacology and Therapeutics, 40*, 165–171.

Heft, M. W., Gracely, R. H., Dubner, R., & McGrath, P. A. (1980). A validation model for verbal descriptor scaling of human clinical pain. *Pain, 9*, 363–373.

Heft, M. W., & Parker, S. R. (1984). An experimental basis for revising the graphic rating scale. *Pain, 19*, 153–161.

Holroyd, K. A., Holm, F. J., Keefe, F. J., et al. (1992). A multi-center evaluation of the McGill Pain Questionnaire: Results from more than 1700 chronic pain patients. *Pain, 48*, 297–300.

Jaffe, J. H. (1975). Opioid analgesics and antagonists. In *The pharmacological basis of therapeutics* (p. 248). L. S. Goodman & A. Gilman (Eds.), New York: Macmillan.

Janal, M. N., Clark, W. C., & Carroll, J. D. (1991). Multidimensional scaling of painful and innocuous electrocutaneous stimuli: Reliability and individual differences. *Perception & Psychophysics, 50*, 108–116.

Janal, M. N., Clark, W. C., & Carroll, J. D. (1993). Multidimensional scaling of painful electrocutaneous stimulation: INDSCAL dimensions, signal detection theory indices, and the McGill Pain Questionnaire. *Somatosensory and Motor Research, 10*, 31–39.

Jensen, M. P., Karoly, P., & Braver, S. (1986). The measurement of clinical pain intensity: A comparison of six methods. *Pain, 27*, 117–126.

Jensen, M. P., & McFarland, C. A. (1993). Increasing the reliability and validity of pain intensity measurement in chronic pain patients. *Pain, 55*, 195–203.

Jones, B. (1980). Algebraic models for integration of painful and nonpainful electric shocks. *Perception and Psychophysics, 28*, 572–576.

Kaufman, E., Chastain, D. C., Gaughan, A. M., & Gracely, R. H. (1992). Staircase assessment of the magnitude and timecourse of 50% nitrous oxide analgesia. *Journal of Dental Research, 71*, 1598–1603.

Kruger, L., & Kenton, B. (1973). Quantitative neural and psychophysical data for cutaneous mechanoreceptor function. *Brain Research, 49,* 1–24.

Kwilosz, D. M., Gracely, R. H., & Torgerson, W. S. (1984). Memory for post-surgical dental pain. *Pain* (Suppl. 2), 426.

Kwilosz, D. M., Schmidt, E. A., & Gracely, R. H. (1986). Assessment of clinical pain: Cross-validation of parallel forms of the descriptor differential scale. *American Pain Society,* (Abstracts) *6,* 81.

LaMotte, R. H., & Cambell, J. N. (1978). Comparison of responses of warm and nociceptive C-fiber afferents in monkey with human judgments of thermal pain. *Journal of Neurophysiology, 41,* 509–528.

Lenz, F. A., Gracely, R. H., Hope, E. J., Baker, F. H., Rowland, L. H., Dougherty, P. M., & Richardson, R. T. (1994). The sensation of angina pectoris can be evoked by stimulation of the human thalamus. *Pain, 59,* 119–125.

Lenz, F. A., Seike, M., Richardson, R. T., Lin, Y. C., Baker, F. H., Khoja, I., Yeager, C. J., & Gracely, R. H. (1993). Thermal and pain sensations evoked by microstimulation in the area of human ventrocaudal nucleus (Vc). *Journal of Neurophysiology, 70,* 200–213.

Levine, F. M., DeSimone, L. L. (1991). The effect of experimenter gender on pain report in male and female subjects. *Pain, 44,* 69–72.

Malone, M. D., Kurts, R. M., & Strube, M. J. (1989). The effects of hypnotic suggestion on pain report. *American Journal of Clinical Hypnosis, 31,* 221–230.

Marks, L. E. (1974). *Sensory processes: The new psychophysics.* New York: Academic Press.

Max, M. B., & Laska, E. M. (1991). Single-dose analgesic comparisons. In M. B. Max, R. K. Portenoy, & E. M. Laska, (Eds.), *Advances in pain research and therapy,* (Vol. 18, pp. 55–95). The Design of Analgesic Clinical Trials. New York: Raven Press.

McArthur, D. L., Cohen, M. J., & Schandler, S. L. (1989). A philosophy for measurement of pain. In C. R. Chapman and J. D. Loeser, (Eds.), *Issues in pain measurement* (pp. 37–49). New York: Raven Press.

McGrath, P. A., Gracely, R. H., Dubner, R., & Heft, M. W. (1983). Non-pain and pain sensations evoked by tooth pulp stimulation. *Pain, 15,* 377–388.

Melzack, R. (1975). The McGill Pain Questionnaire: Major properties and scoring methods. *Pain, 1,* 277–299.

Melzack, R., & Katz, J. (1992). The McGill Pain Questionnaire: Appraisal and current status. In D. C. Turk & R. Melzack (Eds.), (pp. 152–165). *Handbook of pain assessment* New York: Guilford Press.

Melzack, R., & Torgerson, W. S. (1971). On the language of pain. *Anesthesiology, 34,* 50–59.

Merskey, H., & Bodduk, N. (Eds.) (1994). *Classification of chronic pain* (2nd ed.) Seattle: IASP Press.

Meyer, R. A., & Campbell, J. N. (1981). Myelinated nociceptive afferents account for the hyperaglesia that follows a burn to the hand. *Science, 213,* 1527–1529.

Mountcastle, V. B. (1967). The problem of sensing and the neural code of sensory events. In G. C. Quarton, T. Melnechuck, & F. O. Schmitt (Eds.), *The neurosciences* (pp. 393–408). New York: Rockefeller University Press.

Mower, G. (1976). Perceived intensity of peripheral thermal stimuli is independent of internal body temperature. *Journal of Comparative Physiology and Psychology, 90,* (176), 1152–1155.

Mower, G., Mair, R., & Engen, T. (1977). Influence of internal factors on the perception of olfactory and gustatory stimuli. In M. R. Kare & O. Maller (Eds.), *The chemical senses and nutrition* (pp. 104–118). New York: Academic Press.

Parducci, A. (1974). Contextual effects: A range-frequency analysis. In E. C. Carterette & M. P. Friedman (Eds.), *Handbook of perception* (Vol. 2, pp. 127–141). New York: Academic Press.

Pollack, I. (1965). Iterative techniques for unbiased rating scales. *Quarterly Journal of Experimental Psychology, 17,* 139–148.

Poulton, E. C. (1979). Models for biases in judging sensory magnitude. *Psychological Bulletin, 86,* 777–803.

Price, D. D. (1988). *Psychological and neural mechanisms of pain.* New York: Raven Press.

Price, D. D., Bush, F. M., Long, S., & Harkins, S. W. (1994). A comparison of pain measurement characteristics of mechanical and simple numerical rating scales. *Pain.*

Price, D. D., Hu, J. W., Dubner, R., & Gracely, R. H. (1977). Peripheral suppression of first pain and central summation of second pain evoked by noxious heat pulses. *Pain, 3,* 57–68.

Price, D. D., McGrath, P. A., Rafii, A., & Buckingham, B. (1983). The validation of visual analogue scales as ratio scale measures in for chronic and experimental pain. *Pain, 17,* 45–56.

Price, D. D., Barrell, J. J., & Gracely, R. H. (1980). A psychophysical analysis of experiential factors that selectively influence the affective dimension of pain. *Pain 8,* 137–149.

Rollman, G. B. (1977). Signal detection theory measurement of pain: A review and critique. *Pain, 3,* 187–211.

Stevens, S. S. (1975). *Psychophysics: Introduction to its perceptual, neural and social prospects.* New York: Wiley.

Swets, J. A. (1964). *Signal detection and recognition by human observers.* New York: Wiley.

Syrjala, K. (1989). Multidimensional versus unidimensional scaling: Little to debate, much to determine. In C. R. Chapman & J. D. Loeser (Eds.), *Issues of pain measurement* (pp. 327–335). New York: Raven Press.

Teghtsoonian, R. (1971). On the exponents in Steven's law and the constant in Ekman's law. *Psychological Review, 78,* 71–80.

Thurstone, L. I. (1959). *The measurement of values.* Chicago: University of Chicago Press.

Torgerson, W. D. (1958). *Theory and methods of scaling.* New York: John Wiley and Sons.

Torgerson, W. S., BenDebba, M., & Mason, K. J. (1988). Varieties of pain. In R. Dubner et al. (Eds.), *Proceedings of the Vth World Congress on Pain* (pp. 368–374). New York: Elsevier.

Tursky, B. (1974). Physical physiological and psychological factors that affect pain reaction to electric shock. *Psychophysiology, 11,* 95–112.

Tursky, B. (1976). The development of a pain perception profile: a psychophysical approach. In M. Weisenberg, & B. Tursky (Eds.), *Pain: New perspectives in therapy and research* (pp. 171–194). New York: Plenum Press.

Wallenstein, S. L. (1991). The VAS relief scale and other analgesic measures: Carryover effect in parallel and crossover studies. In M. Max, R. Portenoy, & E. Laska (Eds.), *Advances in pain research and therapy* (Vol. 18, pp. 97–103). The Design of Analgesic Clinical Trials. New York: Raven Press.

Walther, D. J., & Gracely, R. H. 91986). Ratio scales of pleasantness and unpleasantness affective descriptors. *American Pain Society Abstracts, 6,* 34.

Warren, R. M., & Warren, R. P. (1963). A critique of SS Stevens' "new psychophysics." *Perceptual and Motor Skill, 16,* 797–810.

Weiffenbach, J. M., Cowart, B. J., & Baum, B. J. (1986). Taste intensity perception in aging. *Journal of Gerontology, 41,* 460–468.

Wolskee, P. J., Gracely, R. H., Dorros, C. M., Schilling, R. M., & Taylor, F. (1983). Fentanyl reduces thermal pain intensity: Effect of stimulus range on response method. *American Pain Society Abstracts, 4,* 74.

Woodworth, R. S., & Schlosberg, H. (1954). *Experimental psychology.* New York: Holt Rinehart and Winston.

Yang, J. C., Clark, W. C., Ngai, S. H., Berkowitz, B. A., & Spector, S. (1979). Analgesic action and pharamcokinetics of morphine and diazepam in man: An evaluation by sensory decision theory. *Anesthesiology, 51,* 495–502.

Zinnes, J. L. (1969). Scaling. *Annual Review of Psychology, 20,* 447–478.

Pathological Pain

C. Richard Chapman
Mark Stillman

I. INTRODUCTION

Broadly defined, pathological pain is severe persisting pain or moderate pain of long duration that disrupts sleep and normal living, ceases to serve a protective function, and instead degrades health and functional capability. Its mechanisms encompass sensitization of peripheral afferents and pathways of spinal transmission, exaggeration of normal neurophysiological mechanisms for nociception, and aberrant central processing of minimally noxious or nonnoxious signals as severe pain. Pathological pain may arise from abnormalities in the transduction of tissue trauma into neural signals, from abnormal peripheral or central transmission or modulation of such signals or from pathological changes at higher levels of the central nervous system (CNS).

When severe, pathological pain is a singularly compelling, all-consuming awareness. Scarry (1985), writing from a humanities perspective, contended that pain's qualities include extreme aversiveness, an ability to annihilate complex thoughts and other feelings, an ability to destroy language, and a strong resistance to objectification. Everyday clinical observations support her contention. Bonica (1990), writing not only as a physician and scholar but also as a chronic pain sufferer, described pain as a malefic force.

Pain and Touch

In this chapter we address the mechanisms of pathological pain and describe several common pain syndromes. We attempt to bridge knowledge gleaned from research to that obtained from clinical observation and treatment. Current scientific understanding of the basic mechanisms of pain derives principally from animal and human laboratory research and has a narrow sensory focus. However, clinical manifestations of pathological pain are complex, colored by the affective status of the patient, and complicated by the patient's general health, psychosocial well-being, and overall psychosocial adjustment. Consequently, knowledge gleaned by laboratory research cannot as yet account adequately for clinical pain states. Our goal in discussing pathological pain is to offer a limited but integrated perspective on pathological pain that draws from clinical medicine, basic science, and behavioral science. Table 1 defines some of the terminology associated with pathological pain.

TABLE 1   Terminology Associated with Pathological Pain

Allodynia	Pain due to a stimulus that normally does not provoke pain.
Anesthesia dolorosa	Pain in an area or region that is anesthetic.
Causalgia	A syndrome of sustained burning pain, allodynia, and hyperpathia after a traumatic nerve lesion, often combined with vasomotor and sudomotor dysfunction and later trophic changes.
Central pain	Pain associated with a lesion of the central nervous system.
Deafferentation Pain	Pain due to loss of sensory input into the central nervous system, as occurs with avulsion of the brachial plexus or other types of lesions of peripheral nerves or due to pathology of the central nervous system. This clinical term encompasses anethesia dolorosa, causalgia, neurogenic, and neuropathic pain.
Dysesthesia	An unplesant abnormal sensation, whether spontaneous or evoked.
Hyperesthesia	Increased sensitivity to stimulation, excluding special senses.
Hyperalgesia	An increased response to a stimulus that is normally painful.
Hyperpathia	A painful syndrome, characterized by increased reaction to a stimulus, especially a repetitive stimulus.
Neuralgia	Pain in distribution of nerve or nerves.
Neurogenic inflammation	The release of peptides that foster inflammation via antidromic, efferent activity in primary afferents.
Neuropathy	A disturbance of function or pathologic change in a nerve; in one nerve, mononeuropathy; in several nerves, mononeuropathy multiplex; if symmetrical and bilateral, polyneuropathy.
Nociceptor	A receptor preferentially sensitive to a noxious stimulus or to a stimulus that would become noxious if prolonged.
Noxious stimulus	A noxious stimulus is one that is potentially or actually damaging to body tissue.
Referred pain	Pain localized to an area adjacent to or distant from the site of its cause.

II. BIOLOGICAL FUNCTIONS AND FEATURES OF NORMAL PAIN

Pain is a complex human perception with marked emotional as well as sensory aspects (Chapman & Gavrin, 1993). The sensory aspect of pain identifies the location and nature of the trauma, whereas its negative emotional component defines the biological significance of the trauma in consciousness. In everyday life pain performs an essential protective function by signaling tissue injury.

A. Protective Functions

Pain protects us from injurious stimuli in the physical environment by facilitating withdrawal from harmful objects and escape from damaging situations. Wall (1985) proposed three periods of normal response to acute injury: the immediate, secondary, and tertiary phases. These phases indicate changes in biological priorities over time. In the first phase, the time of immediate injury, escape or victory over an opponent is more important than attending to wounds, as is summoning help from others by expressing pain. At the second phase, one must cope with the injury and prepare for recovery. In the third phase, by limiting activity and conserving resources, pain contributes to recuperation.

The first phase can involve puzzling phenomena. Sometimes pain fails to occur at the time of significant injury. Soldiers in battle, athletes hurt during intensive contact sports, and vehicular accident victims often experience injury unaccompanied by pain. Clearly, such spontaneous analgesia serves a biological purpose because in emergency situations pain would interfere with actions that survival might require. In other cases, pain sometimes occurs too late to protect us from immediate injury, as in the case of sunburn or overuse of muscles. In such instances pain restricts us from repeating potentially injurious activities in the near future.

B. Pain and Disease

The biological role of pain as a signal of disease raises many perplexing questions. Often its "warning" is too late to assist coping or survival. Nonetheless, speculation helps clarify the nature of pain associated with disease.

Sometimes it is a diffuse tenderness (hyperalgesia) rather than a frank and focused pain that calls disease to our attention. Hyperalgesic changes that portend the onset of infectious disease (e.g., a sore throat) foster recuperative, energy-conserving behaviors. In the case of intestinal infection, hyperalgesia or diffuse abdominal pain warns one that an abnormal and potentially threatening process is underway and discourages further oral intake via its autonomically mediated sequelae, such as nausea and vomiting. When a patient suffers acute coronary insufficiency, chest pain forces him or her to diminish ongoing activity, thus minimizing the heart's work load.

Unfortunately, in many circumstances pain fails to occur when its presence might, at least in contemporary times, provide life-saving notice of a pathological process. Most cancers, for example, are silent and remain undetected for years, and their late discovery often precludes medical rescue from fatal disease. Thus, although pain is essential for our survival, during disease it functions peripatetically as a warning system for pathological processes in tissue.

III. SOME PERCEPTUAL PUZZLES IN PATHOLOGICAL PAIN

Pain is complex and sometimes involves perplexing and intriguing perceptual abberations. It may appear at sites distant from the lesion that causes it, or it may occur in association with an illusory, phantom-like perception of a missing body part. In still other cases, pain involves the entire body and seems to convey no somatosensory information.

A. Pain Referral

Pain signaling tissue trauma in deep tissues or visceral structures sometimes manifests in part, or altogether, as a pattern of hyperalgesia at some area on the surface of the body distant from the diseased organ (Cervero, 1991; Jänig, 1987). Patients suffering myocardial infarction, for example, often feel pain referred to the inner aspect of the left arm and hand or to the throat and jaw. Kidney stones can cause pain in a testicle, the perineum, or the penis. Pain referral patterns tend to be consistent for specific organs and many are well known to physicians, but referred pain from one disease can simulate the pain of another, thus complicating diagnosis. Occasional radical departures from normal pain reference patterns occur and create significant diagnostic challenges.

Referred pain, although strange, is neither a perceptual illusion nor mental "displacement" of pain at the level of the brain. It has clear somatic manifestations, such as reflex skeletal muscle spasm in the area of pain referral, sympathetic hyperactivity (e.g., sweating, piloerection), and a zone of hyperalgesia that patients commonly describe as tenderness. Spinal reflexes appear to contribute to referred pain by creating muscle contraction (Fields, 1987, p. 82). In addition, sympathetic changes may compromise microcirculation, thus producing some degree of ischemia or contributing to chemical changes in the vicinity of nociceptors.

Several investigators have offered explanations for how visceral referred

somatic structure

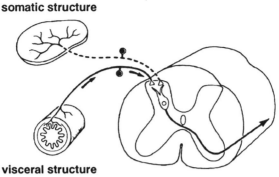

visceral structure

FIGURE 1   The Visceral-Somatic Convergence model for referred pain. Signals from an inflamed visceral organ converge on projection neurons at the dorsal horn of the spinal cord. Sensory input from a distant somatic structure converges on the same neurons. This results in misattribution of pain to the healthy somatic structure. (Adapted from Fields, 1987. Reprinted with permission from McGraw-Hill Companies.)

pain can occur (see Fields, 1987, and Bonica, 1990, pp. 169–174 for reviews). The Visceral-Somatic Convergence model provided by Ruch (1965) provides the best explanation to date. Figure 1 illustrates this concept. Small myelinated and unmyelinated visceral and cutaneous afferent fibers converge onto second-order neurons in lamina V of the spinal and medullary dorsal horn. Considerable supporting evidence has emerged since the model's introduction (see Bonica, 1990). In addition, the model accounts for the observation that most pain referral patterns involve a somatic region innervated by the same spinal segments that innervate the diseased organ.

Gebhart (1993) noted that afferents from multiple viscera often converge on the same spinal cord neurons. This explains why patterns of pain referral from various organs overlap and tend to confound diagnosis. For example, Wiener (1993) indicated that pain referred to the right shoulder could come from acute cholecystitis or perforation of a duodenal ulcer. Pain in the left shoulder could signal myocardial infarction, pericarditis, perforation of a gastric ulcer, or perforation of an ulcerative carcinoma of the stomach, as well as spontaneous rupture of the spleen. Other sources of shoulder pain include hepatic tumor, pleuritis, and subclavian-axillary arterial aneurysm.

Referred pain patterns demonstrate that pain performs its sensory duties inconsistently, with poor correspondence between localization of pathology and the focus of pain experience.

B. Postamputation Phantoms and Pain

After surgical removal of a limb or other body part (nose, tongue, breast, anus, teeth), patients may perceive a phantom limb that seems a part of the body. This occurs universally in limb amputation and variably with other body parts (Loegor, 1990; Melzack, 1989, 1992; Roth & Sugarbaker, 1980). The more distal the severed part, the more vivid the phantom sensation. Phantom-like experiences can occur with any event that causes abrupt cessation of peripheral nerve activity, even if the cessation is temporary as with local anesthetic block. Over time most patients with surgical amputations report that the phantom fades away or telescopes into the stump. Pain accompanies phantoms infrequently, but in some cases it can persist indefinitely. Figure 2 illustrates the phantom pain pattern of a patient who lost his left leg in a war injury. The pain, burning and lancinating in character, persisted for decades (Gross, 1984).

Because phantom limbs seem so vivid, they can cause bizarre experiences. Some patients report that the phantom has a fixed position in space with a definite length and girth. The position may be socially inappropriate. For example a man with a phantom arm extended straight from the left shoulder will refuse to sleep on his left side, tend to enter doorways sideways to allow for the arm to pass through freely, and generally behave as though a real limb extends in space where he perceives the phantom. Many report that their phantom can move, and some find themselves attempting to use the missing limb from force of habit to maintain balance or perform simple tasks. Still others experience their phantoms as distorted in position, sometimes with a distressing sense of torsion or tension. As a phantom telescopes gradually over time, the phantom body part may change grotesquely. A patient may report that his phantom hand is attached to the stump, the rest of the arm having receded already.

The incidence of phantom limb pain remains poorly defined for the

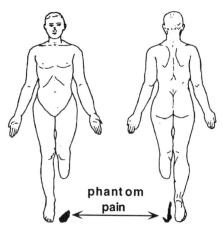

FIGURE 2   Distribution of chronic phantom limb pain in a 60-year-old man. The black area indicated the focus of the pain. (Adapted from Gross, 1984. Reprinted with permission from Lippincott-Raven Press Publishers.)

population as a whole. Shukla, Sahu, and Tripath (1982) observed that 86% of 72 amputees recovering from surgery had a phantom pain that differed qualitatively from the pain of their surgical trauma. A study of 63 patients in the first month after surgery yielded a 97% prevalence for painful phantoms (Roth & Sugarbaker, 1980). A follow-up conducted over several years determined that phantom limb pain eventually resolved in 20% of these patients. Sherman, Sherman, and Parker (1984) gave questionnaires to 5000 patients with amputations related to military service and obtained responses from 55%. Of these, 78% reported that they still experienced phantom limb pain even though the average time since amputation was 26 years. Clearly, the incidence and nature of the syndrome varies widely with the population investigated.

Further amputation or denervation never relieves, but can exacerbate, phantom limb pain (Melzack, 1992). The causes of this pain problem remain at issue. The dominant explanation identifies phantom pain as a special case of central pain combined with the perception of a phantom. The CNS apparently changes in response to loss of neural feedback from normal afferentation and movement, perhaps at the dorsal horn (see Fields, 1987, for a discussion of how peripheral deafferentation can cause hyperactivity in spinal transmission neurons) and perhaps at higher centers as well.

The phenomena of painful phantom body parts demonstrates that pain is a complex perception, intimately linked to central somatosensory representations of body integrity. The brain can impute pain to any part of the body, even when a part on the body no longer exists.

1. L'homme Douloureux: Fibromyalgia

The concept of the "painful person" dates to antiquity. Today, we diagnose many patients in "total body pain" with fibromyalgia syndrome. Patients with this disorder suffer a generalized body pain characterized by diffuse muscle soreness and stiffness that makes every experience or activity painful (Harvey, Cadena, & Dunlap, 1993; Moldofsky, 1993). Moreover, fatigue accompanies the pain, which is continuous and relentless, and patients complain of nonrestorative sleep. This disorder tends to appear in the wake of one or more traumatic events, acute febrile illness, an inflammatory rheumatic disease, or prolonged psychological stress. It occurs five times as often in women as men. Psychological stress, cold, and damp weather exacerbate the condition as do underuse or overexercise of muscles. No satisfactory treatment exists.

The causes of fibromyalgia syndrome remain poorly defined. Work by Pedersen–Bjergaard et al. (1990) suggests that neurogenic inflammation produced by a combination of Substance P (SP), calcitonin gene-related peptide (CGRP), and neurokinin A may play a causal role in the muscle pain that characterizes this syndrome. Zimmerman (1991) hypothesized multiple coexisting mechanisms: nociceptive, neuropathic, and CNS dysregulation. Griep, Boersma, and de Kloet, (1993) demonstrated a significantly greater adrenocorticotropic hormone (ACTH) release in women with fibromyalgia than in normal controls. This suggests that altered adrenal responsiveness within the hypothalamo-pituitary-adrenocortical axis may play a role in the development and persistence of this condition and supports hypothesized links between stress and fibromyalgia.

The pain of this condition appears to serve no biological purpose because it has no somatic focus. It identifies no body part to protect, nor does the pain seem to signal tissue trauma, because the sore muscle tissue appears histologically normal. The presence of pain discourages activity and fosters withdrawal and rest, but sleep is never refreshing for these patients. Thus, pain in this syndrome appears to represent a severe and debilitating perceptual disorder.

IV. BASIC MECHANISMS OF NONPATHOLOGICAL PAIN

Awareness of tissue trauma requires sensory end organs that can detect tissue injury or pathological conditions in tissue and generate signals of such conditions. In addition it necessitates pathways of central transmission that convey such signals faithfully and higher CNS processes that can integrate such signals into ongoing perception and produce the experience of pain. Figure 3 summarizes the central mechanisms that generate pain under normal conditions.

A. Transduction and Transmission

1. Transduction

How does pain occur? Under normal conditions, pain results from the transduction of tissue trauma into neural signals by sensory end organs known as nociceptors (Besson & Chaouch, 1987; Heppelmann, Messlinger, Schaible, & Schmidt, 1991; Willis, 1993). The free nerve endings of thinly myelinated Aδ fibers function as thermal and/or mechanical nociceptors, conducting at 4–44 m/sec. In addition, certain unmyelinated C-fibers that conduct slowly (roughly .5–1 m/sec) act as polymodal nociceptors, responding to various high-intensity mechanical, chemical, and thermal stimuli. Both types of fibers distribute widely in skin and in deep tissue. Recent research indicates that some primary afferents act as "silent nociceptors." They fail to respond to any normal sensory stimuli, but noxious events can sensitize them so that they function thereafter as nociceptors (McMahon & Koltzenburg, 1990; Willis, 1993).

Nociceptors exist in skin, muscle, fascia, joints, tendons, blood vessels,

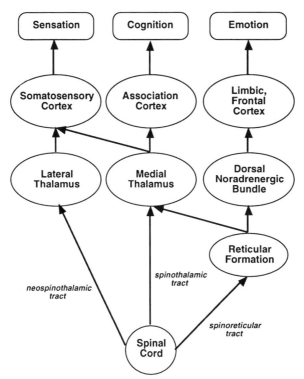

FIGURE 3   A schematic overview of central nervous system processing of nociceptive signals.

and visceral organs. From a sensory perspective, these tissues group into cutaneous, deep, and visceral types. Nociception appears to serve somewhat different functions in the three types of tissues, and the qualities of the pain that ensues from their activation varies across types. Cutaneous pain tends to be well localized, sharp, pricking, or burning. Aδ fibers produce sharp, pricking pain sensations of discreet, brief duration, whereas C-fibers tend to generate burning sensations. Deep tissue pain usually seems diffuse and dull or aching in quality, although deep tissues can give rise to bright, sharp pains under certain conditions (e.g., muscle rupture). Visceral pain is very diffuse, often referred to the body surface, perseverating, and frequently associated with queasiness so that patients tend to describe is as "sickening." Severe visceral pain typically produces an accompaniment of profuse sweating, nausea, and vomiting.

The adequate stimuli for nociception differ across tissue types. Cutaneous receptors detect injurious stimuli from the surrounding environment, and so they respond to severe mechanical and thermal events, such as cutting, burning, or freezing. Nociceptors in deep tissue such as muscle detect overuse strain, deep mechanical injury like tearing and contusion, spasm or cramping, and ischemia. Their function resembles that of nociceptors in cutaneous tissue, but their responses may be more intimately linked to flexor reflexes than are those of their counterparts in skin. Muscle pain tends to beget muscle stiffness and splinting, which serves a protective function by bracing or supporting injured muscle. Visceral nociceptors do not respond to cutting or burning injury like their counterparts in cutaneous tissue but instead fire in response to pathological change (Gebhart & Ness, 1991). A hollow viscus needs to identify and transduce distention, stretch, and isometric contraction. A solid organ needs signal distension of the capsule that contains it and inflammation. Gebhart (1993) listed the following as naturally occurring visceral stimuli: distension of hollow organs, ischemia, inflammation, muscle spasm, and traction. Thus, the peripheral origins of pain vary markedly, depending on whether the nociceptors involved lie in superficial or deep tissues.

The fundamental transduction mechanism(s), the means by which diverse chemical, thermal, and mechanical stimuli depolarize free sensory endings, remains an enigma. Apparently, several distinct tissue trauma transduction mechanisms exist because the threshold of response of polymodal nociceptors to one type of stimulus can change without altering the threshold to others (Jessell & Kelly, 1991). Moreover, responses are poorly graded with regard to severity of tissue trauma. Although psychophysicists can demonstrate relationships between the intensity of noxious but harmless laboratory stimuli and pain report, clinical observations demonstrate that little or no relationship exists between the severity of tissue trauma as defined by size of lesion and pain intensity (Bonica, 1990).

2. Transmission

How do signals of tissue trauma ascend to higher levels of the CNS? Nociceptive afferents enter the spinal cord primarily through the dorsal route, terminating principally in lamina I (the marginal zone) but also in laminae II (the substantia gelatinosa) and V of the dorsal horn (Craig, 1991). The spinal and medullary dorsal horns are much more than simple relay stations; these complex structures participate directly in sensory processing, performing local abstraction, integration, selection, and appropriate dispersion of sensory impulses (Bonica, 1990, 1:28–94; Dubner, 1991; Jänig, 1987; Perl, 1984; Willis, 1988a,b). Upon entry, nociceptive afferents synapse with projection neurons that convey information to higher centers, facilitory interneurons that relay input to projection neurons, and inhibitory interneurons that modulate the flow of nociceptive signals to higher centers (Jessell & Kelly, 1991). Similar neural processing occurs in the spinal cord in the medullary dorsal horn.

Two principal types of projection neurons exist: nociceptive specific and multireceptive or wide-dynamic range (WDR) neurons (Jänig, 1987). The former convey only tissue trauma signals; the latter react to stimuli of increasing intensity. Ascending tracts include spinothalamic, spinoreticular, spinomesencephalic, spinocervical, and postsynaptic dorsal cord tracts. Detailed discussion of this exceeds the scope of this chapter, but Willis and others (1988a,b and Besson & Chaouch, 1987) provide useful reviews.

B. Brain Mechanisms

The principal ascending tracts are the spinothalamic and spinoreticular. We propose that sensory and affective processes subserving pain share common input from afferent sources, injury-sensitive Aδ and C primary afferents. Differentiation of sensory and affective processing begins at the dorsal horn of the spinal cord with sensory transmission following spinothalamic pathways and affective transmission taking place in spinoreticular pathways.

The spinothalamic tract delivers noxious signals to medial and lateral thalamus. These structures in turn activate areas in primary and secondary somatosensory cortex. Detailed reviews of spinothalamic processing appear in Bonica, 1990; Craig, 1991; Fields, 1987; Peschanski and Weil-Fugacza, 1987; and Willis, 1985. The processes associated with these structures equip the individual with a capability for determining the nature of the traumatic event, its location, its duration, and to some extent its severity. Spinothalamic processing plays an important role in the perception of injurious cutaneous events that one can escape, but it performs poorly for trauma in deep tissues and visceral structures.

The spinoreticular tract, which has received far less attention to date, also participates in nociceptive centripetal transmission (Villanueva, Bing, Bouhassira, & Le Bars, 1989; Villanueva, Cliffer, Sorkin, & Willis, 1990). Spinoreticular axons possess receptive fields that resemble those of spinothalamic tract neurons projecting to medial thalamus, and, like their spinothalamic counterparts, they transmit tissue injury information (Bonica, 1990; Fields, 1987; Villanueva et al., 1990). Most spinoreticular neurons carry nociceptive information, and many of them respond preferentially to noxious input (Abou-Samra, 1987; Bing, Villanueva, & Le Bars, 1990; Bowsher, 1976; Willis, 1985). We speculate that it conveys nociceptive signaling to higher central nervous structures that undertake affective (in contradistinction to sensory) processing of those signals. These higher structures are primarily noradrenergic. We emphasize them here because (a) most of the literature on pain overlooks them; (b) they implicate limbic structures in pain perception; (c) the emotional aspect of pain plays a greater role in clinical pain problems than its sensory counterpart; and (d) these pathways link pain and neuroendocrine responses.

1. Nociception and Central Noradrenergic Processing

Processing of nociceptive signals to produce the emotional aspect of pain commences in reticulocortical pathways. Several extrathalamic afferent pathways to neocortex exist (Foote & Morrison, 1987), but clearly the noradrenergic pathways originating in the locus coeruleus (LC) link most closely to negative emotional states (Gray, 1987, 1991). The structures receiving projections from this complex and extensive network, including the hypothalamus, corresponds to the classic definition of the limbic brain (Gray, 1987; Isaacson, 1982; MacLean, 1990; Papez, 1937). Although other processes governed predominantly by other neurotransmitters very likely play important roles in the complex experience of emotion during pain, we emphasize here the role of noradrenergic processing. Two central noradrenergic pathways encompass the key structures: the dorsal and ventral noradrenergic bundles (see Figure 4).

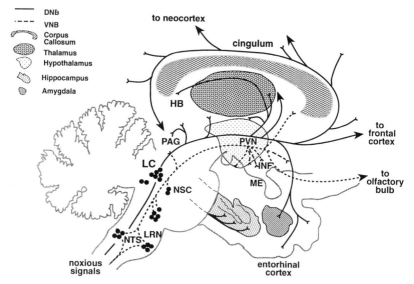

FIGURE 4   The dorsal and ventral noradrenergic bundles. Central corticopetal noradrenergic transmission in a primate brain (parasagittal view). The cell bodies of neurons that produce norepinephrine appear as black circles. The major projections of these cell bodies are the dorsal noradrenergic bundle (DNB) and the ventral noradrenergic bundle (VNB). Tissue trauma signals from spinoreticular pathways excite the primarily noradrenergic locus to coeruleus (LC), activating the DNB, which extends throughout the limbic brain and to neocortex. ME: median eminence; PAG: periaqueductal gray; HB: habenula; NSC: nucleus subcoeruleus; LRN: lateral reticular nucleus; NTS: nucleus tractus solitarius; INF: infundibulum; PVN: paraventricular nucleus. (From Chapman, 1993. Reprinted with permission from Lippincott-Raven Press Publishers.)

2. Locus Coeruleus and the Dorsal Noradrenergic Bundle

The pontine nucleus, LC, resides bilaterally near the wall of the fourth ventricle. It has three major projections: ascending, descending, and cerebellar: The ascending projection, the dorsal noradrenergic bundle (DNB), is the most extensive and important (Fillenz, 1990). It reaches from the LC throughout limbic brain and to all of neocortex, accounting for about 70% of all brain norepinephrine (NE) (Svensson, 1987; Watson, Khachaturian, Lewis, & AK.1, 1986). The LC gives rise to the majority of central noradrenergic fibers in spinal cord, hypothalamus, thalamus, hippocampus (Aston-Jones, Foote, & Segal, 1985; Levitt & Moore, 1979), and its projections extend to limbic cortex and all of neocortex.

The LC responds to sensory stimuli that potentially threaten, or signal injury to, that integrity. Nociception inevitably and reliably increases activity in neurons of the LC, and LC excitation appears to be an inevitable response to nociception (Korf, Bunney, & Aghajanian, 1974; Morilak, Fornal & Jacobs, 1987; Stone, 1975; Svensson, 1987). This does not require cognitively mediated attentional control because it occurs in anesthetized animals. Foote, Bloom, and Aston-Jones (1983) reported that slow, tonic spontaneous activity at LC in rats changed under anesthesia in response to noxious stimulation. Experimental electrical stimulation of the LC causes alarm and apparent fear in primates (Charney, et al., 1990; Redmond & Huang, 1979), and lesions of the LC eliminate normal heart rate increases to threatening stimuli (Redmond, 1977).

How does this relate to tissue trauma? The LC responds consistently, although not exclusively, to tissue injury. However, increased LC activity also follows nonpainful threatening events such as strong cardiovascular stimulation (Elam, Svensson, & Thoren, 1985; Morilak et al., 1987) and certain distressing visceral events such as distension of the bladder, stomach, colon, or rectum (Elam, Svensson, & Thoren, 1986b; Svensson, 1987). Thus, although it reacts to nociception, the LC is not a nociception-specific nucleus. It responds to events that represent biological threat, and tissue trauma is such an event.

Studies of negative emotion and vigilance behavior implicate the DNB as the largest and the most important LC projection for emotional processing of nociception. The DNB makes possible vigilance, and orientation to threatening and novel stimuli can occur because of the DNB; it also regulates attentional processes and facilitates responses (Calogero, Bernadini, Gold, & Chrousos, 1988; Elam, Svensson, & Thoren, 1985, 1986a; Foote & Morrison, 1987; Gray, 1987; Svensson, 1987). Direct activation of the DNB and/or associated limbic structures produces sympathetic nervous system response and releases patterns of emotional behaviors in animals such as defensive threat, fright, enhanced startle, freezing, and vocalization (McNaughton & Mason, 1980). In normal circumstances activity in this pathway increases alertness. Fundamentally, tonically enhanced LC and DNB discharge corresponds to hypervigilance and emotionality (Butler, Weiss, Stout, & Nemeroff, 1990; Foote et al., 1983). The LC and DNB foster survival by making possible global vigilance for threatening and harmful stimuli.

We speculate that the affective dimension of pain shares central mechanisms with vigilance, a biologically important process. Vigilance, intensified by tissue trauma signals, threats from the environment, or a combination of these can progress to hypervigilance and beyond it to panic. Extrapolated to subjective experience, the emotional aspect of pain corresponds to the emotional awareness of threat.

3. The Ventral Noradrenergic Bundle and the Hypothalamo-Pituitary-Adrenocortical Axis

The ventral noradrenergic bundle (VNB) enters the medial forebrain bundle (see Figure 4) and links neurons in the medullary reticular formation to the hypothalamus (Bonica, 1990; Sumal, Blessing, Joh, Reis, & Pickel, 1983). Sawchenko and Swanson (1982) identified two VNB-linked noradrenergic and adrenergic pathways to paraventricular hypothalamus in the rat and described them using the Dahlström and Fuxe (1964) designations: the A1 region of the ventral medulla (lateral reticular nucleus, LRN), and the A2 region of the dorsal vagal complex (the nucleus tractus solitarius, NTS), which receives visceral afferents. These medullary neuronal complexes supply 90% of catecholaminergic innervation to the paraventricular hypothalamus via the VNB (Assenmacher, Szafarczyk, Alonso, Ixart, Barbanel, 1987). Regions A5 and A7 make comparatively minor contributions to the VNB.

The VNB is important for emotion research because it innervates the hypothalamus. The noradrenergic axons in the VNB respond to noxious stimulation (Svensson, 1987), as does the hypothalamus (Kanosue, Nakayama, Ishikawa, & Imai-Matsumura, 1984). Moreover, nociception-transmitting neurons at all segmental levels of the spinal cord project to medial and lateral hypothalamus and several telencephalic regions (Burstein, Cliffer, & Giesler, 1988). These considerations suggest that threatening events, and particularly tissue trauma, excite the hypothalamoadrenocortical (HPA) axis via several routes. As the HPA axis controls the stress response, its reactions to tissue trauma are important corollaries of the pain state and may contribute to the emergence of pathological pain (Griep et al., 1993).

The hypothalamic paraventricular nucleus (PVN) coordinates the HPA axis. Neurons of the PVN receive afferent information from several reticular areas, including ventrolateral medulla, dorsal raphe nucleus, nucleus raphe magnus, LC, dorsomedial nucleus, and the nucleus tractus solitarius (Lopez, Young, Herman, Akil, & Watson, 1991; Peschanski & Weil-Fugacza, 1987; Sawchenko & Swanson, 1982). Still other afferents project to the PVN from the hippocampus and amygdala. Nearly all hypothalamic and preoptic nuclei send projections to PVN.

The PVN responds to potentially injurious or tissue-traumatizing stimuli by initiating a complex series of events that prepare the individual to cope powerfully with the threat at hand (Selye, 1978). Cannon (1929) described this "flight or fight" capability as an emergency reaction. Described another way, such responses constitute stress. The PVN must integrate these signals and coordinate a response.

The hypothalamus contributes to autonomic nervous system reactivity (Panksepp, 1986). Psychophysiologists have long considered diffuse sympathetic arousal to reflect, albeit imperfectly, negative emotional arousal (Lacey & Lacey, 1970). The PVN invokes autonomic arousal through neural as well as hormonal pathways. It sends direct projections to the sympathetic intermediolateral cell column in the thoracolumbar spinal cord and the parasympathetic vagal complex, sources of preganglionic autonomic outflow (Krukoff, 1990). In addition, it prompts the release of epinephrine and NE from the adrenal medulla. These considerations implicate the HPA axis in the neuroendocrinologic and autonomic manifestations of affective changes during pain.

In addition to controlling neuroendocrine and autonomic nervous system reactivity, the HPA axis coordinates emotional arousal with behavior (Panksepp, 1986). Direct stimulation of hypothalamus can elicit well-organized patterns of behavior, including defensive threat behaviors, accompanied by autonomic manifestations (Hess, 1954; Jänig, 1985a,b; Mancia & Zanchetti, 1981). The existence of demonstrable behavioral subroutines suggests that the hypothalamus plays a key role in matching behavioral reactions and bodily adjustments to challenging circumstances or threatening stimuli. The HPA system appears to coordinate behavioral readiness with physiological capability, awareness, and cognitive function. The acute stress response serves this purpose.

The stress response probably interacts with pain, and when pain is limited in duration, it may ameliorate it. Glucocorticoids released by the HPA axis during stress response diminish inflammation and block the sensitization of nociceptors in injured tissue. At the same time, HPA arousal releases ACTH and other pro-opiomelanocortin-derived peptides, including beta-endorphin, into the bloodstream. Moreover, stress hormones, especially glucocorticoids, may affect central emotional arousal, lowering startle thresholds and influencing cognition (Sapolsky, 1992). Saphier (1987) observed that cortisol altered the firing rate of neurons in limbic forebrain. When pain persists, however, the stress response may progress to a "burn-out" of physiological coping resources, a condition Selye (1978) labeled *distress*. Disturbed circadian rhythm (sleep, appetite) and fatigue can ensue. Griep et al. (1993) hypothesized a link between HPA axis dysfunction and the chronic pain of primary fibromyalgia.

V. MECHANISMS OF PATHOLOGICAL PAIN: NOCICEPTION

What can turn pain from a necessary evil to a malefic force? Changes in traumatized tissue, alterations in central pathways of transmission, and damage to the nervous system can convert pain from a protective resource to a source of distress, dysfunction, and disability.

A. Changes in Traumatized Tissue

Nociceptors in specific tissues do not always respond in the same way to the same stimuli; their response properties and sensitivities can alter (Alexander & Black, 1992). As pain becomes pathological, pain thresholds diminish (allodynia), and painful responses to subsequent noxious stimuli increase (hyperalgesia). Such alterations may reflect changes in the transduction process, central changes that facilitate the transmission of noxious messages, or both. Sensitization of nociceptors can result from either repetitive stimulation or inflammation. Enhanced sensitivity is adaptive because it promotes recuperation and repair, minimizing further injury by discouraging all contact rather than just contact with innocuous stimuli.

Repeated application of a noxious stimulus can sensitize a nociceptor, lowering its threshold for firing. This phenomenon, termed sensitization, has some peculiar features (Campbell & LaMotte, 1983; Fields, 1987; Woolf, 1991). If one repeatedly subjects an Aδ high-threshold mechanoreceptor to noxious heat, the receptor changes in two ways: it gives progressively larger response to a constant stimulus, and it lowers its threshold to other temperature stimuli. However, the receptor does not change its threshold to mechanical stimulation. Repeated noxious mechanical stimulation of the receptor does not cause it to reduce its threshold for firing in response to mechanical stimuli, but nearby nociceptors of the same type that were previously unresponsive to mechanical stimulation become sensitized to it. C-fiber polymodal nociceptors also sensitize with repeated stimulation; however, they tend to develop an ongoing background discharge after they become sensitized.

After tissue trauma, inflammation develops in, and at the margin of, the injured area, changing the chemical environment of the nociceptors (Cohen & Perl, 1990; Fields, 1987; Handwerker, 1991). Principal signs of inflammation include heat, redness, edema, and pain. Inflammation produces chemical mediators that decrease the firing thresholds of nociceptors or activate nociceptors directly. Figure 5 illustrates how the "soup" of inflammation can affect nociceptors.

Chemical substances that accumulate around nerve endings during inflammation come from at least three sources: (a) arachidonic acid, histamine, potassium, and other substances leak out from damaged cells; (b) they can be synthesized locally by enzymes from substrates released by damage; and (c) activity in the nociceptor itself may release them (Fields, 1987). The third point is the mechanism of neurogenic inflammation (Pedersen-Bjergaard et al., 1990). The antidromic, efferent activity of nociceptive afferents can release peptides that alter the chemical surroundings of the nociceptors. CGRP, and neurotensin A, acting in combination, can create neurogenic inflammation. SP fosters the spread of edema and hyperalgesia by vasodilatation and by releasing histamine from mast cells, which then acts directly on sensory endings (Fields, 1987). The other peptides may enhance the release of SP.

The chemical factors that cooperate to sensitize or activate nociceptors are many. They include histamine, which excites polymodal nociceptors, adenosine triphosphate (ATP), acetylcholine and serotonin, which act alone or in combination to sensitize nociceptors to other substances, as well as prostaglandins PGE_2 and PGI_2, (cyclooxygenase metabolites of arachidonic acid), which sensitize nociceptors and produce hyperalgesia (Handwerker & Reeh, 1992). Bradykinin, a peptide, appears to activate directly both Aδ and C nociceptors (Dray & Perkins, Koltzenburg, Kress, & Aghajanian, 1992;

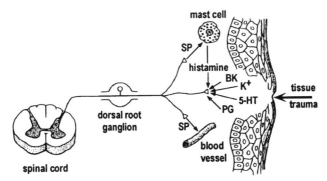

FIGURE 5   Chemical substances associated with inflammation can sensitize or directly activate nociceptors. They include bradykinin (BK), potassium (K+), serotonin (5-HT), and prostaglandins (PG). Nociceptors excited in this way antidromically release Substance P (SP), which acts on nearby mast cells and blood vessels. The mast cells release histamine, which directly excites the nociceptors. The blood vessels dilate in response to SP, producing edema and further release of BK. In this way the chemical surround of nociceptors in inflamed tissue can sustain persistent activity. (Modified from Jessell & Kelly, 1991. Reprinted with permission of the publisher.)

1993), and it increases synthesis and release of prostaglandins from nearby cells. When the chemical surround alters the response properties of nociceptors, previously injured tissues generate intense pain in response to minor noxious stimulation.

We know most about hyperalgesia in skin because study of deep and visceral tissue proves more difficult, but presumably related phenomena occur in all tissues. Figure 6 reproduces a phenomenon observed by LaMotte (1992) following the injection of capsacin into human skin. This procedure activates C nociceptors to produce severe and spontaneous burning pain as well as hyperalgesia to mechanical and heat stimuli at the site of the injection. However, the intradermal injury creates two zones of hyperalgesia: one at the injury site (primary hyperalgesia) and another in the surrounding uninjured tissues (secondary hyperalgesia). Although primary hyperalgesia involves increased sensitivity to both mechanical and thermal stimuli, secondary hyperalgesia is purely mechanical (Treede, Meyer, Raja, & Campbell, 1992). Secondary hyperalgesia appears to result from abnormal central response because no one has been able to demonstrate changes in nociceptors in this region. This differentiation carries at least two important implications for clinical disorders: first, hyperalgesic states probably have both primary peripheral mechanisms and secondary central mechanisms, and second, the secondary hyperalgesia will always appear at some distance from the area of trauma and sensitized primary afferents.

B. Sympathetically Mediated Pain

Some hyperalgesic or allodynic states may arise as an exaggerated response to NE at the sympathetic postganglionic neuron terminal. Campbell, Meyer, Davis, and Raja (1992) described sympathetically maintained pain, SMP, as a disease in which the alpha-1 adrenergic receptor develops a capability to evoke pain when activated. The triggering mechanism for this process ap-

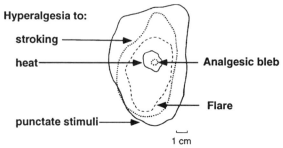

FIGURE 6   Primary and secondary hyperalgesia following intradermal injection of 100 μg of capsaicin at the volar forearm of a volunteer. (Modified from LaMotte, 1992. Reprinted with permission.)

pears to be circulating NE (Ecker, 1989). Levine, Taiwo, and Heller (1992) proposed that sympathetic postganglionic neuron terminals in injured tissues develop sensitivity to NE, which acts at alpha-adrenergic receptors. They elicit activity from primary afferent nociceptors by producing prostaglandins (PGE_2 AND PGI_2) that directly excite nociceptors. The alpha-adrenergic antagonist phentolamine can block this process. Evidence also exists that bradykinin acts indirectly on nociceptors, requiring a similar release of directly acting prostaglandins by sympathetic postganglionic neuron terminals. Clinically, this hypothesis accounts for causalgia and other reflex-sympathetic dystrophies (see below), the pain associated with neuromas that form after the severance of a nerve, and the pain of some inflammatory diseases.

C. Alterations in Central Pathways of Transmission

Pathological changes in central transmission have received less attention from investigators to date then peripheral factors. However, some useful hypotheses exist that link spinal cord mechanisms to pathological pain.

Price (1991) proposed that three aspects of spinal cord nociceptive processing can, when exaggerated, foster pathological pain: radiation, after-response, and slow temporal summation. Radiation refers to the spread of pain across dermatomes when a focused noxious event occurs. Pressing with one's thumb on the infraspinatus muscle on the surface of the scapula, for example, causes a pain that radiates through the shoulder and down the arm. Price attributes such spread of location to propriospinal interconnections, implicating WDR neurons because they possess large receptive fields and rostrocaudal organization. As a stimulus increases in intensity, it recruits more WDR neurons, which extend across an increasing rostrocaudal distance within the cord. This produces a sense of spreading pain. This phenomenon may be related to referred pain (see below).

After-response is the persistence of the disagreeable sensation elicited by a brief noxious stimulus beyond the duration of that stimulus. As nociceptor responses do not outlast a noxious stimulus, the sensation must reflect spinal cord processing, and indeed the subjective duration of the after-discharge phenomenon parallels the discharge activity of nociceptive dorsal horn neurons. Thus, under certain conditions a brief noxious event can generate a painful sensation that can last for several seconds.

Slow temporal summation is specific to "second pain" (i.e., that mediated by C-fibers). Repetitive noxious stimulation occurring at a rate of less than one noxious event per 3 sec results in a gradual increase in the perceived magnitude of the pain (sometimes called "wind up"). Price's (1991) review implicates SP in slow temporal summation. C-polymodal nociceptors release SP at the dorsal horn, where it slows depolarizations and produces slowed temporal summation and ultimately the perception of increasing pain sensation.

VI. MECHANISMS OF PATHOLOGICAL PAIN: NEURAL DYSFUNCTION

Pain can arise in cases where tissue trauma has not activated nociceptors. It originates in dysfunction of the PNS or CNS. Some writers refer to this as neurogenic pain, whereas others prefer the term *neuropathic*. Broadly, neurogenic pain tends to appear with delayed onset following the causative event, its qualities are dysesthetic (burning, "pins and needles," "electricity-like," and perhaps paroxysmal), and its somatic reference tends to follow patterns of sensory loss.

Bowsher (1991) estimated that patients with pain of neurogenic origin make up at least one-quarter of the patients attending most pain clinics and have a prevalence of about 1% in the general population. Here we classify such pains according to two groups: peripheral and central. Some chronic pain syndromes (see above) are a mixture of the two.

A. Peripheral Neuropathic Pain

Some neuropathic conditions result from nerve compression. For example, the carpal tunnel syndrome, caused by entrapment of the median nerve at the wrist, can cause pain that radiates distally into the hand or proximally up the arm and into the shoulder. Other neuropathies ensue from disease such as diabetes, nutritional disorders, severation of a nerve and formation of neuromas (knotted tangles of axon sprouts), chemical injury such as that caused by certain anticancer chemotherapeutic drugs, invasion of a nerve by tumor, or events that tear a major neural structure, such as avulsion of the brachial plexus. Damage to the myelin sheath of a nerve, compression injuries of a nerve, inflammation of a nervous structure (neuritis), and frank deafferentation can all transform peripheral nerves or dorsal roots into sources of abnormal electrical activity. Not all or even most peripheral neuropathies cause pain, but several severe pain problems of this type exist, as the following descriptions illustrate.

1. Sympathetically Maintained Pain

As we noted above, the sympathetic nervous system can affect, and cause, pain. Chronic pain states that subside immediately in response to interruption of sympathetic function are termed sympathetically maintained pains (SMPs). Controversy exists about whether and how sympathetic mechanisms contribute to chronic pain and which syndromes they control. We believe it likely that some element of sympathetic activation exists in many chronic pain states, and that in some pain syndromes they play a major role because chemical sympathectomy (nerve block) or infusion of phentolamine, an alpha-adrenergic blocker, reliably and dramatically relieves pain. To illustrate an SMP related to neuropathy, we describe causalgia, a sequela of high-velocity injury to a peripheral nerve. This syndrome is but one example of a wide range of pain problems generally grouped as reflex sympathetic dystrophies (for further discussion see Bonica, 1990).

Causalgia typically appears after a gunshot, shrapnel, or knife injury that has damaged a major nerve in a limb (Bonica, 1990). Most patients experience superficial pain of a burning quality immediately in the periphery of the injured extremity, and they develop hyperhydrosis and edema in the affected area. The pain worsens and evolves into a constant hyperesthesia and allodynia. With time, the pain spreads proximally and eventually involves the entire limb. Temperature changes, light touch, friction from clothing, air movement, movement of the limb, and any stimulus that affects the patient's emotional state can exacerbate the pain. Minor events like the cry of a child, the rattling of a newspaper, or watching a television program can provoke intense pain. Consequently, patients suffer greatly, become reclusive and withdrawn, and utterly incapacitated by the pain.

Sympathetic blockade achieved with local anesthetic, when repeated, can achieve complete and long-lasting relief of the hyperalgesia and hyperpathia in up to half of causalgia patients (Bonica, 1990). This suggests that the pain involves a self-sustaining vicious circle of reflex response that will continue indefinitely until something interrupts it. Intravenous sympathetic blocks performed with tourniquets and antiadrenergic drugs also produce favorable outcomes. Currently, physicians tend to use intravenous phentolamine infusion as a diagnostic maneuver. Relief of pain with phentolamine implicates alpha-adrenergic mechanisms.

B. Central Pain

Pain syndromes in this group divide into those resulting from spinal cord injury and those caused by injury within the brain (typically stroke). They tend to manifest in the patient's awareness as somatically based pains, and few therapeutic options exist to relieve them. These phenomena demonstrate unequivocally that pain, once established, can persist independently of primary afferent transmission.

1. Spinal Cord Injury

Pain is a common complication of paraplegia or quadriplegia resulting from spinal cord trauma. Davidoff and Roth (1991), drawing on Donovan et al. (1982), described such pain problems as including musculoskeletal pain at or above the zone of injury (probably due to overuse), radicular pain from trauma to the spinal nerve root, visceral pain (a seemingly phantom pain in quadriplegics), dysesthetic (burning, piercing, radiating) pain at or below the zone of injury (probably 5–10% of all pain complaints in this patient population), lesional back pain near the area of impact, sympathetically maintained pain, and painful compressive mononeuropathies of the upper extremities from overuse. The majority of traumatic spinal cord injury patients have a significant pain problem.

2. Stroke

Cerebrovascular lesions can cause painful conditions, although the incidence is far lower than that associated with cord injury. The incidence of post-stroke pain is about 1 in 15,000. Infarction or hemorrhage within the brain can damage structures in spinothalamocortical pathways, such as the thalamus or brain stem (Boivie & Leijon, 1991). The pain comes on gradu-ally and with delayed onset that can range from one to several months poststroke. In most cases the pain is constant and limited to one side of the body, paradoxically well localized in body parts in the patient's subjective experience, and burning, aching, or bizarre in quality. Some patients report that the pain is intolerable but for most it is moderate and it can be intermit-tent. External stimuli and events arousing emotion can exacerbate the pain.

C. Neuropathic Pain of Mixed Origin

Some pain problems appear to involve a combination of peripheral and central mechanisms. As we have already noted, peripheral pathology may cause pathological central changes. *Tic douloureux,* a severe facial pain syn-drome that occurs largely in older persons, illustrates this point.

Patients with tic douloureux present with repeating unilateral facial par-oxysms, or tics (Bonica, 1990). The paroxysms are produced by severe pain that has electrical shock-like features or stabbing qualities. Tic patients have triggers: that is, small areas on the face or scalp or in the mouth that, when touched or otherwise stimulated, cause a paroxysm. Triggers respond to gentle stimuli, such as light touch or wind, cold, chewing, or swallowing. Consequently, patients with this disease suffer enormously, and some can-not maintain normal hygiene or social habits. Fortunately, the syndrome tends to emerge intermittently, allowing weeks or months between attacks in most cases.

Disagreement exists about mechanisms, but the dominant explanation at present points to mechanical compression of the trigeminal nerve by an artery (Janetta, 1967). As the artery pulses, presumably it irritates or even damages the nerve resting against it. Surgery to free the nerve can some-times relieve the condition, but the mainstay of treatment is pharmacology. Many patients respond well to anticonvulsant drugs such as phenytoin or carbamazepine. The mechanisms of pharmacological pain relief remain poorly defined, but the efficacy of centrally acting medications implicates central mechanisms in this peculiar disorder.

Another less common form of trigeminal neuralgia (TN) occurs in patients with multiple sclerosis, a classic CNS disease (Kurtzke, 1970; Moulin, Foley, & Ebers, 1988). These patients who are younger than the normal TN population develop first unilateral TN, which is identical in its presentation, to the more usual idiopathic form of disease. They subsequently develop TN pain on the contralateral side of the face. Pathological and magnetic resonance imaging studies have demonstrated CNS demyelination in the pons either adjacent to, or involving, the trigeminal root entry zone.

VII. SUMMARY AND CONCLUSIONS

Pain serves a protective function in everyday life: it warns of biologically threatening events and promotes escape. In addition, it reminds us of existing tissue trauma and fosters recuperation by limiting activity. However, in these respects pain functions rather crudely. It provides some degree of information about the nature and intensity of an injury to the skin, but it tells us much less about trauma to deep tissue or viscera, often misrepresenting location. Moreover, the sensory end organs that generate noxious signals tend to adjust their thresholds as their chemical environment changes, and central transmission mechanisms can change as well. Consequently, pain severity is often a poor indicator of the severity of tissue trauma. When the nervous system itself suffers damage, pain may appear in the absence of trauma in the tissues with which it is associated.

Pain has both sensory and emotional features. Sensory mechanisms identify location, duration, and sometimes stimulus properties, but the emotional intensity of pain represents the biological importance of a tissue-traumatizing event in consciousness. In clinical situations, the affective aspect of pain is more important than the sensory because it determines distress.

Pain may degrade the health and threaten the well-being of an individual if its intensity greatly exceeds the biological importance of the tissue trauma that generates it of if it persists beyond the normal healing time of an injury. Exaggeration of normal mechanisms that sensitize nerve endings or spinal transmission and direct damage to the nervous system itself can provoke such conditions. When this occurs, pathological pain exists. Hyperalgesia and/or allodynia, emotional distress, fatigue, and somatic preoccupation characterize patients with pathological pain. Unrelieved pathological pain can disturb sleep and appetite and other circadian rhythms, and its prevents normal behavior and gainful activity. When severe and relentless, pain can become an all-consuming focus of awareness and a source of great suffering.

References

Abou-Samra, A. B. (1987). Mechanisms of action of CRF and other regulators of ACTH release in pituitary corticotrophs. In W. F. Ganong, M. F. Dallman, & J. L. Roberts (Eds.), *The hypothalamic-pituitary-adrenal axis revisited. Annals of the New York Academy of Sciences, 512,* 67–84.

Alexander, J., & Black A. (1992). Pain mechanisms and the management of neuropathic pain. *Current Opinions in Neurological Neurosurgery, 5*(2), 228–234.

Assenmacher, I., Szafarczyk, A., Alonso, G., Ixart, G., & Banbanel, G. (1987). Physiology of neuropathways affecting CRH secretion. In W. F. Ganong, M. F. Dallman, J. L. Roberts (Eds.), *The hypothalamic-pituitary-adrenal axis revisited. Annals of the New York Academy Sciences, 512,* 149–161.

Aston-Jones, G., Foote, S. L., & Segal, M. (1985). Impulse conduction properties of noradrenergic locus coeruleus axons projecting to monkey cerebrocortex. *Neuroscience, 15,* 765–777.

Besson, J. M., & Chaouch, A. (1987). Peripheral and spinal mechanisms of nociception. *Physiology Review, 67,* 67–186.

Bing, Z., Villaneuva, L., & Le Bars, D. (1990). Ascending pathways in the spinal cord involved in the activation of subnucleus reticularis dorsalis neurons in the medulla of the rat. *Journal of Neurophysiology, 63,* 424–438.

Boivie, J., & Leijon, G. (1991). Clinical findings in patients with central poststroke pain. In K. L. Casey (Ed.), *Pain and central nervous system disease: The central pain syndromes* (pp. 65–75). New York: Raven Press.

Bonica, J. J. (Ed.). (1990). *The management of pain* (2nd ed.). Philadelphia: Lea & Febiger.

Bowsher, D. (1976). Role of the reticular formation in responses to noxious stimulation. *Pain, 2,* 361–378.

Bowsher, D. (1991). Neurogenic pain syndromes and their management. *British Medical Bulletin, 47*(3), 644–666.

Burstein, R., Cliffer, K. D., & Giesler, G. J. (1988). The spinohypothalamic and spinotelecephalic tracts: Direct nociceptive projections from the spinal cord to the hypothalamus and telencephalon. In R. Dubner, G. F. Gebhart, & M. R. Bond (Eds.), *Proceedings of the 5th World Congress on Pain* (pp. 548–554). New York: Elsevier.

Butler, P. D., Weiss, J. M., Stout, J. C., & Nemeroff, C. B. (1990). Corticotropin-releasing factor produces fear-enhancing and behavioral activating effects following infusion into the locus coeruleus. *Journal of Neurosciences, 10,* 176–183.

Calogero, A. E., Bernardini, R., Gold, P. W., & Chrousos, G. P. (1988). Regulation of rat hypothalamic corticotropin-releasing hormone secretion in vitro: Potential clinical implications. *Advances in Experimental Medicine & Biology, 245,* 167–181.

Campbell, J. N., & LaMotte, R. H. (1983). Latency to detection of first pain. *Brain Research, 266,* 203–208.

Campbell, J. N., Meyer, R. A., Davis, K. D., & Raja, S. N. (1992). Sympathetically maintained pain: A unifying hypothesis. In W. D. Willis, Jr. (Ed.), *Hyperalgesia and allodynia* (pp. 141–149). New York: Raven Press.

Cannon, W. B. (1929). *Bodily changes in pain, hunger, fear, and rage* (2nd ed.). New York: Appleton.

Cervero, F. (1991). Mechanisms of acute visceral pain. *British Medical Bulletin, 47*(3), 549–560.

Chapman, C. R. (1993). The emotional aspect of pain. In C. R. Chapman & K. M. Foley (Eds.), *Current and emerging issues in cancer pain: Research and practice* (pp. 83–98). New York: Raven Press.

Chapman, C. R., & Gavrin, J. (1993). Suffering and its relationship to pain. *Journal of Palliative Care Medicine, 9,* 5–13.

Charney, D. S., Woods, S. W., Nagy, L. M., Southwick, S. M., Krystal, J. H., & Heniger, G. R. (1990). Noradrenergic function in panic disorder. *Journal of Clinical Psychiatry, 51*(Suppl. A), 5–11.

Cohen, E. H., & Perl, E. R. (1990). Contribution of arachidonic acid derivatives and substance P to the sensitization of cutaneous nociceptors. *Journal of Neurophysiology, 64*(2), 457–464.

Craig, A. D. (1991). Supraspinal pathways and mechanisms relevant to central pain. In K. L. Casey (Ed.), *Pain and central nervous system disease: The central pain syndromes* (pp. 157–170). New York: Raven Press.

Dahlström, A., & Fuxe, K. (1964). Evidence for the existence of monoamine-containing neurons in the central nervous system. *Acta Physiologica Scandinavica, 62*, 1–55.

Davidoff, G., & Roth, E. J. (1991). Clinical characteristics of central (dysesthetic) pain in spinal cord injury patients. In K. L. Casey (Ed.), *Pain and central nervous system disease: The central pain syndromes* (pp. 77–83). New York: Raven Press.

Donovan, W. H., Dimitrijevic, M. R., Dahm, L., & Dimitrijevic, M. (1982). Neurophysiological approaches to chronic pain following spinal cord injury. *Paraplegia, 20*, 135–146.

Dray, A., & Perkins, M. (1993). Bradykinin and inflammatory pain. *Trends in Neuroscience, 16*(3), 99–104.

Dubner, R. (1991). Neuronal plasticity in the spinal and medullary dorsal horns: a possible role in central pain mechanisms. In K. L. Casey (Ed.), *Pain and central nervous system disease: The central pain syndromes* (pp. 143–155). New York: Raven Press.

Ecker, A. (1989). Norepinephrine in reflex sympathetic dystrophy: An hypothesis. *Clinical Journal of Pain, 5*(4), 313–315.

Elam, M., Svensson, T. H., & Thoren, P. (1985). Differentiated cardiovascular afferent regulation of locus coeruleus neurons and sympathetic nerves. *Brain Research, 358*, 77–84.

Elam, M., Svensson, T. H., & Thoren, P. (1986a). Locus coeruleus neurons and sympathetic nerves: activation by cutaneous sensory afferents. *Brain Research, 366*, 254–261.

Elam, M., Svensson, T. H., & Thoren, P. (1986b). Locus coeruleus neurons and sympathetic nerves: Activation by visceral afferents. *Brain Research, 375*, 117–125.

Fields, H. L. (Ed.). (1987). *Pain*. New York: McGraw-Hill Book Company.

Fillenz, M. (Ed.). (1990). *Noradrenergic neurons*. Cambridge UK: Cambridge University Press.

Foote, S. L., & Morrison, J. H. (1987). Extrathalamic modulation of corticofunction. *Annual Review of Neuroscience, 10*, 67–95.

Foote, S. L., Bloom, F. E., & Aston-Jones, G. (1983). Nucleus locus ceruleus: New evidence of anatomical and physiological specificity. *Physiology Review, 63*, 844–914.

Gebhart, G. F. (1993). Visceral pain mechanisms. In C. R. Chapman & K. M. Foley (Eds.), *Current and emerging issues in cancer pain: Research and practice* (pp. 99–111). New York: Raven Press.

Gebhart, G. F., & Ness, T. J. (1991). Central mechanisms of visceral pain. *Canadian Journal of Physiology and Pharmacology, 69*(5), 627–634.

Gray, J. A. (Ed.). (1987). *The psychology of fear and stress* (2nd ed.). Cambridge, UK: Cambridge University Press.

Gray, J. A. (1991). *The psychology of fear and stress* (2nd ed.). Cambridge, UK: Cambridge University Press.

Griep E. N., Boersma, J. W., & de Kloet, E. R. (1993). Altered reactivity of the hypothalamic-pituitary-adrenal axis in the primary fibromyalgia syndrome [see comments]. *Journal of Rheumatology, 20*(3), 469–474.

Gross, D. (1984). Contralateral local anesthesia in the treatment of phantom and stump pain. In C. Benedetti, C. R. Chapman & G. Moricca (Eds.), *Advances in pain research and therapy* (Vol. 7, pp. 331–338). New York: Raven Press.

Handwerker, H. O. (1991). Electrophysiological mechanisms in inflammatory pain. *Agents and Actions, 32,* 91–99.

Handwerker, H. O., & Reeh, P. W. (1992). Nociceptors. Chemosensitivity and sensitization by chemical agents. In W. D. Willis, Jr. (Ed.), *Hyperalgesia and allodynia* (pp. 107–115). New York: Raven Press.

Harvey, C. K., Cadena, R., & Dunlap, L. (1993). Fibromyalgia. Part I. Review of the literature. *Journal of the American Podiatrist Medical Association, 83*(7), 412–415.

Heppelmann, B., Messlinger, K., Schaible, H. G., & Schmidt, R. F. (1991). Nociception and pain. *Current Opinions in Neurobiology, 1(2),* 192–197.

Hess, W. R. (Ed.). (1954). *Diencephalon: Autonomic and extrapyramidal functions.* New York: Grune & Stratton.

Isaacson, R. L. (Ed.). (1982). *The limbic system* (2nd ed.). New York: Plenum Press.

Jannetta, P. J. (1967). Arterial compression of the trigeminal nerve at the pons in patients with trigeminal neuralgia. *Journal of Neurosurgery, 26,* 159.

Jänig, W. (1985a). The autonomic nervous system. In R. F. Schmidt (Ed.), *Fundamentals of neurophysiology* (pp. 216–269). New York: Springer-Verlag.

Jänig, W. (1985b). Systemic and specific autonomic reactions in pain: Efferent, afferent and endocrine components. *European Journal of Anaesthesia, 2,* 319–346.

Jänig, W. (1987). Neuronal mechanisms of pain with special emphasis on visceral and deep somatic pain. *Acta Neurochirurgica* (Suppl.) *38,* 16–32.

Jessel, T. M., & Kelly, D. D. (1991). Pain and analgesia. In E. R. Kandel, J. H. Schwartz, & T. M. Jessell (Eds.), *Principles of neural science* (3rd ed.) (pp. 385–399). New York: Elsevier.

Kanosue, K., Nakayama, T., Ishikawa, Y., & Imai-Matsumura, K. (1984). Responses of hypothalamic and thalamic neurons to noxious and scrotal thermal stimulation in rats. *Journal of Thermobiology, 9,* 11–13.

Koltzenburg, M., Kress, M., & Reeh, P. W. (1992). The nociceptor sensitization by bradykinin does not depend on sympathetic neurons. *Neuroscience, 46*(2), 465–473.

Korf, J., Bunney, B. S., & Aghajanian, G. K. (1974). Noradrenergic neurons: Morphine inhibition of spontaneous activity. *European Journal of Pharmacology, 25,* 165–169.

Kurkoff, T. L. (1990). Neuropeptide regulation of autonomic outflow at the sympathetic preganglionic neuron: Anatomical and neurochemical specificity. *Annals of the New York Academy of Sciences, 579,* 162–167.

Kurtzke, J. F. (1970). Clinical manifestations of multiple sclerosis. In P. J. Vinkin & G. E. Bruyn (Eds.). *Handbook of neurology* (Vol. IX, pp. 161–216). Amsterdam: North Holland Press.

Lacey, J. I., & Lacey, B. C. (1970). Some autonomic-central nervous system interrelationships. In P. Black (Ed.), *Physiological correlates of emotion* (pp. 205–227). New York: Academic Press.

LaMotte, R. H. (1992). Neurophysiological mechanisms of cutaneous secondary hyperalgesia in the primate. In W. D. Willis, Jr. (Ed.), *Hyperalgesia and allodynia* (pp. 175–185). New York: Raven Press.

Levine, J. D., Taiwo, Y. O., & Heller, P. H. (1992). Hyperalgesic pain: Inflammatory and neuropathic. In W. D. Willis, Jr. (Ed.), *Hyperalgesia and allodynia* (pp. 117–123). New York: Raven Press.

Levitt, P., & Moore, R. Y. (1979). Origin and organization of the brainstem catecholamine innervation in the rat. *Journal of Comparative Neurology, 186,* 505–28.

Loeser, J. D. (1990). Pain after amputation: Phantom limb and stump pain. In J. J. Bonica (Ed.), *The management of pain* (2nd ed.) (pp. 244–256). Philadelphia: Lea & Febiger.

Lopez, J. F., Young, E. A., Herman, J. P., Akil, H., & Watson, S. J. (1991). Regulatory biology of the HPA axis: An integrative approach. In S. C. Risch (Ed.), *Central nervous system peptide mechanisms in stress and depression* (pp. 1–52). Washington, DC: American Psychiatric Press.

MacLean, P. D. (Ed.) (1990). *The triune brain in evolution: Role in paleocerebral functions.* New York: Plenum Press.

Mancia, G., & Zanchetti, A. (1981). Hypothalamic control of autonomic functions. In J. P. Morgane & J. Panksepp (Eds.), *Handbook of the hypothalamus: behavioral functions of the hypothalamus* (Vol. 3, pp. 147–202). New York: Dekker.

McMahon, S. B., & Koltzenburg, M. (1990). The changing role of primary afferent neurones in pain. *Pain, 43,* 269–272.

McNaughton, N., & Mason, S. T. (1980). The neuropsychology and neuropharmacology of the dorsal ascending noradrenergic bundle—a review. *Progress in Neurobiology, 14,* 157–219.

Melzack, R. (1989). Phantom limbs, the self and the brain. The D. O. Hebb Memorial Lecture. *Canadian Psychology, 30,* 1–16.

Melzack, R. (1992). Phantom limbs. *Science in America, 266,* 120–126.

Moldofsky, H. (1993). Fibromyalgia, sleep disorder and chronic fatigue syndrome. *Ciba Foundation Symposia, 173,* 262–271.

Morilak, D. A., Fornal, C. A., & Jacobs, B. L. (1987). Effects of physiological manipulations on locus coeruleus neuronal activity in freely moving cats. II. Cardiovascular challenge. *Brain Research, 422,* 24–31.

Moulin, D. E., Foley, K. M., & Ebers, G. C. (1988). Pain syndromes in multiple sclerosis. *Neurology, 38,* 1830–1834.

Panksepp, J. (1986). The anatomy of emotions. In R. Plutchik & H. Kellerman (Eds.), *Emotion: theory, research and experience* (Vol. 3, pp. 91–124). Orlando: Academic Press.

Papez, J. W. (1937). A proposed mechanism of emotion. *Archives of Neurology & Psychiatry, 38,* 725–743.

Pedersen-Bjergaard, U., Nielsen, L. B., Jensen, K., Edvinsson, L., Jansen, I., & Olesen, J. (1990). Calcitonin gene-related peptide, Neurokinin A and Substance P: Effects on nociception and neurogenic inflammation in human skin and temporal muscle. *Peptides, 12,* 333–337.

Perl, E. R. (1984). Characterization of nociception and their activation of neurons in the superficial dorsal horn: first steps for the sensation of pain. In L. Kruger & J. C. Liebeskind (Eds.), *Advances in pain research and therapy* (Vol. 6, pp. 23–52). New York: Raven Press.

Peschanski, M., & Weil-Fugacza, J. (1987). Aminergic and cholinergic afferents to the thalamus: Experimental data with reference to pain pathways. In J. M. Besson, G. Guilbaud, & M. Paschanski (Eds.), *Thalamus and pain* (pp. 127–154). Amsterdam: Excerpta Medica.

Price, D. D. (1991). Characterizing central mechanisms of pathological pain states by sensory testing and neurophysiological analysis. In K. L. Casey (Ed.), *Pain and central nervous system disease: The central pain syndromes* (pp. 103–115). New York: Raven Press.

Redmond, D. E., Jr. (1977). Alteration in the functions of the nucleus locus coeruleus: A possible model for studies of anxiety. In I. Hannin & E. Usdin (Eds.), *Animal models in psychiatry and neurology* (pp. 293–306). New York: Pergamon Press.

Redmond, D. E., Jr., & Huang, Y. G. (1979). Current concepts. II. New evidence for a locus coeruleus-norepinephrine connection with anxiety. *Life Sciences, 25,* 2149–2162.

Roth, Y. F., & Sugarbaker, P. H. (1980). Pains and sensations after amputation: Character and clinical significance (Abstract). *Archives of Physical Medicine & Rehabilitation, 61,* 490.

Ruch, T. C. (1965). Pathophysiology of pain. In T. C. Ruch & H. D. Patton (Eds.), *Physiology and biophysics* (pp. 345–363). Philadelphia: Saunders.

Saphier, D. (1987). Cortisol alters firing rate and synaptic responses of limbic forebrain units. *Brain Research Bulletin, 19,* 519–524.

Sapolsky, R. M. (Ed.) (1992). *Stress, the aging brain, and the mechanisms of neuron death.* Cambridge, MA: The MIT Press.

Sawchenko, P. E., & Swanson, L. W. (1982). The organization of noradrenergic pathways from the brain stem to the paraventricular and supraoptic nuclei in the rat. *Brain Research Reviews, 4,* 275.

Scarry, E. (1985). *The body in pain: The making and unmaking of the world.* New York: Oxford University Press.

Selye, H. (Ed.) (1978). *The stress of life.* New York: McGraw-Hill.

Sherman, R. A., Sherman, C. J., & Parker, L. (1984). Chronic phantom and stump pain among American veterans; results of a survey. *Pain, 18,* 83–95.

Shukla, G. T., Sahu, S. C., & Tripathi, R. P. (1982). Phantom limb: A phenomenological study. *British Journal of Psychiatry, 141,* 454–458.

Stone, E. A. (1975). Stress and catecholamines. In A. J. Friedhoff (Ed.), *Catecholamines and behavior* (Vol. 2, pp. 31–72). New York: Plenum Press.

Sumal, K. K., Blessing, W. W., Joh, T. H., Reis, D. J., & Pickel, V. M. (1983). Synaptic interaction of vagal afference and catecholaminergic neurons in the rat nucleus tractus solitarius. *Journal of Brain Research, 277,* 31–40.

Svensson, T. H. (1987). Peripheral, autonomic regulation of locus coeruleus noradrenergic neurons in brain: Putative implications for psychiatry and psychopharmacology. *Psychopharmacology, 92,* 1–7.

Treede, R. D., Meyer, R. A., Raja, S. N., & Campbell, J. N. (1992). Peripheral and central mechanisms of cutaneous hyperalgesia. *Progress in Neurobiology, 38*(4), 397–421.

Villanueva, L., Bing, Z., Bouhassira, D., & Le Bars, D. (1989). Encoding of electrical, thermal, and mechanical noxious stimuli by subnucleus reticularis dorsalis neurons in the rat medulla. *Journal of Neurophysiology, 61,* 391–402.

Villanueva, L., Cliffer, K. D., Sorkin, L. S., Le Bars, D., & Willis, W. D., Jr. (1990). Convergence of heterotopic nociceptive information onto neurons of caudal medullary reticular formation in monkey (*Macaca fascicularis*). *Journal of Neurophysiology, 63,* 1118–1127.

Wall, P. D. (1985). Pain and no pain. In C. W. Coen (Ed.), *Functions of the brain* (pp. 44–66). Oxford: Clarendon Press.

Watson, S. J., Khachaturian, H., Lewis, M. E., & Akil, H. (1986). Chemical neuroanatomy as a basis for biological psychiatry. In P. A. Berger & H. K. H. Brodie (Eds.), *Biological psychiatry* (Vol. 8, pp. 4–33). [Arieti, S., ed. *American handbook of psychiatry,* 2nd ed.]. New York: Basic Books.

Wiener, S. L. (1993). *Differential diagnosis of acute pain by body region.* New York: McGraw Hill, Inc.

Willis, W. D., Jr. (Ed.). (1985). *The pain system: The neurobasis of nociceptive transmission in the mammalian nervous system.* New York: Karger.

Willis, W. D. (1988a). Anatomy and physiology of descending control of nociceptive responses of dorsal horn neurons: Comprehensive review. *Progress in Brain Research, 77,* 1–29.

Willis, W. D. (1988b). Dorsal horn neurophysiology of pain. *Annals of the New York Academy of Sciences, 531,* 76–89.

Willis, W. D., Jr. (1993). Mechanisms of somatic pain. In C. R. Chapman & K. M. Foley (Eds.), *Current and emerging issues in cancer pain: Research and practice* (pp. 67–81). New York: Raven Press.

Woolf, C. J. (1991). Generation of acute pain: central mechanisms. *British Medical Bulletin, 47*(3), 523–533.

Zimmerman, M. (1991). Pathophysiological mechanisms of fibromyalgia. *Clinical Journal of Pain, 7*(Suppl 1), S8–S15.

CHAPTER 8

Control of Pathological Pain

Russell K. Portenoy

I. INTRODUCTION

In most circumstances, the ubiquitous human experience of pain is biologically adaptive. Pain is an acute signal of impending or actual tissue damage, which impels protection of an injured site and other behaviors that foster healing and functional restoration. Pain as a symptom, therefore, plays an essential physiological role.

In contrast, the experience of pain exerts no apparent biological advantage when it persists beyond the healing of an acute injury, continues as a major manifestation of a chronic injury, or occurs as the sole presentation of a physiological or psychological disturbance. From this perspective, long-standing pain is most aptly conceptualized as a disease, which may become far more disabling than any tissue injury associated with its onset. This clinical problem is highly prevalent and extraordinarily diverse. Every discipline encounters patients who describe persistent pain as either a primary condition (e.g., headache) or a distressing complication of an underlying medical disorder (e.g., arthritis or cancer).

Although the diversity of patients with chronic pain has profound implications for management, it is nonetheless true that many of the skills required to treat such patients cross specific disease entities. For most patients, the clinical approach requires a comprehensive assessment and a critical review of a multimodal treatment strategy that targets both comfort and function. In this chapter, the nature of these general skills is emphasized and reference is made to an abundant medical literature that variably focuses on pains associated with particular diseases (e.g., cancer), sites (e.g., headache or low back pain), or inciting events (e.g., surgery). The specific issues surrounding each of these entities are not addressed herein, but are covered in recent textbooks, which may be consulted for more detail (Bonica, 1990; Cousins, 1989; Portenoy & Kanner, 1996; Wall & Melzack, 1989).

II. DEFINITION AND CLASSIFICATION OF PAIN

According to the International Association for the Study of Pain, pain is "an unpleasant sensory and emotional experience associated with actual or potential tissue damage, or described in terms of such damage" (Merskey & Bogduk, 1984, p. 210). This definition emphasizes the lack of a straightforward relationship between tissue damage and pain. Pain is a perception linked at some level to tissue damage, but this linkage is neither uniform nor predictable. An understanding of this complexity is fundamental to the management of chronic pain.

A. Nociception and Pain

Pain is inherently subjective, and given the rarity of factitious complaints (and malingering) in medical practice, the clinician is best served in most cases by a clinical stance that accepts all pain complaints as reflecting the true experience of the patient. This perspective does not imply that the pain is predominantly caused by organic processes, that the patient's disability does not exceed any evidence of tissue damage (or indeed, exceed the reported severity of pain), or that any particular treatment is indicated. Rather, it merely encourages the clinician to avoid the intellectual trap in which patients are evaluated to determine whether or not pain is "real." It is more fruitful to assume that the pain is real, that is, experienced by the patient, and then move to the next element in the assessment, namely the development of inferences about the processes that have incited and sustain the pain.

From the latter perspective, a careful assessment by an experienced clinician may suggest that the pain reported by the patient is appropriate for the evident degree of tissue damage, or either more or less than expected for the degree of tissue damage extant. This inference can be clarified by the distinctions among nociception, pain, and suffering (Loeser & Egan, 1989).

Nociception is the activity produced in the afferent nervous system by potentially tissue-damaging stimuli (Besson & Chaouch, 1987; Willis, 1985). In animal models, objective indicators of nociception are induced by controlled mechanical, thermal, or chemical stimuli, then isolated and measured. These indicators may be molecular, biochemical, neurophysiological, or behavioral. In the clinical arena, such indicators are not available, or in the case of behavioral responses, are unreliable (see below). Hence, the degree of nociception in humans can never be more than an inference based on an appraisal of the extent of tissue damage.

Pain is not equivalent to nociception. Rather, pain may be conceptualized as the perception of nociception, which may or may not be proportionate to the conscious sensory experience of tissue injury. Similar to other perceptions, pain is determined by more than the activity induced in the sensorineural apparatus by external or internal stimuli. Clinical observation of patients immediately following trauma, who may experience no pain whatsoever, provide compelling evidence of this conclusion.

Pain may be perceived by the clinician to be disproportionately greater than the degree of nociception inferred to exist during clinical evaluation. As discussed more fully below, this "nonnociceptive" pain may have any of numerous explanations, some of which indicate an understanding of the pain as predominantly driven by organic factors (e.g., pain sustained by neuropathic processes that become independent of ongoing tissue damage) and others as a manifestation of psychological disorders. In many cases, the clinician perceives that pain reports are disproportionate to tissue damage, but there is insufficient evidence to infer either a predominating organic or psychological pathophysiology; in the absence of such evidence, it is best to label the pain as "idiopathic" (Arner & Meyerson, 1988), a neutral term that does not suggest more knowledge of pain pathophysiology than actually exists.

The foregoing discussion highlights one of the most important challenges inherent in the comprehensive pain assessment. To appropriately target a multimodality therapy, it is necessary to draw relevant inferences about the nociceptive and nonnociceptive factors that may be sustaining the pain. The failure to acquire such an understanding can lead to misguided therapeutic efforts, and presumably, poorer outcome for the patient. For example, interventions that solely attempt to lessen nociception but neglect nonnociceptive contributions to the pain (e.g., psychological factors) can lead to an outcome characterized by reduced tissue damage without symptomatic improvement.

Several models have been proposed in an effort to clarify the complex factors that may contribute to the experience of pain. For example, pain has been postulated to have three broad dimensions—sensory-discriminative, motivational-affective, and cognitive-evaluative—that may be functionally distinct and subserved by different neural systems (Melzack & Casey, 1968). In this heuristic model, the sensory-discriminative dimension reflects an adaptive function that provides the individual with information about the nature of the noxious stimulus. Most relevant to the occurrence of acute pain, the neural systems that underlie this function have the capability to characterize the spatial, temporal, and quantitative aspects of a noxious stimulus. The rapidly conducting afferent pathways well characterized in animal models of nociception (Besson & Chaouch, 1987; Willis, 1985) could subserve the transmission of this type of information.

In contrast, the sensorineural substrate of the other dimensions of the pain experience are more speculative. The motivational-affective dimension, which reflects the reactive component of pain and is highly relevant in both acute and chronic settings, could be mediated by polysynaptic afferent pathways that are known to interconnect with brain stem reticular neurons and the limbic system. The cognitive-evaluative dimension, which reflects the observation that the meaning of pain can profoundly alter its sensory experience, presumably requires input into the cortex, but little is known of these processes.

B. Suffering

Suffering is a more global construct that is, like pain itself, both inherently subjective and multidimensional. Although difficult to define from a medical perspective, a clinical understanding of suffering and its relationship to pain is highly relevant to the assessment and management of painful disorders (Figure 1). Efforts to define suffering have characterized it as a perceived threat to the patient as person (Cassel, 1982) or as "total pain" (Saunders, 1984). It may also be likened to overall impairment in quality of life (Portenoy, 1990b, 1992); the latter construct includes both positive and negative aspects of physical, psychological, and social functioning.

Suffering may be determined by any of numerous aversive perceptions. These may include symptoms other than pain, loss of physical function, social isolation, familial dissolution, or financial concerns. Independent psychological disorders, which may or may not directly contribute to the experience of pain, may be profound determinants. These may include depression, anxiety, and premorbid character pathology. In some cases, spiritual issues (e.g., the meaning of one's life, past mistakes, lack of religious feeling) may be salient to the larger experience.

Clinical inferences related to the degree of suffering and the nature of the contributing factors may be viewed as another component of pain assessment. Just as the treatment of nociception may not reduce pain that is sustained by other factors, a therapeutic approach that addresses comfort alone may not substantively benefit a patient whose suffering is caused by disturbances other than pain.

As a final point, it is important to recognize that the literature that refers

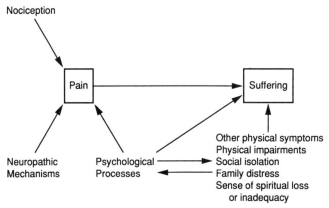

FIGURE 1   Schematic representing the complex interactions between pain and suffering, both of which are multidimensional phenomena that are potentially influenced by numerous physical and psychosocial factors (From Portenoy, 1992, Figure 1, p. 1026.)

to these fundamental issues may use varying nomenclature depending on the clinical perspective. In the context of cancer pain and palliative care, the terms *suffering* and *quality of life* are commonly used and widely understood. In contrast, the extensive literature pertaining to the multidisciplinary management of chronic nonmalignant pain tends to discuss issues related to "function" using more specific language. Reference is often made to the ability to work, engage in meaningful social interactions, contribute to family life, or experience pleasure. Both function and quality of life are mentioned in the literature that discusses symptomatic management of painful medical disorders, such as arthritis. The lesson relevant to pain assessment in all these writings is similar and refers to the need to incorporate a specific understanding of the impact of pain on a broad spectrum of individual perceptions and life activities. This type of comprehensive assessment is a necessary foundation for a therapeutic strategy designed to improve comfort and enhance quality of life.

C. Definition of Chronic Pain

Definitions of chronic pain based solely on symptom duration do not capture the complexity of the clinical situation and are not optimal. A recent definition that incorporates earlier ones and highlights the variability that may be subsumed by the term *chronic* states that pain is chronic if it persists for a month beyond the usual course of an acute illness or a reasonable duration for an injury to heal, if it is associated with a chronic pathologic process, or if it recurs at intervals for months or years (Bonica, 1990).

D. Categories of Patients with Pain

A simple classification of the population of patients who seek treatment for pain may help clarify the extraordinary heterogeneity encountered in the clinical setting. Each of these representative categories subsumes extremely diverse subgroups of patients.

1. Acute Monophasic Pain

Acute pains are short-lived or are anticipated to be short-lived given the natural history of the underlying pathological process. Acute pains that require clinical intervention are typically associated with surgery, major trauma, and burns. Although these pains are physiologically and psychologically complex (Cousins, 1989), treatment tends to be quite uniform, emphasizing the provision of comfort through the administration of opioid drugs. The potential efficacy of opioid pharmacotherapy is very high, but undertreatment is common (Edwards, 1990; Perry & Heidrich, 1982).

2. Recurrent Acute Pains

Pain syndromes characterized by recurrent acute pains include headache, dysmenorrhea, and pains associated with sickle-cell anemia, inflammatory bowel disease, and some arthritides or musculoskeletal disorders (e.g., hemophilic arthropathy). Although these prevalent syndromes share pain-free periods, they have been labeled as chronic pain (Bonica, 1990). In some cases, this designation is apropos, because management imperatives may duplicate those usually applied to pains that are more constant. In other cases, however, the most efficient approach is one that repeatedly applies treatments appropriate for acute pain. The optimal management strategy, and hence the most informative label (recurrent acute pain vs. chronic pain), should be determined by the degree to which the patient manifests the physical, psychosocial, and behavioral characteristics that parallel the disturbances observed in chronic pain syndromes (see below).

3. Chronic Pain Associated with Cancer

A comprehensive assessment of the patient with chronic cancer pain will usually identify an underlying organic lesion (Gonzales, Elliott, Portenoy, & Foley, 1991). Through this process, an opportunity for antineoplastic therapy may be discovered and primary analgesic therapies can be combined with other appropriate treatments. Like acute monophasic pain, opioid therapy is the major analgesic approach in patients with cancer pain (Foley, 1985; Health and Public Policy Committee, 1983; McGivney & Grooks, 1984; Portenoy, 1993; Swerdlow & Stjernsward, 1982; World Health Organization, 1986; World Health Organization, 1990).

4. Chronic Pain Due to Progressive Medical Diseases Other Than Cancer

Chronic pain may develop in association with progressive medical diseases other than cancer, including AIDS, sickle-cell anemia, hemophilia, and some connective tissue diseases. The management of these patients, like those with cancer, requires ongoing consideration of the extent and course of the underlying disease. The potential value of treatments targeted at the underlying pathology must be weighed repeatedly. The importance of disease-related factors, however, should not obscure the potential value of symptomatic therapies and treatments that address psychosocial and rehabilitative concerns. Similar to chronic nonmalignant pain of other types, the therapy should generally have dual goals, namely, comfort and rehabilitation.

5. Chronic Pain Associated with a Nonprogressive Medical Disorder

Many chronic pain syndromes are associated with an organic lesion that is neither rapidly progressive nor life threatening. Included in this category are numerous musculoskeletal pain syndromes (e.g., treatment-refractory osteoporosis and spondylolisthesis) and neuropathic pain syndromes (e.g., postherpetic neuralgia, painful polyneuropathy, central pain, and reflex sympathetic dystrophy). These patients require a careful assessment that characterizes the underlying organic contribution to the pain, after which therapeutic focus can be shifted to symptom control and rehabilitation. Reevaluation of the underlying lesion is indicated if the nature of the symptoms change or progress.

6. Chronic Nonmalignant Pain Syndrome

A comprehensive assessment of the patient with chronic pain may lead to the conclusion that the pain is excessive for the degree of evident organic pathology and is attributable, in part, to psychological disturbances, or that the level of psychological and physical disability presented by the patient is greater than can be explained by the symptoms or underlying pathology. The nomenclature applied to these patients is confusing and tends to vary with the discipline of the practitioner. A site-specific term that has gradually acquired psychological and behavioral implications is often used (e.g., atypical facial pain, failed low back syndrome, chronic tension headache, or chronic pelvic pain of unknown etiology). Idiopathic pain is a newer term that may be less stigmatizing. If positive evidence of a psychiatric disorder, such as psychogenic pain or somatization disorder, exists, it may be diagnosed (American Psychiatric Association, 1994). Speaking more generally, specialists in pain management commonly refer to a chronic nonmalignant pain syndrome or intractable pain syndrome, which again suggests psychological and behavioral disturbances. To a large extent, the multidisciplinary approach to pain management evolved to address the complex physical and psychological needs of these patients.

III. PRINCIPLES OF PAIN ASSESSMENT

A comprehensive assessment is an essential element in the management of acute and chronic pain. This assessment clarifies the organic and psychological contributions to the pain and characterizes the range of problems that may require treatment. As indicated by the useful clinical constructs of suffering or quality of life, these problems may involve symptoms other than pain; disturbances in physical, psychological, or social functioning; or any of numerous other problems that may become associated with the pain complaint.

A useful way of conceptualizing the goal of this assessment is the development of a pain-oriented problem list. In addition to the pain itself, this problem list might include any or all of the related problems that together compromise the comfort and function of the patient. The development of such a list encourages the clinician to prioritize problems, consider their treatability, and develop a multimodality treatment strategy that efficiently addresses the most pressing concerns.

The information required to develop the pain-oriented problem list derives from the history, physical examination, and selected laboratory and radiographic procedures. A review of previous medical records is usually needed. The history of the pain must include reference to temporal features (onset, course, and daily pattern), location, severity, quality, and factors that provoke or relieve it. Specific pain-related impairment in function should be characterized, including sleep disturbance and activities of daily living. The patient's activities during the day should be enumerated to help clarify the degree of physical inactivity and social isolation. A psychosocial assessment should characterize premorbid psychiatric disease or personality disorder, substance abuse history, coping styles demonstrated during earlier episodes of physical disease or psychological stress, current psychological state with particular reference to anxiety and depression, current resources (social, familial, and financial), and present functional status.

The history of present illness should also evaluate any related and unrelated medical conditions. For the patient with cancer, for example, the assessment must acquire information about the tumor type, known extent of disease, types and results of the most recent radiographic procedures, prior antineoplastic therapies, and plans for future primary therapy. The history should also elicit details about a past history of persistent pain, prior pain treatments, and previous use of analgesics and other drugs (both licit and illicit).

The physical examination of patients with chronic pain should aim to clarify the underlying organic contributions to the pain. Based on this examination, clinical hypotheses about the nature of this underlying pathology can be generated. If further evaluation is necessary to assess these hypotheses, appropriate laboratory and imaging procedures should be obtained. The clinician who performs this comprehensive assessment must continually balance the need for objective data and the need to avoid unnecessary testing. It may be appropriate in some cases of low back pain, for example, to exclude treatable pathology in the spine with sophisticated imaging on one or two occasions, then avoid further testing indefinitely while efforts are made to address the patient's disability through rehabilitation; in other cases, the comprehensive assessment of the pain suggests an overriding organic contribution, and the absence of an explanatory radiographic finding impels repeated testing on a regular basis.

This discussion emphasizes that the mere discovery of a lesion should not immediately be construed as sufficient explanation for the pain. A competent evaluation identifies potentially treatable organic conditions and clarifies the degree to which pain and disability can be ascribed to these factors or to other identifiable pathology, including psychological disturbances.

The type of information obtained from the comprehensive pain assessment may be clarified by considering several clinically relevant aspects of the pain. The clinician is usually able to integrate the data provided by the history, examination, review of records, and laboratory or radiographic evaluation into an understanding of these temporal, topographical, syndromic, etiological, and pathophysiological aspects of the pain.

A. Temporal Features

As discussed previously, the distinction between acute and chronic pain highlights important differences in pain presentation and suggests alternative approaches to therapy. Acute pain usually has a well-defined onset, a readily identifiable cause (e.g., surgical incision), and a duration anticipated to be no more than the time required for an injury to heal, usually less that several weeks. There is an association between acute pain and both specific pain behaviors (e.g., moaning, grimacing, and splinting of the painful part) and signs of sympathetic hyperactivity (including tachycardia, hypertension, and diaphoresis).

Chronic pain is usually characterized by an ill-defined onset and a duration that has either already exceeded the healing period of the inciting lesion or is anticipated to continue indefinitely into the future. Importantly, pain behaviors associated with acute pain are typically absent, and there are no signs of sympathetic hyperactivity. The latter observation is often poorly appreciated by clinicians and deserves emphasis: Although the patient with chronic pain can exhibit behaviors or autonomic signs commonly associated with acute pain during intermittent exacerbations of pain, these features are usually absent during periods of sustained pain. The patient does not usually appear to be in pain. In contrast, sleep disturbance is very common, and some patients develop other vegetative signs such as lassitude or anorexia. These observations are equally true for patients with overt tissue injury and those whose pain is suspected to have a primary psychological etiology.

The affective concomitants of pain are highly variable. Although classic teaching suggests that anxiety accompanies acute pain and depression accompanies chronic pain, the clinical reality is more diverse. The level of anxiety associated with an acute pain appears to be influenced by many factors, including premorbid psychological traits and state, the meaning of the pain (e.g., the difference between a predictable postoperative pain and sudden chest pain of unknown cause), the severity of the pain, the availability of relief, and the presence and reactions of others in the environment. Given the variability in these factors, it would be expected that some patients experience intense anxiety, whereas others report little or none. Similarly, the variation in the development of depression associated with chronic pain has been observed (Romano & Turner, 1985), which may also be attributed to heterogeneity in the many potential determinants of this disorder.

It is important to recognize that varying temporal profiles exist across patients categorized as acute or chronic. For example, most patients with chronic pain actually have fluctuating pain with some pain-free intervals. Even those with continuous pain usually experience large fluctuations in the intensity of pain or acute episodes of severe exacerbation that the patient perceives as distinct from the baseline pain. In a survey of patients with cancer, almost two-thirds experienced transitory flares of pain, which varied greatly in duration, frequency, and quality (Portenoy & Hagen, 1990). Information about the specific temporal profiles may be important in the development of a therapeutic approach to the pain.

B. Topographic Features

Some topographic features of the pain have great clinical relevance. The term *focal* may refer to a single site of pain or to a pain location that is superficial to the underlying organic lesion. The latter phenomenon occurs commonly, and it is accepted clinical practice to evaluate the tissues in proximity to the area of pain when pursuing a diagnosis of an underlying organic process. More challenging is the evaluation of pains that are determined to have no propinquituous pathology sufficient to explain the symptom. In this situation, the clinician must consider the possibility that the pain is referred from a remote site. Thus, knowledge of pain referral patterns is needed to adequately assess the underlying etiology of the pain and target appropriate assessment procedures.

The variability of referred pain is exemplified by the number of subtypes associated with neurological lesions. Pain may be referred anywhere along the course of an injured peripheral nerve (e.g., pain in the foot from a sciatic nerve lesion). Damage to a peripheral nerve can also refer pain outside of the dermatomal distribution; for example, shoulder pain may be referred from compression of the median nerve at the wrist (Torebjörk, Ochoa, & Schady, 1984). Similarly, pain may be referred anywhere along the course of the nerve supplied by a damaged nerve root. This referral pattern, known as radicular pain, is easily recognized when pain radiates down the dermatome supplied by the nerve root; diagnosis is more difficult if the patient reports a single site of pain somewhere in that dermatomal distribution. Pain can also be referred anywhere in the body innervated by that part of the central nervous system (CNS) involved by a lesion. For example, highly variable and nondermatomal patterns of pain can be described below a lesion involving the spinal cord or in the hemibody contralateral to a cerebral lesion.

Referral of pain can also occur following injury to nonneural tissues (Kellgren, 1939). Injury to viscera, for example, produces well-recognized cutaneous referral patterns (e.g., shoulder pain from irritation of the ipsilateral diaphragm).

In addition to these diagnostic considerations, the topography of the pain may have implications for therapy. For example, the distinctions among focal pain (meaning pain in an isolated site), multifocal pain, and generalized pain may determine the potential utility of some treatments, such as nerve blocks and cordotomy, that depend on the specific location and extent of the pain.

C. Etiologic Features

For most patients with chronic pain, the relationship between identifiable organic pathology and both the pain and pain-related disability is complex and often varies over time. Elucidation of underlying pathology is a key component of the pain assessment, which must be accompanied by consideration of the larger interactions among physical disease, symptoms, and functional compromise.

The importance of the effort to characterize underlying pathology is most clear in patients with cancer pain and those with pain related to other progressive medical diseases. In these patients, identification of a lesion may present options for primary therapy, which could have analgesic consequences, or shift the prognosis of the disease. A recent survey of patients with cancer pain, for example, noted that the pain evaluation identified previously unsuspected lesions in 63% of patients, approximately 20% of whom received primary therapy as a result (Gonzales et al., 1991).

D. Syndromic Features

Many patients with chronic pain present a constellation of symptoms, signs, and findings on ancillary tests that allow a syndrome diagnosis. From the clinical perspective, such a diagnosis may suggest pathophysiological processes that had not been adequately considered, guide additional evaluation or the selection of therapies, or indicate prognosis. The development of criteria for syndrome identification has been one of the major goals of the taxonomy for pain that has recently been developed by the International Association for the Study of Pain (Merskey & Bogduk, 1994).

E. Pathophysiologic Features

Basic research has yielded extraordinary insights into the anatomy and physiology of nociception, and the pathological disturbances in nociceptive systems that may be responsible for human pain. These advances have encouraged clinicians to attempt a classification of pain on the basis of inferred mechanisms. The highly tentative nature of this effort is apparent: Pain is not equivalent to nociception, is presumably determined by multiple physiological and psychological processes that interact in complex ways, and there is no way to independently confirm that any particular mechanism is operating in the clinical setting. Indeed, the clinical labels are best considered heuristic constructs, rather than a description of actual pathophysiology. Nonetheless, these constructs have been observed to offer a useful guide to the evaluation and treatment of patients and are now widely applied by clinicians.

1. Nociceptive Pain

Pains that are inferred to have a predominating organic contribution have been termed "nociceptive" or "neuropathic." Nociceptive pain is perceived to be commensurate with the degree of ongoing tissue damage from an identifiable peripheral lesion that involves either somatic or visceral structures. Although neural mechanisms that underlie normal nociception are presumed to be involved in the maintenance of this type of pain, the speculative nature of this conclusion is again underscored by the known complexity of these neural systems and the evidence obtained from animal models that persistent tissue injury of any type induces ongoing changes in the nervous system (Besson & Chaouch, 1987; Dubner, 1991; Hammond, 1985; Willis, 1985).

Nociceptive pain originating from somatic structures is termed somatic pain and is typically described as aching, stabbing, throbbing, or pressure-like. Nociceptive pain due to injury to viscera is also known as visceral pain and has been observed to differ depending on the specific structures involved: obstruction of hollow viscus usually causes a gnawing or crampy pain, whereas distention or torsion of organ capsules or other mesentery produces pain typically described as aching. Clinical experience suggests that nociceptive pains can usually be ameliorated through interventions that improve the peripheral nociceptive lesion. For example, joint replacement can relieve refractory arthritic pain, and radiotherapy reduces pain due to neoplasm. Likewise, it has been generally observed that nociceptive pains can be diminished or eliminated be interruption of afferent sensory pathways between the site of tissue injury and the CNS, a result that can be accomplished clinically through the use of nerve blocks or surgical neurolysis.

2. Neuropathic Pain

The term *neuropathic* is applied to those pains inferred to be sustained by aberrant somatosensory processes in the peripheral nervous system (PNS) or CNS. Several major subtypes can be identified. One potentially useful classification first posits the existence of a predominating "generator" for the pain that resides either in the CNS or in the PNS. Although hypothetical, this division gains support from some clinical and experimental observations (Portenoy, 1991). For example, an overriding peripheral etiology can be inferred in a subgroup of patients with pain following peripheral nerve trauma, who are cured by resection of a neuroma.

A subclassification of neuropathic pains flows from this division (Figure 2). Again, the broad groupings have value clinically but, presumably, do not accurately reflect the complex mechanisms that are actually responsible for the pain. For example, deafferentation pains, which are inferred to have a predominating central "generator" could be sustained by processes as diverse as denervation hypersensitivity of central neurons, changes in receptive fields of neurons that innervate regions in proximity to denervated areas, or alteration in central inhibitory processes. In similar fashion, it can be proposed that each of the recognized subtypes of neuropathic pain may actually involve multiple distinct mechanisms. Indeed, it can be speculated that more than one mechanism usually underlies a specific neuropathic pain syndrome (and there are many syndromes subsumed under a broader rubric, such as deafferentation pain). Overlapping patterns of specific mechanisms, some of which may be associated with a particular phenomenology (such as hyperalgesia or autonomic dysregulation), may account for the similarities across neuropathic pain syndromes that are fundamentally distinct.

Identification of a neuropathic mechanism is extremely important in

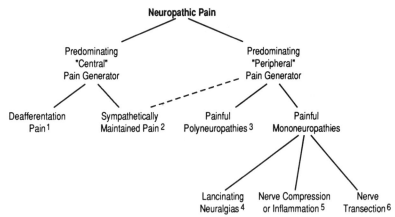

FIGURE 2   Classification of neuropathic pains on the basis of putative mechanism. (1) Response to either peripheral or central nervous system injury. (2) Associated with focal autonomic dysregulation (e.g., edema, vasomotor disturbances), involuntary motor responses, and/or trophic changes that may improve with sympathetic nerve block. (3) Multiple mechanisms probably involved. (4) The patterns of peripheral activity, or peripheral and central interaction, that yield the lancinating quality of these pains are unknown. (5) Nociceptive nervi nervorum (small afferents that innervate larger nerves) may account for neuropathic pain accompanying nerve compression or inflammation. (6) Injury to axons may be followed by neuroma formation, a source of aberrant activity likely to be involved in pain. (Adapted from Portenoy, 1991, Figure 2, p. 396.)

clinical management, because a number of specific therapies may be particularly useful in this situation (see below). In the specific case of sympathetically maintained pain, the correct diagnosis suggests a clinical strategy, namely, the use of sympathetic nerve blocks, that would otherwise not be considered (see below).

3. Idiopathic Pain

Most patients who lack the criteria to fulfill a pathophysiological diagnosis of nociceptive or neuropathic pain can be said to have an idiopathic pain syndrome. As discussed previously, the nomenclature applied to these patients is imprecise and often problematic. Pain specialists often use a site-specific term (e.g., atypical facial pain) or the general label, nonmalignant pain syndrome, particularly if the pain is associated with a relatively high level of disability.

Regrettably, this terminology may suggest the existence of profound psychiatric disease and thereby sustain a bias that can lessen the clinician's enthusiasm for a comprehensive physical and psychological assessment. The use of the term, idiopathic pain, may avoid this stigma and is preferred unless a patient meets criteria for a defined psychiatric syndrome.

4. Other Pathophysiologic Conceptualizations

A recent effort to extend this concept of a pathophysiologic classification of chronic pain patients has attempted to integrate physical and psychosocial dimensions of the pain. Specifically, an approach has been proposed (Turk & Rudy, 1990) in which a multiaxial psychological and behavioral assessment of pain is combined with a medical classification, such as the taxonomy of pain developed by the International Association for the Study of Pain (Merskey & Bogduk, 1994). The multiaxial groupings are based on an empirically derived psychosocial and behavioral classification that categorizes patients into three groups ("dysfunctional," "interpersonally distressed," and "adaptive copers") (Turk & Rudy, 1988). The objective is to develop a pathophysiologic classification that could be useful for targeting a multimodality therapeutic approach based on the character of these various disturbances. This approach has yet to be validated in clinical practice, but it affirms the complexity of chronic pain patients and the need to assess the various processes contributing to both the expression of pain and the patient's response to it.

IV. MANAGEMENT OF CHRONIC PAIN

The spectrum of treatment approaches that may be appropriate for pain is as broad as the interpatient variability encountered in the clinical setting. Within the very large and diverse population with chronic pain, for example, the clinical presentations range from those patients who function well but experience highly distressing pain to those who present a degree of dysfunction that appears to far exceed both the pain and the organic pathology evident to the clinician. This interpatient variability is a fundamental consideration in the management of pain. It indicates that no one approach can be considered a standard that meets the needs of all patients, and underscores the profound responsibility of the primary clinician, who must perform an assessment that is able to guide the evaluation and selection of therapies in a manner that is efficient, cost-effective, and therapeutically optimal.

The selection of the specific therapies for the management of a patient with chronic pain is guided by the nature of the problems identified through the comprehensive assessment. Six broad categories of treatment approaches may be explored in fashioning a therapeutic strategy appropriate to the patient's need for comfort or functional restoration (Table 1). Some types of patients, such as those with acute pain (e.g., postoperative pain) or chronic pain associated with minimal disability or psychological impairment, may require a single modality approach targeted to comfort alone. For example, most patients with cancer pain can be well managed by the individual practitioner using opioid pharmacotherapy.

TABLE 1   Approaches Used in the Management of Chronic Pain

Primary analgesic therapy directed against the underlying etiology
Pharmacological approaches
Anesthetic approaches
Surgical approaches
Physiatric approaches
Neurostimulatory approaches
Psychological approaches

Many types of patients with chronic pain, however, can benefit more from a multimodality approach in which comfort and function are emphasized together. Importantly, this may require reeducation of the patient, who may be convinced that the only appropriate goal of therapy is pain relief, and that comfort alone will immediately improve function. Experienced clinicians know that highly disabled patients are unlikely to improve unless function is addressed directly, and an effort must be made from the start of treatment to impose a clinical agenda that includes physical and psychosocial rehabilitation as major goals of therapy.

The recognition that pain relief and the rehabilitation of physical and psychosocial functioning can often be impossible to address together unless clinicians in multiple disciplines work jointly led to the development of the multidisciplinary pain management approach, which is now promulgated through a large number of pain clinics throughout the United States and other countries. This is a labor-intensive approach that is generally considered to be potentially optimal for a subpopulation of patients with chronic nonmalignant pain associated with severe pain-related disability and affective disturbances.

Most pain patients are not referred to a multidisciplinary pain management program. Rather, patients are offered a multimodality strategy, in which a primary caregiver takes personal responsibility for managing one or more of the broad therapeutic areas and coordinates with other clinicians as appropriate. Most primary caregivers focus on drug therapy, and this approach is emphasized herein.

A. Pharmacological Approaches

The drugs used to treat pain can be divided into three broad categories: (a) nonsteroidal anti-inflammatory drugs (NSAIDs), (b) adjuvant analgesics, and (c) opioids. Adjuvant analgesics can be defined as drugs that have primary indications other than pain but are analgesic in selected circumstances.

1. Nonsteroidal Anti-Inflammatory Drugs

The NSAIDs inhibit the enzyme cyclooxygenase, an action that reduces inflammatory mediators that sensitize or activate peripheral nociceptors (Higgs & Moncada, 1983; Vane, 1971) and may also alter central nociceptive processing (Malberg & Yaksh, 1992; Willer, De Broucker, Bussel, Roby-Brami, & Harrewyn, 1989). The central actions of these drugs may explain the marked disproportion between peripheral anti-inflammatory effects and analgesia that occurs with some of these drugs, such as acetaminophen.

The NSAIDs comprise numerous distinct classes, which share several fundamental characteristics (Brooks & Day, 1991; Sunshine & Olson, 1989). NSAID analgesia is characterized by a ceiling dose, beyond which additional increments fail to yield greater pain relief, and by a lack of physical dependence or tolerance. Although the dose–response relationship is inferred from large dose-ranging studies, the minimal effective dose, ceiling dose, and dose associated with adverse effects is unknown in any individual patient and may be higher or lower than the standard recommended dose.

Numerous NSAIDs are commercially available in the United States (Table 2) and other countries. They are useful independently for mild to moderate pain and provide additive analgesia when combined with opioid drugs in the treatment of more severe pain (Ferrer-Brechner & Ganz, 1984). Although most of these agents have been approved as anti-inflammatory drugs rather than analgesics, clinical experience suggests that any can potentially be useful in the treatment of pain. They are typically considered to be first-line agents for pain associated with an underlying inflammatory condition, such as arthritis, and are often given an empirical trial in pains of all other types. The recent approval in the United States of an injectable NSAID, ketorolac, has enlarged the potential clinical indications to acute postoperative pain (O'Hara, Fragen, Kinzer, & Pemberton, 1987).

The most common toxicity associated with the NSAIDs is gastrointestinal. NSAID-induced gastropathy ranges from dyspepsia to life-threatening hemorrhage. The risk of gastric ulceration is increased fivefold over the general population (Loeb, Ahlquist, & Talley, 1992). Although various studies have evaluated the relative risk associated with the different NSAIDs, comparative data are incomplete. The evidence suggests that the risk is relatively greater with piroxicam and relatively less with ibuprofen; the risk is probably negligible with acetaminophen, and may also be relatively low with the nonacetylated salicylates, such as choline magnesium trisalicylate and salsalate. Although it has not been confirmed that the risk of duodenal ulceration is increased from NSAID therapy, the complications that ensue following duodenal ulceration are worse in patients who are receiving these drugs.

Despite the high prevalence and serious nature of NSAID-induced

TABLE 2   Nonsteroidal anti-inflammatory drugs

Class	Examples
p-aminophenol derivative	Acetaminophen
Salicylates	Aspirin
	Diflunisal
	Choline magnesium trisalicylate
	Salsalate
Proprionic acids	Ibuprofen
	Naproxen
	Naproxen sodium
	Fenoprofen
	Ketoprofen
	Flurbiprofen
Acetic acids	Indomethacin
	Tolmetin
	Sulindac
	Diclofenac
	Etodolac
	Ketorolac
Oxicams	Piroxicam
Fenamates	Mefenamic acid
	Meclofenamate
Naphthylalkanones	Nabumetone
Pyrazoles	Phenylbutazone

gastropathy, the use of prophylactic agents remains controversial. Misoprostol, a prostaglandin E1 analogue, is the only established prophylactic treatment for NSAID-induced gastric ulceration (Graham, Agrawal, & Roth, 1988), but a recent study suggests that the cost may be excessive in patients with no prior episode of gastrointestinal hemorrhage (Edelson, Tosteson, & Sax, 1990). This drug also produced distressing gastrointestinal side effects in a high proportion of patients. A study in healthy volunteers suggests that cimetidine, an H_2-blocker, can reduce the risk of NSAID-induced gastropathy (Frank et al., 1989), but controlled trials have failed to demonstrate benefit during long-term NSAID therapy (Robinson et al., 1989; Roth, Bennett, Mitchell, & Hartman, 1987). Sucralfate reduces symptoms, but not the incidence of ulceration (Caldwell et al., 1987). Omeprazole, a new hydrogen-potassium ATP-ase pump inhibitor blocks gastric acid production and may be protective, but has not been adequately tested.

Given the limited data, it is reasonable to consider prophylactic treatment with misoprostol in those patients who receive long-term NSAID therapy and are at relatively high risk for gastric ulceration. Based on epidemiology data, this high risk includes the elderly (greater than age 60), those with rheumatoid arthritis, those concurrently receiving a corticosteroid, those with recent upper abdominal pain, and those with distressing gastrointestinal side effects from NSAIDs (Fries et al., 1989). Patients with a prior history of gastric ulceration and those who would be at particularly high risk should an ulcer develop, such as those receiving anticoagulants, might also be considered for this therapy. On theoretical grounds, patients at high risk who cannot tolerate misoprostol might be considered for treatment with a combination of sucralfate and a H2 blocker (Roth, 1988).

All NSAIDs should be used cautiously in patients with renal insufficiency. Acetaminophen is the preferred drug in this setting, notwithstanding recent data establishing its potential for renal toxicity (Sandler et al., 1989). Acetaminophen may also be much safer in patients with a bleeding diathesis. The risk of a bleeding complication in a patient receiving an NSAID, which derives from the effects of these drugs onn platelet aggregation, is also presumably less with the nonacetylated salicylates, such as choline magnesium salicylate or salsalate, which do not impair platelet aggregation at usual clinical doses (Danesh, Saniabadi, Russell, & Lowe, 1987).

All NSAIDs, except acetaminophen, may cause or exacerbate encephalopathy and may be problematic in patients at risk from volume overload. They must therefore be administered cautiously to patients with preexisting encephalopathy and those with congestive heart failure, peripheral edema, or ascites. NSAIDs can also cause hepatotoxicity. The pyrazole subclass, including phenylbutazone, is more likely to do this, as is acetaminophen, which must be administered cautiously in those with preexisting liver disease.

Although the selection and administration of an NSAID is commonly believed to be empirical, a growing experience with these drugs and new information about their pharmacology may provide a foundation for rational dosing guidelines (Brooks & Day, 1991; Ingham & Portenoy, 1993) (Table 3). The drug-selective toxicities discussed previously must be considered in choosing a drug in a specific clinical setting. Other considerations in drug selection include favorable prior experience, concerns about compliance (which may suggest the value of a drug with once daily dosing), and cost.

Given the observation that the standard recommended dose of an NSAID may or may not be appropriate for the individual patient, it is prudent to consider the potential value of dose titration, particularly in those patients who may be predisposed to adverse effects. By initiating treatment with a dose lower than the standard recommended dose, then increasing the dose at intervals while monitoring effects, the minimal effective dose, ceiling dose, and toxic dose may be identified. The risk of serious

TABLE 3   Guidelines for Nonsteroidal Anti-Inflammatory Drugs Therapy in Pain Management[a]

1. Comprehensive assessment
   Define pain syndrome, functional status, psychosocial disturbances, and concurrent diseases.
   Consider efficacy of NSAIDs in the defined pain syndrome and the role of this treatment in a multimodality approach.
2. Drug selection
   Avoid NSAIDs, if possible, in patients with gastroduodenopathy, bleeding diathesis, renal insufficiency, hypertension, severe encephalopathy, and cardiac failure; avoid acetaminophen in patients with liver disease.
   If NSAID strongly indicated in patient with relative contraindication (except liver disease), consider acetaminophen; if anti-inflammatory effects desirable, consider NSAID with good safety profile and long clinical experience (e.g., ibuprofen or a nonacetylated salicylate).
   In all patients, consider drug-selective differences in toxicity (e.g., aspirin less well tolerated than other NSAIDs).
   Consider the effects of concurrent drugs with possible pharmacokinetic and pharmacodynamic interactions.
   Consider individual differences (note prior treatment outcomes) and patient preference.
   Be aware of available preparations (e.g., oral, intravenous, topical).
   Be aware of cost differences.
3. Route selection
   Use the least invasive route possible.
4. Dosing and dose titration
   Begin with low dose and adjust weekly or less often.
   Increase dose until adequate analgesia occurs, ceiling dose is identified, or maximal recommended dose is reached.
   Be aware that several weeks may be necessary to fully assess efficacy.
   In timing the dose consider pharmacokinetic properties and, if necessary, adjust regimen to circadian rhythms in pain and inflammation.
   Review therapy at regular intervals to avoid unnecessarily long treatment.
5. Trials of alternative NSAIDs
   Noting individual differences in the response to various NSAIDs, consider trial of another NSAID following treatment failure.
6. Toxicity monitoring
   Monitor potential gastrointestinal, renal, and hepatic toxicity regularly; increase frequency of monitoring in the elderly and those with concurrent disease.
7. Prophylaxis against adverse gastroduodenal events
   Consider in patients >60 years old and those with a history of gastroduodenal disease (especially bleeding), significant gastrointestinal pain, high risk of serious adverse outcome if bleeding occurs, or coagulopathy; misoprostol is preferred for prophylaxis.

[a]From Ingham & Portenoy (1993, Table 1, p. 839).

toxicity will be reduced by the use of the lowest dose capable of providing meaningful therapeutic effects.

The existence of serious dose-related toxicities, which may occur without warning (including gastric ulceration and renal failure), combined with the paucity of information available about the long-term safety of high doses, suggests that dose exploration should be limited by an empirical maximum dose. A reasonable guideline limits upward dose titration to 1.5–2 times the standard starting dose. Long-term therapy above the standard dose should be carefully monitored with more frequent tests for occult fecal blood, urinalysis, and serum tests of renal and hepatic function.

Extensive clinical experience has demonstrated that failure with one NSAID may be followed by effective analgesia with another. Given this intraindividual variability in drug response, it is reasonable to consider sequential trials of NSAIDs in patients with refractory pain.

2. Adjuvant Analgesics

The adjuvant analgesics comprise many drug classes, within which numerous specific compounds have been discovered to have analgesic effects (Table 4).

a. Antidepressants

Antidepressants are analgesic in diverse chronic pain states (Butler, 1984; Getto, Sorkness & Howell, 1987; Max et al., 1991; Portenoy, 1990c). Of those drugs available in the United States, the analgesic efficacy of the tertiary amine tricyclic antidepressants (TCAs), amitriptyline, doxepin, imipramine, and clomipramine, and the secondary amine TCAs, desipramine and nortriptyline, has been established through favorable controlled trials in populations with diabetic neuropathy, postherpetic neuralgia, headache, myofascial pain, arthritis, and psychogenic pain (Carette, McCain, Bell, & Fam, 1986; Couch, Ziegler, & Hassanein, 1976; Diamond & Baltes, 1971; Gingras, 1976; Hameroff et al., 1982; Kishore-Kumar, Max, Schafer, Gaughan, Smoller, Gracely, & Dubner, 1990; Kvinesdal, Molin, Froland, & Gram, 1984; Langohr, Stohr, & Petruch, 1982; Max et al., 1987; Max et al., 1991; Okasha, Ghaleb, & Sadek, 1973; Pilowsky, Hallet, Bassett, Thomas, & Penhall, 1982; Watson & Evans, 1985; Watson et al., 1982). Studies of some of the "newer" antidepressants, specifically maprotiline and paroxetine, have also been favorable (Eberhard et al., 1988; Sindrup, Gram, Brosen, Eshoj, & Mogensen, 1990; Watson, Chipman, Reed, Evans, & Birkett, 1992); results with trazodone have been equivocal (Davidoff, Guarracini, Roth, Sliwa, & Yarkony, 1987; Ventafridda, 1987). The monoamine oxidase inhibitors, which also increase the availability of central monoamines, may also have analgesic effects (Anthony & Lance, 1969; Lascelles, 1966), but possess a less favorable safety profile than other antidepressant drugs.

This abundant literature suggests that antidepressants can be considered nonspecific analgesics that may be potentially beneficial in most types of chronic pain. This primary analgesic effect was initially believed to be

TABLE 4   Adjuvant Analgesics

Class	Examples
Antidepressants	
Tricyclic antidepressants	Amitriptyline
	Doxepin
	Imipramine
	Nortriptyline
	Desipramine
	Clomipramine
"Newer" antidepressants	Fluoxetine
	Paroxetine
	Trazodone
	Maprotiline
Monoamine oxidase inhibitors	Phenelzine
Anticonvulsants	Carbamazepine
	Phenytoin
	Valproate
	Clonazepam
	Gabapentin
Oral local anesthetics	Mexiletine
	Tocainide
Neuroleptics	Fluphenazine
	Haloperidol
	Methotrimeprazine
	Pimozide
Muscle relaxants	Orphenadrine
	Carisoprodol
	Methocarbamol
	Chlorzoxazone
	Cyclobenzaprine
Antihistamines	Hydroxyzine
Psychostimulants	Caffeine
	Methylphenidate
	Dextroamphetamine
Corticosteroids	Dexamethasone
	Prednisone
	Methylprednisolone
Sympatholytic drugs	Prazosin
	Phenoxybenzamine
Calcium channel blockers	Nifedipine
	Verapamil
Miscellaneous	Baclofen
	Clonidine
	Capsaicin
	Calcitonin

related to the reversal of depression (Evans, Gensler, Blackwell, & Galbrecht, 1973), but strong evidence, including the demonstration of antinociceptive effects in animal models (Spiegel, Kalb, & Pasternak, 1983), contradicts this conclusion (Couch et al., 1976; Kishore-Kumar et al., 1990;

Max et al., 1987; Watson, et al., 1982). The primary analgesic action of TCA drugs may relate to their effects on central monoamines in endogenous pain modulating pathways (Besson & Chaouch, 1987; Hammond, 1985), and current evidence suggests that TCA drugs with primary effects on norepinephrine synapses or both norepinephrine and serotonin synapses are most likely to be analgesic. TCAs also bind to many other receptors (Charney, Menkes, & Heninger, 1981; Richelson, 1979), including some implicated in the process of pain modulation (Besson & Chaouch, 1987; Sosnowski & Yaksh, 1990), and the potential for alternative analgesic mechanisms must be considered.

A trial of a TCA should be considered in virtually all chronic pain syndromes, unless contraindicated by specific medical conditions, including significant cardiac arrhythmias, symptomatic prostatic hypertrophy, or narrow angle glaucoma. Given the abundant data from controlled studies, amitriptyline is the preferred drug for an initial trial. Patients unable to tolerate this drug should undergo a trial with an alternative; the secondary amine TCAs and some of the newer antidepressants, such as paroxetine, usually produce fewer adverse effects.

Based on clinical experience, guidelines for TCA administration should include low initial doses and gradual dose escalation. The existence of a therapeutic window for analgesia during treatment with nortriptyline has been suggested anecdotally, but confirmatory data are lacking, and this effect has not been observed with other drugs. In the absence of side effects, most patients should undergo dose escalation into the antidepressant range. This guideline is particularly important if pain is associated with depressed mood. Although the plasma level associated with analgesic effects is unknown for all agents, a plasma level can be measured to assure that noncompliance, poor absorption, or unusually rapid catabolism is not compromising efforts to achieve therapeutic levels.

b. Anticonvulsants

Anticonvulsants have become widely accepted in the management of chronic neuropathic pain, particularly those characterized by lancinating or paroxysmal pains (Swerdlow, 1984). Their mode of analgesic action is not known, but presumably relates to the capacity to suppress paroxysmal discharges, neuronal hyperexcitability, or spread of abnormal discharges (Weinberger, Nicklas, & Berl, 1976). Nerve injury can result in spontaneous electrical activity (Albe-Fessard & Lombard, 1982; Loeser, Ward, & White, 1968; Nystrom & Hagbarth, 1981; Wall & Gutnick, 1974), which may be the target of these drugs.

The analgesic efficacy of phenytoin has been suggested in anecdotal reports and controlled trials that describe varying populations, all of which have neuropathic pain associated with a prominent lancinating component. The populations include those with trigeminal neuralgia, glossopharyngeal neuralgia, tabetic lightning pains, paroxysmal pain in postherpetic neuralgia, thalamic pain, postsympathectomy pain, posttraumatic neuralgia, painful neuropathy from Fabry's disease, and painful diabetic neuropathy (Blom, 1963; Braham & Saia, 1960; Cantor, 1972; Chadda & Mathur, 1978; Green, 1961; Hatangdi, Boas, & Richards, 1976; Lockman, Hunninghake, Krivit, & Desnick, 1973; Raskin, Levinson, Hoffman, Pickett, & Fields, 1974; Swerdlow & Cundill, 1981; Taylor, Gray, Bicknell, & Rees, 1977).

Controlled trials have established the efficacy of carbamazepine in trigeminal neuralgia, the lancinating (but not continuous) pains of postherpetic neuralgia, and painful diabetic neuropathy (Campbell, Graham, & Zilkha, 1966; Killian & Fromm, 1968; Rockcliff & Davis, 1966; Rull, Quibrora, Gonzalez-Miller, & Castaneda, 1969). There have also been numerous anecdotal reports that suggest benefit in glossopharyngeal neuralgia, tabetic lightning pains, paroxysmal pain in multiple sclerosis, postsympathectomy pain, lancinating pains due to cancer and post-traumatic neuralgia (Ekbom, 1972; Elliot, Little, & Milbrandt, 1976; Espir & Millac, 1970; Mullan, 1973; Raskin et al., 1974; Swerdlow & Cundill, 1981; Taylor et al., 1977).

Uncontrolled clinical trials and anecdotal reports have similarly suggested that clonazepam and valproate may be effective in neuropathic pains characterized by lancinating dysesthesias. Clonazepam has been reported to be useful in the treatment of trigeminal neuralgia, paroxysmal postlaminectomy pain, and posttraumatic neuralgia (Caccia, 1975; Martin, 1981; Swerdlow & Cundill, 1981). Valproate has been beneficial in the management of trigeminal neuralgia, postherpetic neuralgia, and other lancinating neuropathic pains (Peiris, Perera, Devendra, & Lionel, 1980; Raftery, 1979; Swerdlow & Cundill, 1981). Gabepentin and lamotrigine are new anticonvulsants that may also have analgesic effects. Both are antinociceptive in animal models of neuropathic pain, and the potential utility of gabapentin has been noted in numerous clinical anecdotes.

Three other drugs that are not classified as anticonvulsants have been demonstrated to have analgesic effects in trigeminal neuralgia. They or their congeners are therefore considered with the aforementioned agents as potential treatments for patients with lancinating or paroxysmal neuropathic pains. Baclofen, which is a gamma–aminobutyric acid (GABA) agonist primarily indicated in the treatment of spasticity, is widely considered to be a second-line agent in trigeminal neuralgia (Fromm, Terence, & Chatta, 1984), and is commonly used to treat lancinating pains of other types. The efficacy of tocainide, an oral local anesthetic chemically related to lidocaine, in trigeminal neuralgia (Lindstrom & Lindblom, 1987) is further justification for empirical trials of a related drug, mexiletine, for neuropathic pain of all types, including pain characterized by a prominent lancinating or paroxysmal component (see below). Finally, pimozide is a neuroleptic drug that has been demonstrated to be analgesic in trigeminal neuralgia (Lechin et al., 1989).

Guidelines for the selection and administration of all these drugs in patients with neuropathic pain are empirical. Although the primary indication is lancinating or paroxysmal neuropathic pain, clinical experience suggests that these drugs can occasionally be useful in the management of continuous dysesthesias, and a trial is reasonable in refractory cases. The administration of each of these drugs must use dosing guidelines appropriate for the primary indication. Both surveys (Swerdlow & Cundill, 1981) and clinical experience indicate that patients may have markedly different analgesic responses to the various drugs, and should one fail, a trial of another must be considered.

c. Oral local anesthetics

Systemically administered local anesthetics can be analgesic in diverse types of acute and chronic pain (Glazer & Portenoy, 1991). The recent introduction of oral local anesthetic drugs for the management of cardiac arrhythmias has spurred the use of this class in the treatment of chronic pain. As noted previously, the efficacy of tocainide in trigeminal neuralgia has been established (Lindstrom & Lindblom, 1987), and other controlled trials have similarly demonstrated the efficacy of mexiletine in other types of neuropathic pain, such as painful diabetic neuropathy (Dejgard, Peterson, & Kastrup, 1988).

On the basis of these data and a growing clinical experience, it is reasonable to view the oral local anesthetic drugs as second-line agents for the treatment of refractory neuropathic pain syndromes of any type. Mexiletine is the safest compound (Kreeger & Hammill, 1987) and should be administered first. Dosing is empirical and should mimic that applied in the treatment of cardiac arrhythmias (Kreeger & Hammill, 1987).

d. Neuroleptics

Numerous anecdotal reports have suggested the value of haloperidol, fluphenazine, perphenazine, thioridazine, chlorprothixene, and chlorpromazine in the management of patients with a variety of painful disorders, the majority of which were neuropathic (Kocher, 1976; Margolis & Gianascol, 1956; Nathan, 1978; Weis, Sriwatanakul, Weintraub, 1982). There have been few confirmatory controlled studies, however. The analgesic efficacy of methotrimeprazine has been established in controlled single-dose studies (Beaver, Wallenstein, Houde & Rogers, 1966; Bloomfield, Simard-Savoie, Bernier, & Tretault, 1964; Lasagna & DeKornfeld, 1961), and as noted previously, pimozide has been demonstrated to be analgesic in trigeminal neuralgia (Lechin et al., 1989). A controlled single-dose trial of another phenothiazine, chlorpromazine, failed to demonstrate analgesic effects in cancer pain (Houde & Wallenstein, 1966).

Given the lack of adequate data and the risks associated with neuroleptic drugs, the most important of which is the capacity to produce refractory movement disorders, it is appropriate to restrict their use to trials in patients with intractable neuropathic pain. Low doses should be administered, and there are no data to support upward dose titration beyond a limited range.

e. Muscle relaxants

Pain in muscles or adjacent soft tissues, which is sometimes associated with trigger points or spasm, is commonplace. The "muscle relaxant" drugs, which include orphenadrine, carisoprodol, chlorzoxazone, methocarbamol, and cyclobenzaprine, are analgesic in a variety of musculoskeletal pains (Bercel, 1977; Birkeland & Clawson, 1968; Gold, 1978) and are frequently used in these settings. Ironically, however, these drugs do not in fact relax skeletal muscle. Although they suppress polysnaptic myogenic reflexes in experimental preparations (Smith, 1965), the clinical relevance of this phenomenon is uncertain and the mechanism of analgesic effects is unknown. Given the chemical distinctions among these agents, varying modes of action are likely. All these drugs have prominent sedative effects, and it is possible that analgesia may be linked to hypnotic or anxiolytic actions, at least for some of them.

The lack of a demonstrable effect on striated muscle indicates that the muscle relaxant drugs should not be used in the management of true spasticity. The latter condition, which is due to a lesion of the CNS, is more effectively managed with baclofen, dantrolene, or diazepam, all of which can diminish muscle contraction.

The selection and administration of the muscle relaxant drugs is largely empirical. There are no comparative data by which to judge the relative efficacy and safety of the various drugs within this class, and neither the risks nor benefits of these agents have been compared to any of the NSAIDs or the opioids. Furthermore, there have been no studies to clarify optimal dosing guidelines. Based on favorable clinical experience, one or another of these drugs is typically administered for a short period as a therapy for an acute musculoskeletal pain.

Regardless of the duration of therapy, the potential sedative and anticholinergic effects of these drugs must be considered prior to treatment, then monitored thereafter. These effects can be additive to other centrally active drugs and may further impair function or increase patient distress. There is no evidence that upward dose titration beyond usually recommended doses produces any other effect than progressive sedation, with an unknown degree of accruing risk.

There are no data by which to judge the safety or potential benefits associated with the long-term administration of the muscle relaxant drugs. Clinicians who treat patients referred to multidisciplinary pain management programs have observed that adverse consequences can result from overuse or misuse of these drugs, and this potential must be recognized in the chronic pain population. Long-term administration, therefore, should be likened to treatment with other potentially abusable centrally acting drugs (e.g., benzodiazepines and opioids): ongoing assessment is needed to ensure that the patient continues to experience symptomatic improvement without functional compromise, adverse pharmacological effects, or aberrant drug-taking behaviors.

f. Antihistamines

Controlled studies have demonstrated the analgesic efficacy of antihistamine drugs (Rumore & Schlichting, 1986), including orphenadrine, diphenhy-dramine, and hydroxyzine (Birkeland & Clawson, 1968; Gold, 1978; Stambaugh & Lance, 1983). The mechanism of this analgesic effect is not known, and its clinical relevance has proved to be limited. With the exception of orphenadrine, which is used as an muscle relaxant, and several analgesic combination products that include an antihistamine, these drugs have not been widely used in the treatment of pain.

g. Alpha-2 adrenergic agonists

Clonidine has established analgesic effects (Max et al., 1988) and has been used to treat chronic and recurrent pains (Shafar, Tallett, Knowlson, 1972; Tan & Croese, 1986). Long-term trials of oral or transdermal clonidine in patients with chronic pain have not yet appeared, but clinical experience suggests that some patients, including some with refractory neuropathic pains, will respond favorably to this drug. In patients with persistent pain and no medical contraindications, a therapeutic trial of oral or transdermal clonidine is reasonable.

h. Benzodiazepines

Studies in the postoperative setting have suggested that benzodiazepine drugs, including diazepam and midazolam, can be analgesic (Miller et al., 1986; Singh, Sharma, Gupta, & Pandey, 1981). Treatment with alprazolam yielded favorable effects in a survey of cancer patients with neuropathic pain (Fernandez, Adams, & Holmes, 1987). Other clinical trials have been negative (Yosselson-Superstine, Lipman, & Sanders, 1985), and a study using an experimental pain paradigm indicated that the analgesic effects produced by diazepam could be ascribed to psychological influences (change in response bias) rather than a shift in sensory discriminability, such as is observed with morphine (Yang, Clark, Ngai, Berkowitz, & Spector, 1979). In clinical practice, some patients report enhanced analgesia following treatment with a benzodiazepine, and it is often impossible to distinguish a primary analgesic action from a favorable secondary outcome resulting from anxiolysis or relaxation of muscle.

The short-term administration of a benzodiazepine drug is widely accepted in the treatment of acute musculoskeletal pain. Concerns about persistent sedation and other adverse cognitive effects, as well as the observation that these drugs can be abused by chronic pain patients, have limited the long-term use of this class as primary analgesics. Some patients do appear to gain persistent benefit from the ongoing administration of a benzodiazepine, and the approach can be considered for highly selected patients who undergo repeated assessment to establish continued benefits without adverse effects. As noted previously, the type of monitoring required is similar to that recommended for other potentially abusable drugs, including opioids.

i. Drugs for sympathetically maintained pain

Patients who have a so-called complex regional pain syndrome (also known as reflex sympathetic dystrophy or causalgia) may have a favorable response to interruption of sympathetic efferent activity to the painful region. If so, the syndrome is known as a sympathetically maintained pain. Patients suspected of having a sympathetically maintained pain are usually offered a series of sympathetic nerve blocks, which may be both diagnostic and therapeutic. Patients who are candidates for such blocks, but are unable to tolerate them have been empirically treated with sympatholytic drugs. Case reports and clinical series have suggested that phenoxybenzamine (Ghostine, Comair, Turner, Kassell, & Azar, 1984), prazosin (Abram & Lightfoot, 1981), oral guanethidine (Tabira, Shibasaki, & Kuroiwa, 1983), and propranolol (Simson, 1974) may be useful in these pains. Other, nonsympatholytic drugs, specifically a corticosteroid (Kozin, Ryan, Carerra, Soin, & Wortmann, 1981) and nifedipine (Prough et al., 1985), have also been successfully administered in anecdotal reports. A controlled trial suggested that calcitonin may be beneficial in reflex sympathetic dystrophy (Gobelet, Waldburger, & Meier, 1982).

j. Miscellaneous drugs

In addition to their empirical use in the treatment of sympathetically maintained pain, corticosteroids are also commonly administered to patients with cancer pain of diverse types (Ettinger & Portenoy, 1988). Empirically, they have been found to be particularly useful for patients with either cancer-related neuropathic pain or bone pain.

Malignant bone pain is often managed with the combination of an opioid and NSAID or corticosteroid. Newer treatments have gained support through both anecdotal reports and clinical trials. These have included the bisphosphonates, calcitonin, and radionuclides, such as strontium-89 (Blomquist, Elomaa, Porkka, Karonen, & Lamberg-Allardt, 1988; Silberstein & Williams, 1985; Holten-Verzantvoort, et al., 1991).

The analgesic effects produced by calcitonin may involve mechanisms other than those related to bone. As noted, a controlled trial has suggested that this compounded may be analgesic in reflex sympathetic dystrophy, and another controlled trial demonstrated analgesic effects in phantom limb pain (Jaeger & Maier, 1992). Thus, there may be a larger role for this drug as an adjuvant analgesic.

Capsaicin is a compound that depletes peptides in small primary afferent neurons, including those, such as substance P, that are putative neurotransmitters of nociceptive processing. There is some evidence for the utility of this drug as a topical therapy for selected neuropathic pains, including postherpetic neuralgia, postmastectomy pain, and painful diabetic neuropathy (Watson, Evans, & Watt, 1988, 1989). Clinical experience has yielded mixed results.

Psychostimulant drugs, including dextroamphetamine, methylphenidate and caffeine, are analgesic (Bruera, Chadwick, Brenneis, MacDonald, 1987; Forrest et al., 1977; Laska et al., 1984). Caffeine is commonly added to combination products used in the treatment of headache, and both dextroamphetamine and methylphenidate are used to reverse opioid-induced sedation and provide co-analgesic effects in patients with cancer pain.

Finally, cannabinoid drugs have been demonstrated to be analgesic (Noyes, Brunk, Avery, & Canter, 1975). Tetrahydrocannabinol is commercially available in the United States and is used as an antiemetic in the cancer population. In part due to the potential for psychotomimetic side effects, its use as an analgesic has not been explored in the clinical setting.

3. Opioid Analgesics

Opioid analgesics are the mainstay therapy for acute pain and chronic cancer pain. Numerous drugs within the agonist–antagonist and pure agonist subclasses are available in the United States (Table 5). Optimal administration of these drugs in acute pain settings has a very high efficacy, and guidelines for chronic administration in the cancer population (Table 6) have been demonstrated to yield favorable results in 70-90% of patients (Foley, 1985; Portenoy, 1993; World Health Organization, 1986; Zech, Grand, Lynch, Hertel, & Lehmann, 1995). Given the striking benefits that can accrue to patients who receive optimal opioid therapy for acute pain or cancer pain, a major issue in opioid pharmacotherapy relates to public health: increasing access to this effective and affordable care for all patients.

The gratifying outcomes in the cancer population have been among the most salient observations in the ongoing controversy surrounding the use of opioids in patients with chronic nonmalignant pain. The long-term administration of opioids to patients with cancer has failed to confirm many concerns about this therapy that were previously taken to be axiomatic (Portenoy, 1990a, 1994). For example, the development of analgesic tolerance, which would be expected to compromise long-term benefits, is rarely

TABLE 5 Opioid Analgesics

Agonist–antagonist opioids—Usually limited to acute pain
Pentazocine
Nalbuphine
Butorphanol
Buprenorphine
Dezocine
Pure agonist opioids—Used orally for moderate acute or chronic pain
Codeine
Propoxyphene
Hydrocodone (combined with aspirin or acetaminophen)
Dihydrocodeine (combined with aspirin or acetaminophen)
Oxycodone (combined with aspirin or acetaminophen)
Pure agonist opioids—Used orally or systemically for severe acute or chronic pain
Morphine
Hydromorphone
Oxycodone
Methadone
Meperidine
Oxymorphone
Fentanyl
Sufentanil
Alfentanil

TABLE 6 Opioid Pharmacotherapy for Cancer Pain

1. Select specific opioid based on usual severity of the pain, favorable prior experience, availability of desired formulation, desirable time–action relationship, cost, or other factors; one of the pure agonists is generally used for chronic administration.
2. Use the oral route unless an alternative route is indicated by the inability to swallow or absorb, the need for very rapid onset of each dose, or a desire to simplify the logistics of drug administration.
3. For continuous or very frequent pain, administer the drug on "around-the-clock" basis.
4. If breakthrough pains occur, coadminister a "rescue dose" using the same drug, if this is feasible and the drug is available in a formulation with a short duration of effect, or an alternative short-acting drug; the "rescue dose" is usually 5–15% of the total daily dose and is offered at an appropriate interval on an "as needed" basis.
5. Dose escalation at an interval appropriate to the severity of the pain and the pharmacology of the drug and route should be performed until a favorable balance between analgesia and side effects is obtained or intolerable and unmanageable adverse effects supervene.
6. Side effects, such as constipation, nausea, and somnolence, should be treated during the process of reevaluation and dose titration.
7. If adverse effects prevent needed dose escalation, consider one or more of the following maneuvers: (a) more intensive treatment of side effects; (b) a trial of a different opioid drug ("sequential opioid trials"); (c) reduction of the systemic requirement for the opioid through a spinal infusional therapy (opioid with or without a local anesthetic) or coadministration of a nonopioid analgesic or an adjuvant analgesic; or (d) use of a nonpharmacological treatment, including neural blockade, surgical neurolysis, a physiatric therapy, or a psychological approach.

observed to be the reason for loss of efficacy; in the absence of progressive disease, opioid doses typically remain stable (Foley, 1991; Kanner & Foley, 1981; Twycross, 1974). Similarly, the perception that opioid side effects, such as somnolence and cognitive impairment, will compromise function during long-term treatment is not affirmed by clinical investigations or experience (Bruera, Macmillan, Selmser, & MacDonald, 1989). These observations have been supplemented by many hundreds of published cases, which together document the existence of a subpopulation with chronic nonmalignant pain that can attain satisfactory partial analgesia from opioids for a long period without the development of adverse pharmacological effects or addiction (France, Urban, & Keefe, 1984; Portenoy & Foley, 1986; Taub, 1982; Tennant, Robinson, Sagherian, & Seecof, 1988; Tennant & Uelman, 1983; Urban, France, Steinberger, & Scott, & Maltbie, 1986; Zenz, Strumpf, & Tryba, 1992).

Addiction continues to be a concern in the use of opioids for cancer pain, as well as nonmalignant pain. Addiction may be defined as a syndrome characterized by psychological dependence and aberrant drug-related behaviors. Commonly applied diagnostic criteria have been derived from experience in the addict population and cannot be simply transferred to medically ill patients who receive the drug as a treatment for pain. In the medically ill population, it is necessary to highlight the cardinal features of the addiction disorder, namely loss of control over drug use, compulsive use, and continued use despite harm, and operationally define the aberrant drug-related behaviors that suggest the diagnosis (Portenoy & Payne, 1992).

Addiction must be fully distinguished from the phenomenon of physical dependence, which is a pharmacological property defined solely by the potential for an abstinence syndrome following abrupt dose reduction or administration of an antagonist drug. Whereas physical dependence should be presumed to exist in all patients who receive opioids on a regular basis for more than a few days, addiction is a rare phenomenon among medical patients with no prior history of substance abuse who are administered opioids for painful disorders (Chapman, 1989; Perry & Heidrich, 1982; Portenoy, 1990a; Porter & Jick, 1980). Addiction is not determined by the mere exposure to a potentially abusable drug, but rather, by a complex and poorly understood interaction between inherently reinforcing properties of the drug and a constellation of genetic, psychological, and social characteristics of the patient. The terms *addict* or *addiction* can be highly stigmatizing and should never be applied to a patient when referring to the potential for withdrawal.

Although clinical experience has highlighted factors that may be important in the development of addiction, such as a prior substance abuse history or severe character pathology, the predictive validity of these factors has not been determined. Among populations with chronic nonmalignant pain,

TABLE 7   Proposed Guidelines for the Management of Long-Term Opioid Therapy for Chronic Nonmalignant Pain[a]

1. Should be considered only after all other reasonable attempts at analgesia have failed.
2. A history of substance abuse, severe character pathology, and chaotic home environment should be viewed as relative contraindications.
3. A single practitioner should take primary responsibility for treatment.
4. Patients should give informed consent before the start of therapy; points to be covered include recognition of the low risk of true addiction as an outcome, potential for cognitive impairment with the drug alone and in combination with sedative/hypnotics, likelihood that physical dependence will occur (abstinence possible with acute discontinuation), and understanding by female patients that children born when the mother is on opioid therapy will likely be physically dependent at birth.
5. After drug selection, doses should be given on an around-the-clock basis; several weeks should be agreed upon as the period of initial dose titration, and although improvement in function should be continually stressed, all should agree to at least partial analgesia as the appropriate goal of therapy.
6. Failure to achieve at least partial analgesia at relatively low initial doses in the nontolerant patient raises questions about the potential treatability of the pain syndrome with opioids.
7. Emphasis should be given to attempts to capitalize on improved analgesia by gains in physical and social function; opioid therapy should be considered complementary to other analgesic and rehabilitative approaches.
8. In addition to the daily dose determined initially, patients should be permitted to escalate dose transiently on days of increased pain; two methods are acceptable; (a) Prescription of an additional 4–6 "rescue doses" to be taken as needed during the month; (b) Instruction that one or two extra doses may be taken on any day, but must be followed by an equal reduction of dose on subsequent days.
9. Initially, patients must be seen and drugs prescribed at least monthly. When stable, less frequent visits may be acceptable.
10. Exacerbations of pain not effectively treated by transient, small increases in dose are best managed in the hospital, where dose escalation, if appropriate, can be observed closely and return to baseline doses can be accomplished in a controlled environment.
11. Evidence of drug hoarding, acquisition of drugs from other physicians, uncontrolled dose escalation, or other aberrant behaviors must be carefully assessed. In some cases, tapering and discontinuation of opioid therapy will be necessary. Other patients may appropriately continue therapy with rigid guidelines. Consideration should be given to consultation with an addiction medicine specialist.
12. At each visit, assessment should specifically address
    a. Comfort (degree of analgesia)
    b. Opioid-related side effects
    c. Functional status (physical and psychosocial)
    d. Existence of aberrant drug-related behaviors
13. Use of self-report instruments may be helpful but should not be required.
14. Documentation is essential and the medical record should specifically address comfort, function, side effects, and the occurrence of aberrant behaviors repeatedly during the course of therapy.

[a]From Portenoy, 1994, Table IV, p. 274–275.

both case selection and practice guidelines remain empirical (Table 7). Controlled prospective studies are needed to evaluate the safety and efficacy of this approach in terms of the dual goals—comfort and functional restoration—that are actively pursued during the treatment of most patients with chronic nonmalignant pain.

B. Anesthetic Approaches

Pain management is an important subspecialty of anesthesiology, and analgesic techniques that are uniquely within the purview of anesthesiologists play an important role in the treatment of acute and chronic pain. In most hospitals, acute pain services are directed by anesthesiologists with advanced training in pain management. In the care of patients with chronic pain, techniques that are specifically performed by these physicians include neural blockade and intraspinal infusional modalities.

Neural blockade encompasses a diverse group of procedures that transiently or more permanently block sympathetic and/or somatic nerves (Cousins, Dwyer, & Gibb, 1988; Raj, 1988). Temporary somatic nerve blocks with local anesthetic may be diagnostic (to elucidate the afferent pathways involved in the experience of pain), prognostic (used prior to a neurolytic procedure), or therapeutic. Repeated local anesthetic blocks are used therapeutically in patients who obtain substantial and fairly prolonged relief after each procedure. For example, repeated blocks of sympathetic nerves or ganglia are a mainstay approach for sympathetically maintained neuropathic pain. More permanent nerve blocks produced by neurolytic solutions (e.g., phenol or alcohol) are an important modality in cancer pain management.

Infusion techniques can be used to provide spinal opioid therapy or more prolonged neural blockade. Epidural or intrathecal opioid infusion is a well-accepted approach that is usually considered for refractory cancer pain. Local anesthetics can also be infused intraspinally, typically into the epidural space. Other infusion techniques have been developed to deliver local anesthetic in the region of selected peripheral nerves.

Although trigger point injections are typically classified as an anesthetic approach, the technique of local anesthetic injection into painful soft tissues is extremely simple and may be considered within the purview of all practitioners. Anecdotal observation suggests that many patients with defined trigger points (Travell & Simons, 1983) or discrete areas of intense soft tissue tenderness without the characteristics of a trigger point gain substantial temporary relief following an injection.

C. Neurostimulatory Approaches

It has been appreciated for some time that stimulation of afferent neural pathways may eventuate in analgesia. The best known application of this principle is transcutaneous electrical nerve stimulation (TENS). Other approaches include counterirritation (systematic rubbing of the painful part), percutaneous electrical nerve stimulation, dorsal column stimulation, deep brain stimulation and acupuncture. Surveys of the techniques that have been used most extensively, specifically TENS, acupuncture, and dorsal column stimulation, have suggested that the majority of patients will achieve analgesia soon after the approach is implemented, but fewer can obtain prolonged relief. A recent controlled trial that compared TENS to a program of stretching exercises in patients with chronic low back pain failed to identify any positive effect from the stimulation (Deyo, Walsh, Martin, Schoenfeld, & Ramamurthy, 1990).

Notwithstanding these ambiguous data, that safety of the noninvasive techniques (TENS and counterirritation) and acupuncture, combined with the observation that some patients benefit greatly from even temporary relief, justifies therapeutic trials in selected patients. Occasionally patients will opt to continue with one or another approach. For TENS, an adequate trial typically involves several weeks, during which the patient should experiment with different electrode placements and stimulation parameters. There is no one correct approach, and it is possible that many patients who could potentially benefit do not as a result of an incomplete trial. The application of other neurostimulatory approaches should be limited to experienced practitioners knowledgeable about the management of chronic pain.

D. Physiatric Approaches

The potential for analgesic effects from physiatric therapies, including the use of orthoses or prostheses, occupational therapy, and physical therapy, is inadequately recognized. Although these approaches are often provided to selected patients with chronic nonmalignant pain as part of an overall rehabilitation strategy, they are uncommonly recommended for analgesic purposes in other clinical contexts. Splinting of a painful limb, for example, may be a valuable treatment for patients with refractory cancer pain. Similarly, the comfort experienced by patients with myofascial pains may be improved by the use of systematic stretching and strengthening techniques.

E. Neurosurgical Approaches

Procedures designed to surgically denervate a painful region of the body have been developed for every level of the nervous system, from peripheral nerve to cortex (Gybels & Sweet, 1989). Other procedures, such as lobotomy or cingulotomy, are not denervating, but may reduce the affective concomitants of the pain. All of these procedures have been extensively applied to the management of cancer pain (Arbit, 1993). In this setting, cordotomy has been the most useful. Surgical modalities are rarely considered for the treatment of nonmalignant pain, because pain often recurs with time and the risks associated with the procedure, including the risk of inducing a new persistent pain, are too great in patients with normal life expectancies. The potential exceptions to this view include resection of painful neuromas and the use of the dorsal root entry zone lesion in patients with avulsion of neural plexus (Young, 1990).

F. Psychological Approaches

For patients with chronic pain, psychological approaches are often an essential component of the overall therapeutic strategy. Specific cognitive and behavioral treatments have become widely accepted (Turk, Meichenbaum, & Genest, 1983). Cognitive approaches comprise numerous procedures, including relaxation training, distraction techniques, hypnosis, and biofeedback, all of which may enhance a patient's sense of personal control and potentially reduce pain. Behavior therapy may be effective in improving the functional capabilities of the patient with chronic pain (Fordyce, Brockway, Bergman, & Spengler, 1986).

These cognitive and behavioral approaches, combined with intensive physiatric therapies, are the foundation of the multidisciplinary pain management clinic. As discussed previously, this model is generally considered to be optimal for a selected subpopulation with chronic pain complicated by high levels of disability and affective disturbance. It is important to recognize, however, that those patients whose level of disability is not sufficient to warrant referral to a multidisciplinary pain program, or who are not amenable to treatment in such a setting, may still benefit from psychological interventions designed to lessen discomfort, improve coping, or enhance function. Although referral to a specialist in this area is often required, some techniques may be within the purview of most practitioners. For example, many patients could probably benefit from simple cognitive techniques, such as relaxation training or distraction. Similarly, a graduated exercise program or a behavioral program designed to improve social interaction can be implemented through the use of an activity diary maintained by the patient.

Occasionally, the comprehensive pain assessment indicates the need for other types of psychological treatments, including individual insight-oriented therapy or family therapy. Again, these approaches may be viewed as part of a multimodality pain-oriented therapy, the goals of which encompass both enhanced comfort and improved function.

V. CONCLUSION

Burgeoning basic and clinical information has propelled the development of a subspeciality focus on pain in numerous disciplines. This has clearly expanded the opportunities for expert care in the management of acute and chronic painful disorders. All health-care providers must have sufficient information to perform a detailed assessment and thereby identify the nature of the pain problem and associated disturbances. This assessment can yield an appropriate referral or provide a foundation for a therapeutic strategy undertaken by the individual clinician. This strategy may integrate a multimodality approach designed to enhance comfort and improve overall quality of life.

References

Abram, S. E., & Lightfoot, R. W. (1981). Treatment of longstanding causalgia with prazosin. *Regional Anesthesia, 6,* 79–81.
Albe-Fessard, D., & Lombard, M. C. (1982). Use of an animal model to evaluate the origin of deafferentation pain and protection against it. In J. J. Bonica, U. Lindblom, & A. Iggo (Eds.), *Advances in pain research and therapy, vol. 5, Proceedings of the IIIrd world congress on pain* (pp. 691–700). New York: Raven Press.
American Psychiatric Association. (1994). *Diagnostic and statistical manual of mental disorders IV* (4th ed.). Washington, DC: American Psychiatric Association.
Anthony, M., & Lance, J. W. (1969). MAO inhibition in the treatment of migraine. *Archives of Neurology, 21,* 263–268.
Arbit, E. (Ed.). (1993). *Management of cancer-related pain.* Mount Kisco, NY: Futura Publishing.
Arner, S., & Meyerson, B. A. (1988). Lack of analgesic effect of opioids on neuropathic and idiopathic forms of pain. *Pain, 33,* 11–23.
Beaver, W. T., Wallenstein, S. M., Houde, R. W., & Rogers, A. (1966). A comparison of the analgesic effects of methotrimeprazine and morphine in patients with cancer. *Clinical Pharmacology & Therapeutics, 7,* 436–446.
Bercel, N. A. (1977). Cyclobenzaprine in the treatment of skeletal muscle spasm in osteoarthritis of the cervical and lumbar spine. *Current Therapy & Research, 22,* 462–468.
Besson, J. -M., & Chaouch, A. (1987). Peripheral and spinal mechanisms of nociception. *Physiology Review, 67,* 67–185.
Birkeland, I. W., & Clawson, D. K. (1968). Drug combinations with orphenadrine for pain relief associated with muscle spasm. *Clinical Pharmacology & Therapeutics, 9,* 639–646.
Blom, S. (1963). Tic douloureux treated with new anticonvulsant. *Archives of Neurology, 9,* 285–290.
Blomquist, C., Elomaa, I., Porkka, L., Karonen, S. L., & Lamberg-Allardt, C. (1988). Evaluation of salmon calcitonin treatment in bone metastasis from breast cancer—a controlled trial. *Bone, 9,* 45–51.
Bloomfield, S., Simard-Savoie, S., Bernier, J., & Tretault, L. (1964). Comparative analgesic activity of levomepromazine and morphine in patients with chronic pain. *Canadian Medical Association Journal, 90,* 1156–1159.
Bonica, J. J. (1990). Definitions and taxonomy of pain. In J. J. Bonica (Ed.), *The management of pain* (pp. 18–27). Philadelphia: Lea & Febiger.
Braham, J., & Saia, A. (1960). Phenytoin in the treatment of trigeminal and other neuralgias. *Lancet, 2,* 892–893.
Brooks, P. M., & Day, R. O. (1991). Nonsteroidal anti-inflammatory drugs—differences and similarities. *New England Journal of Medicine, 24,* 1716–1725.
Bruera, E., Chadwick, S., Brenneis, C., & MacDonald, R. N. (1987). Methylphenidate associated with narcotics for the treatment of cancer pain. *Cancer Treatment Reports, 71,* 67–70.

Bruera, E., Macmillan, K., Selmser, P., & MacDonald, R. N. (1989). The cognitive effects of the administration of narcotic analgesics in patients with cancer. *Pain, 39,* 13–16.

Butler, S. (1984). Present status of tricyclic antidepressants in chronic pain. In C. R. Benedetti, C. R. Chapman, & G. Moricca (Eds.), *Advances in pain research and therapy, vol. 7, Recent advances in the management of pain* (pp. 173–198). New York: Raven Press.

Caccia, M. R. (1975). Clonazepam in facial neuralgia and cluster headache: Clinical and electrophysiological study. *European Journal of Neurology, 13,* 560–563.

Caldwell, J. R., Roth, S. H., Wu, W. C., Semble, E. L., Castell, D. O., Heller, M. D., & Marsh, W. H. (1987). Sucralfate treatment of nonsteroidal anti-inflammatory drug-induced gastrointestinal symptoms and mucosal damage. *American Journal of Medicine, 83*(Suppl. 3B), 74–82.

Campbell, F. G., Graham, J. G., & Zilkha, K. J. (1966). Clinical trial of carbamazepine (Tegretol) in trigeminal neuralgia. *Journal of Neurology, Neurosurgery & Psychiatry, 29,* 265–267.

Cantor, F. K. (1972). Phenytoin treatment of thalamic pain. *British Medical Journal, 2,* 590.

Carette, S., McCain, G. A., Bell, D. A., & Fam, A. G. (1986). Evaluation of amitriptyline in primary fibrositis. *Arthritis and Rheumatism, 29,* 655–659.

Cassel, E. J. (1982). The nature of suffering and the goals of medicine. *New England Journal of Medicine, 306,* 639–645.

Chadda, V. S., & Mathur, M. S. (1978). Double blind study of the effects of diphenylhydantoin sodium in diabetic neuropathy. *Journal of the Association of Physicians of India, 26,* 403–406.

Chapman, C. R. (1989). Giving the patient control of opioid analgesic administration. In C. S. Hill & W. S. Fields (Eds.), *Advances in pain research and therapy, vol. 11, Drug treatment of cancer pain in a drug-oriented society* (pp. 339–352). New York: Raven Press.

Charney, D. S., Menkes, D. B., & Heninger, F. R. (1981). Receptor sensitivity and the mechanism of action of antidepressant treatment. *Archives of General Psychiatry, 38,* 1160–1180.

Couch, J. R., Ziegler, D. K., & Hassanein, R. (1976). Amitriptyline in the prophylaxis of migraine effectiveness and relationship of antimigraine and antidepressant effects. *Neurology, 26,* 121–127.

Cousins, M. (1989). Acute and postoperative pain. In P. D. Wall & R. Melzack (Eds.), *Textbook of pain* (pp. 284–305). Edinburgh: Churchill Livingstone.

Cousins, M. J., Dwyer, B., & Gibb, D. (1988). Chronic pain and neurolytic neural blockade. In M. J. Cousins & P. O. Bridenbaugh (Eds.), *Neural blockade in clinical anesthesia and management of pain* (2nd ed.) (pp. 1053–1084). Philadelphia: J. B. Lippincott.

Danesh, B. J. Z., Saniabadi, A. R., Russell, R. I., & Lowe, G. D. O. (1987). Therapeutic potential of choline magnesium trisalicylate as an alternative to aspirin for patients with bleeding tendencies. *Scottish Medical Journal, 32,* 167–168.

Davidoff, G., Guarrancini, M., Roth, Sliwa, J., & Yarkony, G. (1987). Trazodone hydrochloride in the treatment of dysesthetic pain in traumatic myelopathy: A randomized, double-blind, placebo-controlled study. *Pain, 29,* 151–161.

Dejgard, A., Petersen, P., & Kastrup, J. (1988). Mexiletine for treatment of chronic painful diabetic neuropathy. *Lancet, 1,* 9–11.

Deyo, R. A., Walsh, N. E., Martin, D. C., Schoenfeld, L. S., & Ramamurthy, S. (1990). A controlled trial of transcutaneous electrical nerve stimulation (TENS) and exercise for chronic low back pain. *New England Journal of Medicine, 322,* 1627–1634.

Diamond, S., & Baltes, B. J. (1971). Chronic tension headache—treatment with amitriptyline—a double blind study. *Headache, 11,* 110–116.

Dubner, R. (1991). Neuronal plasticity and pain following peripheral tissue inflammation or nerve injury. In M. R. Bond, J. E. Charlton, & C. J. Woolf (Eds.), *Proceedings of the VIth world congress on pain* (pp. 263–276). Amsterdam: Elsevier.

Eberhard, G., von Knorring, L., Nilsson, H. L., Sundequist, U., Björling, G., Linder, H., Svärd, K. O., & Tysk, L. (1988). A double-blind randomized study of clomipramine versus maprotiline in patients with idiopathic pain syndromes. *Neuropsychobiology, 19,* 25–34.

Edelson, J. T., Tosteson, A. N. A., & Sax, P. (1990). Cost-effectiveness of misoprostol for prophylaxis against nonsteroidal anti-inflammatory drug-induced gastrointestinal tract bleeding. *Journal of the American Medical Association, 264,* 41–47.

Edwards, W. T. (1990). Optimizing opioid treatment of postoperative pain. *Journal of Pain and Symptom Management, 5,* S24–S36.

Ekbom, K. (1972). Carbamazepine in the treatment of tabetic lightning pains. *Archives of Neurology, 26,* 374–378.

Elliot, F., Little, A., & Milbrandt, W. (1976). Carbamazepine for phantom limb phenomena. *New England Journal of Medicine, 295,* 678.

Espir, M. L. E., & Millac, P. (1970). Treatment of paroxysmal disorders in multiple sclerosis with carbamazepine (Tegretol). *Journal of Neurology, Neurosurgery and Psychiatry, 33,* 528–531.

Ettinger, A. B., & Portenoy, R. K. (1988). The use of corticosteroids in the treatment of symptoms associated with cancer. *Journal of Pain and Symptom Management, 3,* 99–103.

Evans, W., Gensler, F., Blackwell, B., & Galbrecht, C. (1973). The effects of antidepressants drugs on pain relief and mood in the chronically ill. *Psychosomatics, 14,* 214–219.

Fernandez, F., Adams, F., & Holmes, V. F. (1987). Analgesic effect of alprazolam in patients with chronic, organic pain of malignant origin. *Journal of Clinical Psychopharmacology, 7,* 167–169.

Ferrer-Brechner, T., & Ganz, P. (1984). Combination therapy with ibuprofen and methadone for chronic cancer pain. *American Journal of Medicine, 77,* 78–83.

Foley, K. M. (1985). The treatment of cancer pain. *New England Journal of Medicine, 313,* 84–95.

Foley, K. M. (1991). Clinical tolerance to opioids. In A. I. Basbaum & J. -M. Besson, (Eds.), *Towards a new pharmacotherapy of pain* (pp. 818–204). Chichester, UK: John Wiley & Sons.

Fordyce, W. E., Brockway, J. A., Bergman, J. A., & Spengler, D. (1986). Acute back pain: A control group comparison of behavioral versus traditional management methods of acute back pain. *Journal of Behavioral Medicine, 9,* 127–140.

Forrest, W. H., Brown, B. W., Brown, C. R., Defalque, R., Gold, M., Gordon, E., James, K. E., Katz, J., Mahler, D. L., Schroff, P., & Teutch, G. (1977). Dextroamphetamine with morphine for the treatment of postoperative pain. *New England Journal of Medicine, 296,* 712–715.

France, R. D., Urban, B. J., & Keefe, F. J. (1984). Long-tern use of narcotic analgesics in chronic pain. *Social Science & Medicine, 19,* 1379–1382.

Frank, W. O., Wallin, B. A., Berkowitz, J. M., Kimmey, M. B., Palmer, R. H., Rockhold, F., & Young, M. D. (1989). Reduction of indomethacin-induced gastroduodenal mucosal injury and gastrointestinal symptoms with cimetidine in normal subjects. *Journal of Rheumatology, 16,* 1249–1252.

Fries, J. F., Miller, S. F., Spitz, P. W., Williams, C. A., Hubert, H. B., & Bloch, D. A. (1989). Toward an epidemiology of gastropathy associated with nonsteroidal anti-inflammatory drug use. *Gastroenterology, 16,* 815–823.

Fromm, G. H., Terence, C. F., & Chatta, A. S. (1984). Baclofen in the treatment of trigeminal neuralgia. *Annal of Neurology, 15,* 240–247.

Getto, C. J., Sorkness, C. A., & Howell, T. (1987). Antidepressants and chronic nonmalignant pain: A review. *Journal of Pain and Symptom Management, 2,* 9–18.

Ghostine, S. Y., Comair, Y. G., Turner, D. M., Kassell, N. F., & Azar, C. G. (1984). Phenoxybenzamine in the treatment of causalgia. *Journal of Neurosurgery, 60,* 1263–1268.

Gingras, M. A. (1976). A clinical trial of Tofranil in rheumatic pain in general practice. *Journal of Internal Medicine Research, 4,* 41–49.

Glazer, S., & Portenoy, R. K. (1991). Systemic local anesthetics in pain control. *Journal of Pain and Symptom Management, 6,* 30–39.

Gobelet, C., Waldburger, M., & Meier, J. L. (1992). The effect of adding calcitonin to physical treatment on reflex sympathetic dystrophy. *Pain, 48,* 171–175.

Gold, R. H. (1978). Treatment of low back syndrome with oral orphenadrine citrate. *Current Therapy & Research, 23,* 271–276.

Gonzales, G. R., Elliott, K. J., Portenoy, R. K., & Foley, K. M. (1991). The impact of a comprehensive evaluation in the management of cancer pain. *Pain, 47,* 141–144.

Graham, D. Y., Agrawal, N. M., & Roth, S. H. (1988). Prevention of NSAID-induced gastric ulcer with misoprostol: Multicentre, double-blind, placebo-controlled trial. *Lancet, 2,* 1277–1280.

Green, J. B. (1961). Dilantin in the treatment of lightning pains. *Neurology,* (Minneap.) *11,* 257–258.

Gybels, J. M., & Sweet, W. H. (1989). *Neurosurgical treatment of persistent pain.* Basel: Karger.

Hameroff, S. R., Cork, R. C., Scherer, K., Crago, R., Neuman, C., Wamble, J. R., & Davis, T. P. (1982). Doxepin effects on chronic pain, depression and plasma opioids. *Journal of Clinical Psychology, 43,* 22–27.

Hammond, D. L. (1985). Pharmacology of central pain-modulating networks (biogenic amines and nonopioid analgesics). In H. L. Fields, R. Dubner, & F. Cervero (Eds.), *Advances in pain research and therapy,* Vol. 9, *Proceedings on the fourth world congress on pain* (pp. 499–513). New York: Raven Press.

Hatangdi, V. S., Boas, R. A., & Richards, E. G. (1976). Postherpetic neuralgia: Management with antiepileptic and tricyclic drugs. In J. J. Bonica, D. Albe-Fessard (Eds.), *Advances in pain research and therapy* (Vol. 1, pp. 583–587). New York: Raven Press.

Health and Public Policy Committee, American College of Physicians (1983). Drug therapy for severe chronic pain in terminal illness. *Annals of Internal Medicine, 99,* 870–873.

Higgs, G. A., & Moncada, S. (1983). Interaction of arachidonate products with other pain mediators. In J. J. Bonica, U. Lindblom, & A. Iggo (Eds.), *Advances in pain research and therapy* (Vol. 5, pp. 617–626). New York: Raven Press.

Houde, R. W., & Wallenstein, S. L. (1966). Analgesic power of chlorpromazine alone and in combination with morphine [Abstract]. *Federation Proceedings, 14,* 353.

Ingham, J. M., & Portenoy, R. K. (1993). Drugs in the treatment of pain: NSAIDs and opioids. *Current Opinion in Anaesthesiology, 6,* 838–844.

Jaeger, H., & Maier, C. (1992). Calcitonin in phantom limb pain: A double-blind study. *Pain, 48,* 21–27.

Kanner, R. M., & Foley, K. M. (1981). Patterns of narcotic drug use in a cancer pain clinic. *Annals of the New York Academy of Science, 362,* 161–172.

Kellgren, J. G. (1939). On distribution of pain arising from deep somatic structures with charts of segmental pain areas. *Clinical Sciences, 4,* 35–46.

Killian, J. M., & Fromm, G. H. (1968). Carbamazepine in the treatment of neuralgia: Use and side effects. *Archives of Neurology, 19,* 129–136.

Kishore-Kumar, R., Max, M. B., Schafer, S. C., Gaughan, A. M., Smoller, B., Gracely, R. H., & Dubner, R. (1990). Desipramine relieves postherpetic neuralgia. *Clinical Pharmacology and Therapeutics, 47,* 305–312.

Kocher, R. (1976). Use of psychotropic drugs for the treatment of chronic severe pain. In J. J. Bonica & D. Albe-Fessard et al. (Eds.), *Advances in pain research and therapy* (Vol. 1, pp. 279–282). New York: Raven Press.

Kozin, F., Ryan, L. M., Carerra, G. F., Soin, L. S., & Wortmann, R. L. (1981). The reflex sympathetic dystrophy syndrome (RSDS). III. scintigraphic studies, further evidence for the therapeutic efficacy of systemic corticosteroids, and proposed diagnostic criteria. *American Journal of Medicine, 70,* 23–29.

Kreeger, W., & Hammill, S. C. (1987). New antiarrhythmic drugs: tocainide, mexiletine, flecainide, encainide, and amiodarone. *Mayo Clinic Proceedings, 62,* 1033–1050.

Kvinesdal, B., Molin, J., Froland, A., & Gram, L. F. (1984). Imipramine treatment of painful diabetic neuropathy. *Journal of the American Medical Association, 251,* 1727–1730.

Langohr, H. D., Stohr, M., & Petruch, F. (1982). An open and double-blind cross-over study on the efficacy of clomipramine (Anafranil) in patients with painful mono- and poly-neuropathies. *European Neurology, 21,* 309–317.

Lasagna, L., & DeKornfeld, T. J. (1961). Methotrimeprazine—a new phenothiazine derivative with analgesic properties. *Journal of the American Medical Association, 178,* 119–122.

Lascelles, R. G. (1966). Atypical facial pain and depression. *British Journal of Psychiatry, 122,* 651–659.

Laska, E. M., Sunshine, A., Mueller, F., Elvers, W. B., Siegel, C., & Rubin, A. (1984). Caffeine as an analgesic adjuvant. *Journal of the American Medical Association, 251,* 1711–1718.

Lechin, F., van der Dijs, B., Lechin, M. E., Amat, J., Lechin, A. E., Cabrera, A., Gomez, F., Acosta, E., Arocha, L., & Villa, S. (1989). Pimozide therapy for trigeminal neuralgia. *Archives of Neurology, 9,* 960–962.

Lindstrom, P., & Lindblom, U. (1987). The analgesic effect of tocainide in trigeminal neuralgia. *Pain, 28,* 45–50.

Lockman, L. A., Hunninghake, D. B., Krivit, W., & Desnick, R. J. (1973). Relief of pain of Fabry's disease by diphenylhydantoin. *Neurology,* (Minneap.)*23,* 871–875.

Loeb, D. S., Ahlquist, D. A., & Talley, N. J. (1992). Management of gastroduodenopathy associated with use of nonsteroidal anti-inflammatory drugs. *Mayo Clinic Proceedings, 67,* 354–364.

Loeser, J. D., & Egan, K. J. (1989). History and organization of the University of Washington Multidisciplinary Pain Center. In J. D. Loeser & K. J. Egan (Eds.), *Managing the chronic pain patient: Theory and practice at the University of Washington Multidisciplinary Pain Center* (pp. 3–20). New York: Raven Press.

Loeser, J. D., Ward, A. A., & White, L. E. (1968). Chronic deafferentation of human spinal cord neurons. *Journal of Neurosurgery, 29,* 48–50.

Malberg, A. B., & Yaksh, T. L. (1992). Hyperalgesia medicated by spinal glutamate or substance P receptors blocked by spinal cyclooxygenase inhibition. *Science, 257,* 1276–1279.

Margolis, L. H., & Gianascol, A. J. (1956). Chlorpromazine in thalamic pain syndrome. *Neurology,* (Minneap.)*6,* 302–304.

Martin, G. (1981). The management of pain following laminectomy for lumbar disc lesions. *Annals of the Royal College of Surgeons of England, 63,* 244–252.

Max, M. B., Culnane, M., Schafer, S. C., Gracely, R. H., Walter, D. J., Smoller, B., & Dubner, R. (1987). Amitriptyline relieves diabetic neuropathy pain in patients with normal or depressed mood. *Neurology, 37,* 589–594.

Max, M. B., Kishore-Kumar, R., Schafer, S. C., Meister, R. H., Gracely, B., Smoller, B., & Dubner, R. (1991). Efficacy of desipramine in painful diabetic neuropathy: A placebo-controlled trial. *Pain, 45,* 3–9.

Max, M. B., Schafer, S. C., Culnane, M., Dubner, R., & Gracely, R. H. (1988). Association of pain relief with drug side effects in post-herpetic neuralgia: A single-dose study of clonidine, codeine, ibuprofen, and placebo. *Clinical Pharmacology and Therapeutics, 43*(4), 363–371.

McGivney, W. T., & Grooks, G. M. (1984). The care of patients with severe chronic pain in terminal illness. *Journal of the American Medical Association, 251,* 1182–1188.

Melzack, R., & Casey, K. L. (1968). Neurophysiology of pain. In R. A. Sternbach (Ed.), *The psychology of pain* (2nd ed) (pp. 1–24). New York: Raven Press.

Merskey, H., & Bogduk, N. (1994). *Classification of chronic pain: Descriptions of chronic pain syndromes and definitions of pain terms* (2nd ed.). Seattle: IASP Press.

Miller, R., Eisenkraft, J. B., Cohen, M., Toth, C., Mora, C. T., & Bernstein, J. L. (1986). Midazolam as an adjunct to meperidine analgesia for postoperative pain. *Clinical Journal of Pain, 2,* 37–43.

Mullan, S. (1973). Surgical management of pain in cancer of the head and neck. *Surgical Clinics of North America, 53,* 203–210.

Nathan, P. W. (1978). Chlorprothixene (Taractan) in postherpetic neuralgia and other severe pains. *Pain, 5,* 367–371.

Noyes, R., Brunk, S. F., Avery, D. H., & Canter, A. (1975). The analgesic properties of delta-9-tetrahydrocannabinol and codeine. *Clinical Pharmacology and Therapeutics, 18,* 84–89.

Nystrom, B., & Hagbarth, K. E. (1981). Microelectrode recordings from transected nerves in amputees in phantom limb pain. *Neuroscience Letter, 27,* 211–216.

O'Hara, D. A., Fragen, R. J., Kinzer, M., Pemberton, D. (1987). Ketorolactromethamine as compared with morphine sulfate for treatment of postoperative pain. *Clinical Pharmacology and Therapeutics, 41,* 556–561.

Okasha, A., Ghaleb, A. A., & Sadek, A. (1973). A double-blind trial for the clinical management of psychogenic headache. *British Journal of Psychiatry, 122,* 181–183.

Peiris, J. B., Perera, G. L. S., Devendra, S. V., & Lionel, N. D. W. (1980). Sodium valproate in trigeminal neuralgia. *Medical Journal of Austria, 2,* 278.

Perry, S., & Heidrich, G. (1982). Management of pain during debridement: A survey of U.S. burn units. *Pain, 13,* 267–280.

Pilowsky, I., Hallet, E. C., Bassett, D. L., Thomas, P. G., & Penhall, R. K. (1982). A controlled study of amitriptyline in the treatment of chronic pain. *Pain, 14,* 169–179.

Portenoy, R. K. (1990a). Chronic opioid therapy in non-malignant pain. *Journal of Pain and Symptom Management, 5,* S46–S62.

Portenoy, R. K. (1990b). Pain and quality of life: Clinical issues and implications for research. *Oncology, 4,* 172–178.

Portenoy, R. K. (1990c). Pharmacologic management of chronic pain. In H. L. Fields (Ed.), *Pain syndromes in neurology* (pp. 257–278). London: Butterworths.

Portenoy, R. K. (1991). Issues in the management of neuropathic pain. In A. Basbaum & J.-M. Besson (Eds.), *Towards a new pharmacotherapy of pain* (pp. 393–416). Chichester, UK: John Wiley & Sons.

Portenoy, R. K. (1992). Cancer pain: Pathophysiology and syndromes. *Lancet, 339,* 1026–1031.

Portenoy, R. K. (1993). Cancer pain management. *Seminars in Oncology, 20,* 19–35.

Portenoy, R. K. (1994). Opioid therapy for chronic nonmalignant pain: current status. In H. L. Fields & J. C. Liebeskind (Eds.), *Progress in pain research and management: Pharmacological approaches to the treatment of chronic pain: New concepts and critical issues* (pp. 247–287). Seattle: IASP Press.

Portenoy, R. K., & Foley, K. M. (1986). Chronic use of opioid analgesics in non-malignant pain: Report of 38 cases. *Pain, 25,* 71–86.

Portenoy, R. K., & Hagen, N. A. (1990). Breakthrough pain: Definition, prevalence and characteristics. *Pain, 41,* 273–281.

Portenoy, R. K., & Kanner, R. M. (1996). *Pain management: Theory and practice.* Philadelphia: F. A. Davis.

Portenoy, R. K., & Payne, R. (1992). Acute and chronic pain. In J. H. Lowinson, P. Ruiz, & R. B. Millman (Eds.), *Substance abuse: A comprehensive textbook* (pp. 691–721). Baltimore: Williams & Wilkins.

Porter, J., & Jick, H. (1980). Addiction rare in patients treated with narcotics. *New England Journal of Medicine, 302,* 123.

Prough, D. S., McLeskey, C. H., Borshy, G. G., Poehling, G. G., Koman, L. A., Weeks,

D. B., Whitworth, T., & Semble, E. L. (1985). Efficacy of oral nifedipine in the treatment of reflex sympathetic dystrophy. *Anesthesiology, 62,* 796–799.

Raftery, H. (1979). The management of postherpetic pain using sodium valproate and amitriptyline. *Journal of the Irish Medical Association, 72,* 399–401.

Raj, P. P. (1988). Prognostic and therapeutic local anesthetic block. In M. J. Cousins & P. O. Bridenbaugh (Eds.), *Neural blockade in clinical anesthesia and management of pain* (2nd ed.) (pp. 899–934). Philadelphia: J. B. Lippincott Company.

Raskin, N. H., Levinson, S. A., Hoffman, P. M., Pickett, J. B. E., & Fields, H. L. (1974). Postsympathectomy neuralgia: Amelioration with diphenylhydantoin and carbamazepine. *American Journal of Surgery, 128,* 75–78.

Richelson, E. (1979). Tricyclic antidepressants and neurotransmitter receptors. *Psychiatric Annals, 9,* 186–194.

Robinson, M. G., Griffin, J. W., Bowers, J., Kogan, F. J., Kogut, D. G., Lanza, F. L., & Warner, C. W. (1989). Effect of ranitidine on gastroduodenal mucosal damage induced by nonsteroidal anti-inflammatory drugs. *Digestive Diseases and Sciences, 34,* 424–428.

Rockliff, B. W., & Davis, E. H. (1966). Controlled sequential trials of carbamazepine in trigeminal neuralgia. *Archives of Neurology, 15,* 129–136.

Romano, J. M., & Turner, J. A. (1985). Chronic pain and depression: Does the evidence support a relationship? *Psychiatry Bulletin, 97,* 18–34.

Roth, S. H. (1988). NSAID and gastropathy: A rheumatologist's review. *Journal of Rheumatology, 15,* 912–919.

Roth, S. H., Bennett, R. E., Mitchell, C. S., & Hartman, R. J. (1987). Cimetidine therapy in nonsteroidal anti-inflammatory drug gastropathy. Double-blind long-term evaluation. *Archives of Internal Medicine, 147,* 1798–1801.

Rull, J. A., Quibrera, R., Gonzalez-Millan, H., & Castaneda, O. L. (1969). Symptomatic treatment of peripheral diabetic neuropathy with carbamazepine (Tegretol): Double blind cross-over trial. *Diabetologia, 5,* 215–218.

Rumore, M. M., & Schlichting, D. A. (1986). Clinical efficacy of antihistamines as analgesics. *Pain, 25,* 7–22.

Sandler, D. P., Smith, J. C., Weinberg, C. R., Buckalew, V. M., Dennis, V. W., Blythe, W. B., & Burgess, W. P. (1989). Analgesic use and chronic renal disease. *New England Journal of Medicine, 320,* 1238–1243.

Saunders, C. (1984). The philosophy of terminal care. In C. Saunders (Ed.), *The management of terminal malignant disease* (pp. 232–241). London: Edward Arnold.

Shafter, J., Tallett, E. R., & Knowlson, P. A. (1972). Evaluation of clonidine in prophylaxis of migraine. *Lancet, 1,* 403–407.

Silberstein, E. B., & Williams, C. (1985). Strontium-89 therapy for the pain of osseous metastases. *Journal of Nuclear Medicine, 26,* 345–348.

Simson, G. (1974). Propranolol for causalgia and Sudek's atrophy. *Journal of the American Medical Association, 227,* 327.

Sindrup, S. H., Gram, L. F., Brosen, K., Eshoj, O., & Morgensen, E. F. (1990). The selective serotonin reuptake inhibitor paroxetine is effective in the treatment of diabetic neuropathy symptoms. *Pain, 42,* 135–144.

Singh, P. N., Sharma, P., Gupta, P. K., & Pandey, K. (1981). Clinical evaluation of diazepam for relief of postoperative pain. *British Journal of Anaesthesia, 53,* 831––836.

Smith, C. M. (1965). Relaxants of skeletal muscle. In W. S. Root & F. G. Hoffman (Eds.), *Physiological pharmacology* (Vol. 2, pp. 2–96). New York: Academic Press.

Sosnowski, M., & Yaksh, T. L. (1990). Spinal administration of receptor-selective drugs as analgesics: New horizon. *Journal of Pain and Symptom Management, 5,* 204–213.

Spiegel, K., Kalb, R., & Pasternak, G. W. (1983). Analgesic activity of tricyclic antidepressants. *Annals of Neurology, 13,* 462–465.

Stambaugh, J. E., & Lance, C. (1983). Analgesic efficacy and pharmacokinetic evaluation of meperidine and hydroxyzine, alone and in combination. *Cancer Investigations, 1,* 111–117.

Sunshine, A., & Olson, N. Z. (1989). Non-narcotic analgesics. In P. D. Wall & R. Melzac (Eds.), *Textbook of pain* (2nd ed.) (pp. 670–685). New York: Churchill Livingstone.

Swerdlow, M. 91984). Anticonvulsan drugs and chronic pain. *Clinical Neuropharmacology, 7,* 51–82.

Swerdlow, M., & Cundill, J. G. (1981). Anticonvulsant drugs used in the treatment of lancinating pains: A comparison. *Anesthesa, 36,* 1129–1132.

Swerdlow, M., & Stjernsward, J. (1982). Cancer pain relief—an urgent problem. *World Health Forum, 3,* 325–330.

Tabira, T., Shibasaki, H., & Kuroiwa, Y. (1983). Reflex sympathetic dystrophy (causalgia) treatment with guanethidine. *Archives of Neurology, 40,* 430–432.

Tan, Y.-M., & Croese, J. (1986). Clonidine and diabetic patients with leg pains. *Annals of Internal Medicine, 105,* 633.

Taub, A. (1982). Opioid analgesics in the treatment of chronic intractable pain of non-neoplastic origin. In L. M. Kitahata & D. Collins (Eds.), *Narcotic analgesics in anesthesiology* (pp. 199–208). Baltimore: Williams and Wilkins.

Taylor, P. H., Gray, K., Bicknell, R. G., & Rees, J. R. (1977). Glossopharyngeal neuralgia with syncope. *Journal of Laryngology and Otology, 91,* 859–868.

Tennant, F. S., Robinson, D., Sagherian, A., & Seecof, R. (1988). Chronic opioid treatment of intractable non-malignant pain. *Pain Management,* Jan/Feb, 18–36.

Tennant, F. S., & Uelman, G. F. (1983). Narcotic maintenance for chronic pain: medical and legal guidelines. *Postgraduate Medicine, 73,* 81–94.

Torebjork, H. E., Ochoa, J. L., & Schady, W. (1984). Referred pain from intraneural stimulation of muscle fascicles in the median nerve. *Pain, 18,* 145–156.

Travell, J. G., & Simons, D. G. (1983). *Myofascial pain and dysfunction: The trigger point manual.* Baltimore: Williams and Wilkins.

Turk, D. C., Meichenbaum, D., & Genest, M. (1983). *Pain and behavioral medicine: a cognitive-behavioral perspective.* New York: Guilford Press.

Turk, D. C., & Rudy, T. E. (1988). Toward an empirically derived taxonomy of chronic pain patients: Integration of psychological assessment data. *Journal of Consultation and Clinical Psychology, 56,* 233–238.

Turk, D. C., & Rudy, T. E. (1990). The robustness of an empirically derived taxonomy of chronic pain patients. *Pain, 43,* 27–35.

Twycross, R. G. (1974). Clinical experience with diamorphine in advanced malignant disease. *International Journal of Clinical Pharmacology and Therapeutics, 9,* 184–198.

Urban, B. J., France, R. D., Steinberger, D. L., Scott, D. L., & Maltbie, A. A. (1986). Long-term use of narcotic/antidepressant medication in the management of phantom limb pain. *Pain, 24,* 191–197.

Van Holten-Verzantvoort, A. T. M., Zwinderman, A. H., Aaronson, N. K., Hermans, J., van Emmenk, B., van Dam, F. S., van den Bos, B., Bijvoet, O. L., & Cleton, F. J. (1991). The effect of supportive pamidronate treatment on aspects of quality of life of patients with advanced breast cancer. *European Journal of Cancer, 27,* 544–549.

Vane, J. R. (1971). Inhibition of prostaglandin synthesis as a mechanism of action for aspirin-like drugs. *Nature New Biology, 231,* 232–235.

Ventafridda, V., Bonezzi, C., Caraceni, A., DeConno, F., Guarise, G., Ramella, G., Saita, L., Silvani, V., Tamburini, M., & Toscani, F. (1987). Antidepressants for cancer pain and other painful syndromes with deafferentation component: Comparison of amitriptyline and trazodone. *Italian Journal of Neurological Sciences, 8,* 579–587.

Wall, P. D., & Gutnick, M. (1974). Ongoing activity in peripheral nerves. 2. The physiology and pharmacology of impulses originating in a neuroma. *Experimental Neurology, 43,* 580–593.

Wall, P. D., & Melzack, R. (1989). *Textbook of pain* (2nd ed.). New York: Churchill Livingstone.

Watson, C. P. N., Chipman, M., Reed, K., Evans, R. J., & Birkett, N. (1992). Amitriptyline versus maprotiline in postherpetic neuralgia: a randomized, double-blind, crossover tiral. *Pain, 48,* 29–36.

Watson, C. P. N., & Evans R. J. (1985). A comparative trial of amitriptyline and zimelidine in postherpetic neuralgia. *Pain, 23,* 387–94.

Watson, C. P. N., Evans, R. J., & Watt, V. R. (1988). Postherpetic neuralgia and topical capsaicin. *Pain, 33,* 333–340.

Watson, C. P. N., Evans, R. J., & Watt, V. R. (1989). The post-mastectomy syndrome and the effect of topical capsaicin. *Pain, 38,* 177–186.

Watson, C. P. N., Evans, R. J., Reed, K., Merskey, H., Goldsmith, L., & Warsh, J. (1982). Amitriptyline versus placebo in postherpetic neuralgia. *Neurology, 32,* 671–673.

Weinberger, J., Nicklas, W. J., & Berl, S. (1976). Mechanism of action of anticonvulsants. *Neurology,* (Minneap.) *26,* 162–173.

Weis, O., Sriwatanakul, K., & Weintraub, M. (1982). Treatment of postherpetic neuralgia and acute herpetic pain with amitriptyline and perphenazine. *South African Medical Journal, 62,* 274–275.

Willer, J.-C., De Broucker, T., Bussel, B., Roby-Brami, A., & Harrewyn, J.-M. (1989). Central analgesic effect of ketoprofen in humans: Electrophysiological evidence for a supraspinal mechanism in a double-blind and cross-over study. *Pain, 38,* 1–8.

Willis, W. D. (1985). *The pain system: The neural basis of nociceptive transmission in the mammalian nervous system.* Basel: Karger.

World Health Organization. (1986). *Cancer pain relief.* Geneva: World Health Organization.

World Health Organization. (1990). *Cancer pain relief and palliative care.* Geneva: World Health Organization.

Yang, J. C., Clark, W. C., Ngai, S. H., Berkowitz, B. A., & Spector, S. (1979). Analgesic action and pharmacokinetics of morphine and diazepam in man: An evaluation by sensory decision theory. *Anesthesiology, 51,* 495–502.

Yosselson-Superstine, S., Lipman, A. G., & Sanders, S. H. (1985). Adjunctive antianxiety agents in the management of chronic pain. *Israel Journal of Medical Science, 21,* 113–117.

Young, R. F. (1990). Clinical experience with radio frequency and laser DREZ lesions. *Journal of Neurosurgery, 72,* 715–720.

Zech, D. F. J., Grond, S., Lynch, J., Hertal, D., & Lehmann, K. A. (1995). *Validation of the World Health Organization guidelines for cancer pain relief: A 10 year prospective study. Pain, 63,* 65–76.

Zenz, M., Strumpf, M., & Tryba, M. (1992). Long-term oral opioid therapy in patients with chronic nonmalignant pain. *Journal of Pain and Symptom Management, 7,* 69–77.

Index